Strength Prediction of Adhesively-Bonded Joints

Strength Prediction of Adhesively-Bonded Joints

Editor

Raul D.S.G. Campilho
Departamento de Engenharia Mecânica
Instituto Superior de Engenharia do Porto
Porto
Portugal

CRC Press
Taylor & Francis Group
Boca Raton London New York

CRC Press is an imprint of the
Taylor & Francis Group, an **informa** business

A SCIENCE PUBLISHERS BOOK

CRC Press
Taylor & Francis Group
6000 Broken Sound Parkway NW, Suite 300
Boca Raton, FL 33487-2742

First issued in paperback 2021

ISBN-13: 978-0-367-78241-2 (pbk)
ISBN-13: 978-1-4987-2246-9 (hbk)

Library of Congress Cataloging-in-Publication Data

Names: Campilho, Raul D. S. G., editor.
Title: Strength prediction of adhesively-bonded joints / editor, Raul D. S.
G. Campilho, Departamento de Engenharia Mecãanica Instituto Superior de
Engenharia do Porto, Porto, Portugal.
Description: Boca Raton, FL : Taylor & Francis Group, CRC Press, [2016] | "A
science publishers book." | Includes bibliographical references and index.
Identifiers: LCCN 2016029025| ISBN 9781498722469 (hardback) | ISBN
9781498722476 (e-book)
Subjects: LCSH: Adhesive joints. | Strength of materials.
Classification: LCC TA492.A3 .S74 2016 | DDC 660/.293--dc23
LC record available at https://lccn.loc.gov/2016029025

Visit the Taylor & Francis Web site at
http://www.taylorandfrancis.com

and the CRC Press Web site at
http://www.crcpress.com

Preface

Adhesively bonded joints are increasingly becoming an alternative to mechanical joints in engineering applications and provide many advantages over conventional mechanical fasteners. The application of adhesively bonded joints in structural components made of fibre reinforced composites has increased significantly in recent years. The traditional fasteners usually result in the cutting of fibres and hence the introduction of stress concentrations, both of which reduce structural integrity. By contrast, bonded joints are more continuous and have potential advantages of strength-to-weight ratio, design flexibility and ease of fabrication. A lack of suitable material models and failure criteria has resulted in a tendency to 'overdesign' adhesive joints. Safety considerations often require that adhesively bonded structures, particularly those employed in primary load-bearing applications, include mechanical fasteners (e.g. bolts) as an additional safety precaution. These practices result in heavier and more costly components. The development of reliable design and predictive methodologies can be expected to result in more efficient use of adhesives. In order to design structural joints in engineering structures, it is necessary to be able to analyse them. This means to determine stresses and strains under a given loading, and to predict the probable points of failure. There are two basic mathematical approaches for the analyses of adhesively bonded joints: closed-form analyses (analytical methods) and numerical methods (i.e. finite element or FE analyses).

The proposed book is divided in 12 chapters, in which the relevant strength prediction tools for the strength prediction of adhesively bonded joints are discussed in detail with examples of application. The first chapter, written by the book editor, aims to give an overview of the available methodologies and enable the reader to focus the attention in the most suited technique/s for a given application, by presenting a critical comparison between the methods. Analytical methods are then discussed, with emphasis to the single-lap joint, including joints in fibre-reinforced composite structures. Numerical methods are the main subject of the book, beginning with the continuum mechanics approach for strength estimation. Cohesive zone models are most likely the most accepted and reliable method to analyse bonded joints and, because of this, 3 contributions are provided on this method: (1) static applications, (2) parameter identification techniques and accuracy issues and (3) fatigue applications. Damage mechanics is another possibility, and the description of this technique includes static and fatigue modelling. Finally, the extended finite element method has been tested for application in bonded joints and it is also discussed in this book. Selected application of design are also presented.

Raul D.S.G. Campilho

Contents

1

Introduction

Raul D.S.G. Campilho

1.1 BACKGROUND

Although adhesives have been used since thousands of years, it was only in the beginning of the 20th century that adhesives based on synthetic polymers were developed. Major improvements were undertaken from the 1940's because of the rapid expansion of the chemical knowledge of polymers, resulting in the improvement of strength and ductility and cost reduction (Adams 2008). Currently, adhesive bonding is a major joining method in several fields. The pioneering use of this bonding method was in the aerospace industry, but other industries such as automotive are changing paradigms and implementing adhesive bonding in the fabrication process of vehicles, which enables the combination of different materials and new designs. Other applications are in the fields of aeronautics, electronics, construction, sports and packaging industries. Among the advantages of this technique are lower structural weight, lower fabrication cost and improved damage tolerance. However, it should be realised that the capacity of a bonded joint to perform as expected in the short and long term depends on different variables such as the joint design, load application and in-service environmental conditions (da Silva et al. 2011a). In view of this, to obtain good results with this joining technique, it is necessary to gather knowledge from different fields like surface chemistry, chemistry and physics of polymers, materials engineering and mechanical engineering (Petrie 2000). There are several reference books dealing with adhesive joints such as those of Kinloch (1987) and da Silva et al. (2011b).

Decades ago, when adhesive joints started to be implemented in the industry, design was based on trial-and-error and extensive experimentation, and optimization techniques were non-existent. With time, driven by the industry and technological demands and made possible by modern computers, different advanced techniques were developed and refined, such that it became possible to evaluate a given solution

Departamento de Engenharia Mecânica, Instituto Superior de Engenharia do Porto, Instituto Politécnico do Porto, Rua Dr. António Bernardino de Almeida, 431, 4200-072 Porto, Portugal.
E-mail: raulcampilho@gmail.com

1

and to optimize it before shifting from design to mass production. Currently, strength prediction techniques abound for static conditions, but other conditions such as impact loadings, vibrations, fatigue and durability are yet not properly addressed. For example, the durability concern is the main limitation for a wider use of adhesives in road vehicles (da Silva and Sato 2013). Actually, it is nowadays difficult to predict the joints' behaviour after long-term exposure to temperature and moisture. Even for the static loading case, the lack of a general-purpose predictive method has resulted in a tendency to 'overdesign' adhesive joints. In some cases, these issues make adhesively-bonded structures, particularly those employed in primary load-bearing applications, dependent on the combination of this joining technique with mechanical fasteners as an additional safety precaution. This results in heavier and more costly components. With the availability of accurate predictive methods, it is expected that the use of adhesives becomes more efficient. These methods are grouped into two main categories: closed-form analyses (analytical methods) and numerical methods (Finite Element or FE analyses). The analytical techniques become highly complex in the presence of composite adherends, plasticization of the adhesive layer or an adhesive fillet at the bonding edges. If any of these conditions is to be considered, a large number of non-linear and non-homogeneous differential equations of high complexity can be obtained. Numerical methods can be a viable alternative to deal with these scenarios. The FE method, the boundary element (BE) method and the finite difference (FD) method are the three major numerical methods for solving partial differential equations in science and engineering. The FE is by far the most common technique used in the context of adhesively-bonded joints and, thus, it is the only method described in this book. Adams and co-workers are among the first to have used the FE method for analysing adhesive joint stresses (Adams and Peppiatt 1974). One of the first reasons for the use of the FE method was to assess the influence of the spew fillet. The joint rotation and the adherends and adhesive plasticity are other aspects that are easier to treat with a FE analysis. Several FE approaches to failure analysis exist: continuum mechanics, fracture mechanics, Cohesive Zone Modelling (CZM), damage mechanics and the more recent eXtended Finite Element Method (XFEM). The use of fracture mechanics in the adhesive bonding context is marginal and, thus, this method is not addressed here.

1.2 OVERVIEW OF STRENGTH PREDICTION TECHNIQUES

Within the context of continuum mechanics, information of the stress or strain distributions inside the adhesive layer of an adhesively-bonded joint is essential for joint strength prediction and joint design. There are two big groups of techniques for the stress and strain analysis of bonded joints, namely analytical and numerical methods (da Silva and Campilho 2012). Analytical methods using closed form algebra employ classical linear theories in which some simplifications are used. Volkersen (1938) was the precursor of these methods applied to bonded joints by using a closed-form solution that considers the adhesive and adherends as elastic and that the adhesive deforms only in shear. The equilibrium equation of a single-lap joint led to a simple governing differential equation with a simple algebraic form. However, as previously stated, the analysis can be highly complex if composite adherends are used, the adhesive deforms plastically or if there is an adhesive fillet. There are many analytical models in the literature for obtaining stress and strain distributions. Many closed-form models are based on modified shear-lag equations, as proposed originally by Volkersen (1938). Reviews of these closed-form theories and their assumptions can be found in the reviews of da Silva et al. (2009a), da Silva et al. (2009b) and Tong and Luo (2011). With the inclusion of factors such as stress variation through the adhesive thickness, plasticity, thermal effects, composite materials and others in the formulations, the analytical equations become more complex and there is a greater requirement to use computing power to attain a solution. Hart-Smith (1973) developed a number of methods used for stress analysis of adhesive joints. Versions of this method have been prepared as Fortran programmes and were largely used in the aerospace industry. Formulations by different authors were implemented in spreadsheets or as a programme to be run in personal computers (PC). Currently, software packages are available to assist in the design of efficient joints. da Silva et al. (2009c) reviews the available commercial PC based analysis/design software packages. The main characteristics and field of application of each software package are compared. It is shown that the available software packages are tailored to specific geometries and most of them only cover few of them. Also, only few of these find application in the field of composites. A common limitation of these softwares, apart from those of da Silva et al. (2009c) and Dragoni et al. (2010), is that the choosing process is dependent on the previous experience of the designer. The FE method gives an approximate solution to problems in various fields of engineering. Ashcroft (2011) described this technique in the context of adhesive joints. The first efforts to use piecewise continuous functions defined over triangular domains are credited to Courant (1943). The aerospace industry and the development of computers in the 1950's and 1960's induced the fast development and computer implementation of the FE method (Turner et al. 1956). To predict the joint strength, one must have the stress or strain distribution and a suitable failure criterion. Actually, in continuum mechanics methods, stress or strain-based criteria are used, in which the maximum stresses and strains in the models are compared with the allowable limiting values for assessing the maximum load (Adams et al. 1997). The stress distribution can be obtained by a closed-form model or a FE analysis. For complex geometries and elaborated material models, the FE method is preferable. Fracture mechanics principles (not addressed in this book, as previously mentioned) can also be used within a FE analysis. This can be based on either the stress intensity factor or energy approaches (Shahin and Taheri 2008).

Another possibility for strength prediction of adhesive joints is the CZM technique, which also allows simulating damage of a material ahead of a crack. The technique is a combination of continuum mechanics and fracture mechanics. CZM do not depend on an initial flaw, unlike conventional fracture mechanics approaches. Usually, CZM are based on spring or cohesive elements (Feraren and Jensen 2004), connecting plane or two-dimensional (2D) or three-dimensional (3D) solid elements of structures. CZM are based on a softening relationship between stresses and relative displacements between the crack faces, which simulates the gradual degradation of material properties. The shape of the softening laws can be adjusted to conform to the behaviour of the material or interface that the CZM is simulating. Thus, they can be adapted to simulate ductile adhesive layers, which can be modelled with trapezoidal laws (Campilho et al. 2008, 2009). Under mixed-mode, energetic criteria are often used to combine pure modes I and II (2D analyses) or I, II and III (3D analyses), thus simulating the typical mixed-mode behaviour inherent to bonded assemblies.

In Damage Mechanics methodologies, a damage parameter is established to modify the constitutive response of materials by the depreciation of stiffness or strength, e.g. for thin adhesive bonds (Khoramishad et al. 2010), or composite delaminations or matrix failure (Daudeville and Ladevèze 1993), to represent the severity of material damage during loading. This state variable can be used in a damage evolution law to model both pre-cracking damage uptake and crack growth. By Damage Mechanics, the growth of damage is defined as a function of the load for static modelling (Raghavan and Ghosh 2005) or the cyclic count for fatigue analyses (Wahab et al. 2001, Imanaka et al. 2003). Generic Damage Mechanics models for bulk materials are available in the literature (Lemaitre and Desmorat 2005), and respective modifications for particular kinds of damage modelling, e.g. damage nucleation from voids and the formation of micro cracks (Gurson 1977, Tvergaard and Needleman 1984, Kattan and Voyiadjis 2005). Damage Mechanics has also been used to numerically model scenarios of constant and variable amplitude fatigue (Bhattacharya and Ellingwood 1998). For bonded joints, little work is published in this field. Compared to fatigue CZM, Damage Mechanics techniques do not provide a clear distinction between fatigue initiation and propagation phases, although they can give a basis for the predictive analysis (Khoramishad et al. 2010). Nonetheless, the evolution of damage prior to macro-crack growth can be simulated. On the other hand, damage modelling with fatigue CZM is restricted to pre-defined crack paths and, for specific applications, Damage Mechanics may be recommended if the damage is more widespread or if the failure path is not known (Shenoy et al. 2010).

The XFEM is a recent improvement of the FE method for modelling damage growth in structures. It uses damage laws for the prediction of fracture that are based on the bulk strength of the materials for the initiation of damage and strain for the assessment of failure, rather than the cohesive tractions and tensile/shear relative displacements used in CZM. XFEM gains an advantage over CZM modelling as it does not require the crack to follow a predefined path. Actually, cracks are allowed to grow freely within a bulk region of a material without the requirement of the mesh to match the geometry of the discontinuities neither remeshing near the crack (Mohammadi 2008). This method is an extension of the FE method, whose fundamental features were firstly presented in the late 90's by Belytschko and Black (1999). The XFEM relies on the concept of partition of unity, which can be implemented in the traditional FE method by the introduction of local enrichment functions for the nodal displacements near the crack to allow its growth and separation between the crack faces (Moës et al. 1999). Due to crack growth, the crack tip continuously changes its position and orientation depending on the loading conditions and structure geometry, simultaneously to the creation of the necessary enrichment functions for the nodal points of the finite elements around the crack path/tip. Works regarding the application of XFEM in bonded joints are very scarce.

1.3 BOOK LAYOUT

This book aims at providing a state-of-the-art assessment between the most relevant strength prediction tools for bonded joints, with all chapters written by internationally renowned specialists in the field. The proposed book is divided in 13 chapters, in which the relevant strength prediction tools for the strength prediction of adhesively-bonded joints are discussed in detail with examples of application. The current chapter, written by the book editor, aims to give an overview of the most relevant methodologies and enable the reader to focus the attention in the most suited technique/s for a given application, by presenting a critical comparison between the methods. Analytical methods are initially discussed in this book, with emphasis to the single-lap joint, including joints in fibre-reinforced composite structures, and also dynamic and impact loads. Numerical methods are the main subject of the book, beginning with the continuum mechanics approach for strength estimation. CZM are the most accepted and reliable method to analyse bonded joints and, because of this, 3 contributions are provided on this method: (1) static applications, (2) parameter identification techniques and accuracy issues and (3) fatigue applications. Damage mechanics is another possibility, also addressed here. Finally, the XFEM has been tested for application in bonded joints and it is also discussed in this book. Selected design applications are also presented: (1) use of CZM for strength prediction of adhesively-bonded joints and repairs and (2) CZM approach for the optimization of mixed adhesive joints for the aerospace industry. Each chapter is considered as a stand-alone document, such that it can be studied independently of the others. Whenever possible, references to review works are given for further details on each topic. Nevertheless, each chapter is documented with a significant amount of references that can be consulted for more detailed information regarding a specific topic.

1.4 REFERENCES

Adams, R. D. and Peppiatt, N. A. 1974. Stress analysis of adhesive-bonded lap joints. The Journal of Strain Analysis for Engineering Design 9: 185-196.

Adams, R. D., Comyn, J. and Wake, W. C. 1997. Structural Adhesive Joints in Engineering. Chapman & Hall, London.

Adams, R. D. 2008. Preface. In: L. F. M. da Silva and A. Öchsner (eds.). Modeling of Adhesively Bonded Joints. Springer, Heidelberg.

Ashcroft, I. A. 2011. Fatigue load conditions. pp. 845-874. In: L. F. M. da Silva, A. Öchsner and R. D. Adams (eds.). Handbook of Adhesion Technology. Springer, Berlin Heidelberg.

Belytschko, T. and Black, T. 1999. Elastic crack growth in finite elements with minimal remeshing. International Journal for Numerical Methods in Engineering 45: 601-620.

Bhattacharya, B. and Ellingwood, B. 1998. Continuum damage mechanics analysis of fatigue crack initiation. International Journal of Fatigue 20: 631-639.

Campilho, R. D. S. G., de Moura, M. F. S. F. and Domingues, J. J. M. S. 2008. Using a cohesive damage model to predict the tensile behaviour of CFRP single-strap repairs. International Journal of Solids and Structures 45: 1497-1512.

Campilho, R. D. S. G., de Moura, M. F. S. F., Pinto, A. M. G., Morais, J. J. L. and Domingues, J. J. M. S. 2009. Modelling the tensile fracture behaviour of CFRP scarf repairs. Composites Part B: Engineering 40: 149-157.

Courant, R. 1943. Variational methods for the solution of problems of equilibrium and vibrations. Bulletin of the American Mathematical Society 49: 1-23.

da Silva, L. F. M., das Neves, P. J. C., Adams, R. D. and Spelt, J. K. 2009a. Analytical models of adhesively bonded joints—Part I: Literature survey. International Journal of Adhesion and Adhesives 29: 319-330.

da Silva, L. F. M., das Neves, P. J. C., Adams, R. D., Wang, A. and Spelt, J. K. 2009b. Analytical models of adhesively bonded joints—Part II: Comparative study. International Journal of Adhesion and Adhesives 29: 331-341.

da Silva, L. F. M., Lima, R. F. T. and Teixeira, R. M. S. 2009c. Development of a computer program for the design of adhesive joints. The Journal of Adhesion 85: 889-918.

da Silva, L. F. M., Öchsner, A. and Adams, R. D. 2011a. Introduction to adhesive bonding technology. pp. 1-11. *In*: L. F. M. da Silva, A. Öchsner and R. D. Adams (eds.). Handbook of Adhesion Technology. Springer, Heidelberg.

da Silva, L. F. M., Öchsner, A. and Adams, R. D. (Eds.). 2011b. Handbook of Adhesion Technology. Springer, Heidelberg.

da Silva, L. F. M. and Campilho, R. D. S. G. 2012. Advances in Numerical Modelling of Adhesive Joints. Springer, Heidelberg.

da Silva, L. F. M. and Sato, C. 2013. Preface. *In*: L. F. M. da Silva and C. Sato (eds.). Design of Adhesive Joints under Humid Conditions. Springer, Heidelberg.

Daudeville, L. and Ladevèze, P. 1993. A damage mechanics tool for laminate delamination. Composite Structures 25: 547-555.

Dragoni, E., Goglio, L. and Kleiner, F. 2010. Designing bonded joints by means of the JointCalc software. International Journal of Adhesion and Adhesives 30: 267-280.

Feraren, P. and Jensen, H. M. 2004. Cohesive zone modelling of interface fracture near flaws in adhesive joints. Engineering Fracture Mechanics 71: 2125-2142.

Gurson, A. L. 1977. Continuum theory of ductile rupture by void nucleation and growth: Part I—Yield criteria and flow rules for porous ductile media. Journal of Engineering Materials and Technology 99: 2-15.

Hart-Smith, L. J. 1973. Adhesive-bonded double-lap joints. NASA Contract Report, NASA CR-112235.

Imanaka, M., Hamano, T., Morimoto, A., Ashino, R. and Kimoto, M. 2003. Fatigue damage evaluation of adhesively bonded butt joints with a rubber-modified epoxy adhesive. Journal of Adhesion Science and Technology 17: 981-994.

Kattan, P. I. and Voyiadjis, G. Z. 2005. Damage Mechanics with Finite Elements. Springer, Heidelberg.

Khoramishad, H., Crocombe, A. D., Katnam, K. B. and Ashcroft, I. A. 2010. Predicting fatigue damage in adhesively bonded joints using a cohesive zone model. International Journal of Fatigue 32: 1146-1158.

Kinloch, A. J. 1987. Adhesion and Adhesives: Science and Technology. Chapman & Hall, London.

Lemaitre, J. and Desmorat, R. 2005. Engineering Damage Mechanics. Springer, Heidelberg.

Moës, N., Dolbow, J. and Belytschko, T. 1999. A finite element method for crack growth without remeshing. International Journal for Numerical Methods in Engineering 46: 131-150.

Mohammadi, S. 2008. Extended Finite Element Method for Fracture Analysis of Structures. Blackwell Publishing, New Jersey.

Petrie, E. M. 2000. Handbook of Adhesives and Sealants. McGraw-Hill, New York.

Raghavan, P. and Ghosh, S. 2005. A continuum damage mechanics model for unidirectional composites undergoing interfacial debonding. Mechanics of Materials 37: 955-979.

Shahin, K. and Taheri, F. 2008. The strain energy release rates in adhesively bonded balanced and unbalanced specimens and lap joints. International Journal of Solids and Structures 45: 6284-6300.

Shenoy, V., Ashcroft, I. A., Critchlow, G. W. and Crocombe, A. D. 2010. Fracture mechanics and damage mechanics based fatigue lifetime prediction of adhesively bonded joints subjected to variable amplitude fatigue. Engineering Fracture Mechanics 77: 1073-1090.

Tong, L. and Luo, Q. 2011. Analytical approach to joint design. pp. 597-627. *In*: L. F. M. da Silva, A. Öchsner and R. D. Adams (eds.). Handbook of Adhesion Technology. Springer, Heidelberg.

Turner, M. J., Clough, R. W., Martin, H. C. and Topp, J. L. 1956. Stiffness and deflection analysis of complex Structures. Journal of the Aeronautical Sciences (Institute of the Aeronautical Sciences) 23: 805-823.

Tvergaard, V. and Needleman, A. 1984. Analysis of the cup-cone fracture in a round tensile bar. Acta Metallurgica 32: 157-169.

Volkersen, O. 1938. Die nietkraftoerteilung in zubeanspruchten nietverbindungen mit konstanten loschonquerschnitten. Luftfahrtforschung 15: 41-47.

Wahab, M. M. A., Ashcroft, I. A., Crocombe, A. D. and Shaw, S. J. 2001. Prediction of fatigue thresholds in adhesively bonded joints using damage mechanics and fracture mechanics. Journal of Adhesion Science and Technology 15: 763-781.

2
CHAPTER

Analytical Modelling for the Single-Lap Joint

Lucas F.M. da Silva[1,*], Marcelo Costa[2],
Guilherme Viana[2] and Raul D.S.G. Campilho[3]

2.1 INTRODUCTION

Adhesive joints are increasingly being used due to their improved mechanical performance and a better understanding of the mechanics of failure. To predict the joint strength, one must have the stress distribution and a suitable failure criterion. The stress distribution can be obtained by a finite element analysis (FEA) or a closed-form model. For complex geometries and elaborate material models, a FEA is preferable. However, for a fast and easy answer, a closed-form analysis is more appropriate. Adhesive joints have been intensively investigated over the past 70 years and numerous analytical models have been proposed. The design engineer is therefore confronted with a long list of models and has difficulty in finding the most appropriate one/model for a particular situation. The objective of this chapter is to review the analytical models available in the literature and discuss the conditions of applicability for each of them in order to facilitate their application.

This analysis starts with the simple classical analyses of Volkersen (1938) and Goland and Reissner (1944), discussing their limitations and describing developments from these analyses. Only lap joints with flat adherends are discussed, but there are analyses for other kind of joints such as those of Lubkin and Reissner (1955), Adams and Peppiatt (1977) and Nemeş et al. (2006) for tubular joints. Two-dimensional linear elastic analyses, two-dimensional elasto-plastic analyses, three-dimensional analyses and mixed adhesive joint (MAJ) analyses are also presented. A summary has been made (see Table 2.1) indicating for each model the assumptions made, the

[1] Departamento de Engenharia Mecânica, Faculdade de Engenharia, Universidade do Porto, Rua Dr. Roberto Frias, 4200-465 Porto, Portugal.
[2] INEGI – Pólo FEUP, Rua Dr. Roberto Frias, s/n, 4200-465 Porto, Portugal.
[3] Departamento de Engenharia Mecânica, Instituto Superior de Engenharia do Porto, Instituto Politécnico do Porto, Rua Dr. António Bernardino de Almeida, 431, 4200-072 Porto, Portugal.
* Corresponding author: lucas@fe.up.pt

Table 2.1 Summary of both linear and nonlinear two-dimensional analytical models available in the literature.

Analytical method	Material linearity						Adherends			Adhesive stresses			Solution	
	Adhesive		Adherend		Isotropic	Composite	Similar	Dissimilar		σ_x	σ_y	τ_{xy}	Closed-form	Numerical
	Linear	Nonlinear	Linear	Nonlinear				Thickness	Material					
Volkersen (1938)	X		X		X		X	X				X	X	
Goland and Reissner (1944)	X		X		X		X				X	X	X	
Wah (1973)	X		X		X	X	X	X				X	X	X
Hart-Smith (1973a, 1973b)	X	X	X		X		X				X	X	X	
Pirvics (1974)	X		X		X		X	X	X	X	X	X		X
Grimes and Greimann (1975)	X	X	X	X	X	X	X	X	X		X	X		X
Renton and Vinson (1975b)	X		X		X	X	X	X	X		X	X		
Srinivas (1975)	X		X		X	X	X	X	X	X	X	X	X	
Allman (1977)	X		X		X	X	X				X	X	X	

Table 2.1 contd....

Analytical method	Material linearity				Adherends					Adhesive stresses			Solution	
	Adhesive		Adherend		Isotropic	Composite	Similar	Dissimilar		σ_x	σ_y	τ_{xy}	Closed-form	Numerical
	Linear	Nonlinear	Linear	Nonlinear				Thickness	Material					
Ojalvo and Eidinoff (1978)	X		X		X		X		X		X	X	X	
Delale et al. (1981)	X	X	X		X	X	X	X	X		X	X	X	
Bigwood and Crocombe (1989)	X		X		X		X	X	X		X	X		X
Bigwood and Crocombe (1990)	X	X	X		X		X	X	X		X		X	
Cheng et al. (1991)	X		X		X		X	X	X	X	X	X		X
Crocombe and Bigwood (1992)	X	X	X	X	X		X	X	X		X			X
Adams and Mallick (1992)	X	X	X		X	X	X	X	X	X	X	X	X	
Tong (1996)	X	X	X		X		X			X	X	X	X	

Table 2.1 contd....

Analytical method	Material linearity				Adherends					Adhesive stresses			Solution	
	Adhesive		Adherend		Isotropic	Composite	Similar	Dissimilar		σ_x	σ_y	τ_{xy}	Closed-form	Numerical
	Linear	Nonlinear	Linear	Nonlinear				Thickness	Material					
Yang and Pang (1996)	X		X		X	X	X	X	X		X	X		X
Frostig et al. (1999)	X		X		X	X	X	X	X		X	X		X
Sawa et al. (2000)	X		X		X		X	X	X	X	X	X		X
Mortensen and Thomsen (2002)	X	X	X		X	X	X	X	X			X		
Adams et al. (1997)	X	X	X	X	X		X						X	
Wang et al. (2003)	X	X	X	X	X	X	X	X	X	X	X	X		X
Smeltzer and Klang (2003)	X	X	X	X	X		X	X	X		X	X		X
Zhao and Lu (2009), Zhao et al. (2014)	X		X		X		X				X	X	X	
Carbas et al. (2014)	X		X		X		X	X			X	X	X	
Luo and Tong (2009)	X	X	X	X	X	X	X	X	X		X	X		X
Sayman (2012)	X		X		X	X	X	X			X	X	X	

stresses they give, and the type of solution (algebraic or numerical). Various models of increasing complexity are compared in terms of time of implementation and accuracy (joint strength prediction) and a comparison between results for different methods is also presented.

This chapter focuses on stress–strain-based models but a mention should be made to fracture mechanics-based approaches (Fernlund et al. 1994, Ripling et al. 1971, Suo 1990). For example, the resistance to crack propagation in an adhesive bondline can be quantified in terms of the critical energy release rate, G_C. However, G_C cannot be considered a unique property of the bondline since it is dependent on the mode ratio of loading of the joint (Modes I, II and III). The plot of G_C as a function of the mode ratio for a given adhesive system is known as the fracture envelope (Fernlund and Spelt 1994). In one implementation of the fracture approach, an adhesive sandwich element is isolated from the surrounding joint structure as a free-body (Fernlund et al. 1994, Papini et al. 1994), in analogy with the sandwich stress–strain model proposed by Bigwood and Crocombe (1989), discussed later in this chapter. The reactions acting at the ends of the free-body sandwich can be determined using analytical or numerical techniques, depending on the complexity of the problem. If the loads in the substrates are less than what is required to yield the substrates, the applied energy release rate, G, and the mode ratio are calculated at the end of the bondline. For the calculated mode ratio, the critical energy release rate, G_C, is then obtained from the experimentally determined fracture envelope. If the calculated applied G is less than G_C, the joint will not fail, otherwise failure by crack propagation in the bondline will occur.

2.2 TWO-DIMENSIONAL LINEAR ELASTIC ANALYSES

2.2.1 Classical Analyses

2.2.1.1 Simplest Linear Elastic Analysis

The simplest analysis considers one of the most common joints that can be found in practice, the single-lap joint (SLJ). In this analysis, the adhesive is considered to deform only in shear and the adherends to be rigid. The adhesive shear stress (τ) is constant over the overlap length (l), as shown in Fig. 2.1, and is given by

$$\tau = \frac{P}{bl} \tag{2.1}$$

where P is the applied load and b is the joint width.

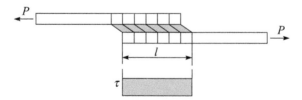

FIG. 2.1 Deformations in loaded SLJ with rigid adherends.

The value for τ can be interpreted as the average shear stress acting on the adhesive layer. This analysis is not very realistic due to many simplifications but is still the basis for quoting adhesive shear strength in many test situations such as ASTM and ISO standards.

2.2.1.2 Volkersen's Analysis

Volkersen's analysis (1938) introduced the concept of differential shear, illustrated in Fig. 2.2. It was assumed that the adhesive deforms only in shear but that the adherends can deform in tension, as can be seen in Fig. 2.3, because they are considered elastic and not rigid. The tensile stress in the upper adherend is maximum at A (see Fig. 2.2) and decreases to zero at B (free surface), so the strain must progressively reduce from A to B. The reduction of the strain in the adherends along the overlap and the continuity of the adhesive/adherend interface cause a non-uniform shear strain (and stress) distribution in the adhesive layer. τ is maximum at the ends of the overlap and is much lower in the middle, as shown in Fig. 2.16.

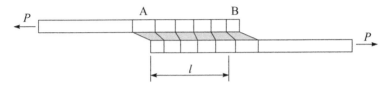

FIG. 2.2 Deformations in loaded SLJ with elastic adherends.

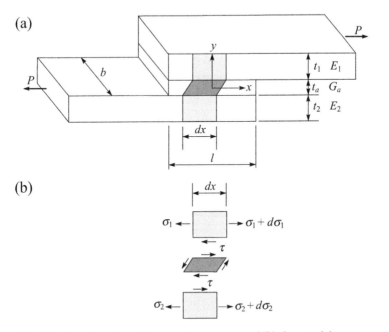

FIG. 2.3 SLJ analysed by Volkersen: (a) geometry and (b) elemental diagram.

The differential equation obtained by Volkersen is

$$\frac{d^2\sigma_2(x)}{dx^2} - \frac{G_a}{t_a}\cdot\left(\frac{1}{E_1 t_1} + \frac{1}{E_2 t_2}\right)\cdot\sigma_2(x) + \frac{P}{t_a b \cdot E_2 t_2 t_1}G_a = 0$$

The equation for τ is

$$\tau = \frac{P}{bl}\frac{w}{2}\frac{\cosh(wX)}{\sinh(w/2)} + \left(\frac{t_1 - t_2}{t_1 + t_2}\right)\frac{w}{2}\frac{\sinh(wX)}{\cosh(w/2)} \tag{2.2}$$

with $X = x/l$ where $-0.5 \leq X \leq 0.5$ and

$$w = \sqrt{\frac{G_a l^2}{E t_1 t_a}\left(1 + \frac{t_1}{t_2}\right)} \tag{2.3}$$

where t_t is the top adherend thickness, t_b the bottom adherend thickness, E the adherend modulus, G_a the adhesive shear modulus, and t_a the adhesive thickness. The origin of the longitudinal co-ordinate x is the middle of the overlap.

However, this analysis does not account for the bending effect caused by the eccentric load path of SLJ. The solution is more representative of a double lap joint (DLJ) than a SLJ since in a DLJ the bending of the adherends is not as significant as in the SLJ.

2.2.1.3 Goland and Reissner's Analysis

The eccentric load path of a SLJ causes a bending moment (M), and a transverse force (V) to be applied to the joint ends in addition to the applied tensile load per unit width ($\bar{P}$), as shown in Fig. 2.4. Because of this bending moment, the joint will rotate, altering the direction of the load line with the tendency of the applied tensile forces to come into line. As the joint rotates, the bending moment will decrease, giving rise to a nonlinear geometric problem where the effects of the large deflections of the adherends must be accounted for.

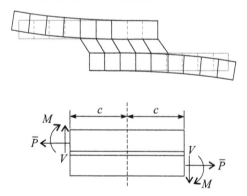

FIG. 2.4 Goland and Reissner's model.

The first to consider these effects were Goland and Reissner (1944). They used a bending moment factor (k) and a transverse force factor (k') that relate $\bar{P}$, M and V at the overlap ends, according to the following:

$$M = k\frac{\bar{P}t}{2} \qquad\qquad V = k'\frac{\bar{P}t}{2} \qquad\qquad (2.4)$$

where t is the adherend thickness ($t_1 = t_2$), and c is half the overlap length. If the joint does not rotate, i.e. for very small applied loads, the factors k and k' will be approximately equal to 1. As the joint rotates with the increase of load, k and k' will decrease and, consequently, the bending moment and the transverse load will decrease too. Goland and Reissner (1944) took into account the effect of large deflections of the adherends, but assumed that the adherends were integral, with an infinitely thin adhesive layer. Their expression for the bending moment factor is

$$k = \frac{\cosh(u_2 c)}{\cosh(u_2 c) + 2\sqrt{2}\sinh(u_2 c)} \qquad\qquad (2.5)$$

with

$$u_2 = \sqrt{\frac{3(1-v^2)}{2}}\frac{1}{t}\sqrt{\frac{P}{tE}} \qquad\qquad (2.6)$$

where v is the adherends Poisson's ratio.

Hart-Smith (1973b) took into account the effect of large deflections but considered the individual deformation of the upper and lower adherends in the overlap, thus not neglecting the adhesive layer. Hart-Smith (1973b) presented an alternative expression for Goland and Reissner's bending moment factor

$$k = \left(1 + \frac{t_a}{t}\right)\frac{1}{1 + \xi c + \frac{(\xi c)^2}{6}} \qquad\qquad (2.7)$$

where $\xi = \sqrt{\bar{P}/D}$ (D is the adherend bending stiffness).

Oplinger (1994) presented a more detailed analysis. Departing from the analysis of Goland and Reissner (1944), he took into account the effects of large deflections both outside and inside the overlap, considering also the individual deformation of the upper and lower adherends in the overlap. Oplinger (1994) found similar results to those of Goland and Reissner for large adherend to adhesive layer thickness ratios, and substantial differences for relatively thin adherends. More recently, Zhao et al. (2010) developed a simpler form of the bending moment factor that is accurate for thick and stiff adherends but has limitations for short overlaps. His version is much simpler and easier to use than those of Hart-Smith or Goland and Reissner, and is

$$k = \frac{1}{1 + \xi c} \qquad\qquad (2.8)$$

After the determination of the loads at the ends of the overlap, Goland and Reissner (1944) calculated τ and σ in the adhesive layer, solving a plane strain problem. Instead of solving a nonlinear geometric problem due to the eccentric load path, they solved a linear problem in the overlap with the loads applied at the ends. In this way, they avoided a more complex problem with the consideration of the geometric nonlinearity effect. The nonlinear geometric problem was solved by the determination of the loads at the ends of the overlap. Two limiting cases were considered for finding the adhesive stresses. In the first, the adhesive layer was assumed relatively inflexible and the overall overlap was treated as a single deformed body with the same material properties as the adherend. In the second, the adhesive was relatively flexible and the joint flexibility was mainly due to the adhesive layer. The first case is typically applicable to thick wood and plastic adherends, and the second is applicable to metal joints as in the case of aircraft structures. In the second case, the adherends were treated as cylindrically bent plates, in which the deformation of the adherends was due only to the longitudinal normal stress (σ_x). As a result of adherend bending, a transverse normal stress through the thickness direction will be induced in the adhesive layer, the so-called peel stress (σ). The adhesive layer was modelled as an infinite number of shear springs with an infinite number of tension/compression springs through the thickness direction, giving rise to shear and transverse direct stresses in the adhesive layer. The longitudinal direct stress in the adhesive layer was neglected. The adhesive layer thickness was considered to be negligible compared to the adherend thickness, so that the stress in the adhesive layer was assumed to be constant through the thickness. This second case is applicable to many metallic joints, provided they satisfy the following conditions

$$\frac{tG_a}{t_aG} < 0.1 \qquad\qquad \frac{tE_a}{t_aE} < 0.1 \tag{2.9}$$

where G is the adherend shear modulus, G_a the adhesive shear modulus and E_a the adhesive Young's modulus. The authors state that, for adhesive joints that satisfy these bounds, the adherend shear and transverse (through the thickness) deformations can be neglected compared with those in the adhesive layer. The distributions of the adhesive shear and peel stresses given by Goland and Reissner (1944) are illustrated in Fig. 2.16 for the second case, and the equations to determine the stresses are

$$\tau(x) = -\frac{\bar{P}}{8c}\left[\frac{\beta c}{t}(1+3k)\frac{\cosh\left(\dfrac{\beta x}{t}\right)}{\sinh\left(\dfrac{\beta c}{t}\right)} + 3(1-k)\right] \tag{2.10}$$

$$\sigma(x) = \frac{\bar{P}t}{\Delta c^2}\left(\left(R_2\lambda^2\frac{k}{2} + \lambda k'\cosh(\lambda)\cos(\lambda)\right)\cosh\left(\frac{\lambda x}{c}\right)\cos\left(\frac{\lambda x}{c}\right)\right.$$
$$\left. + \left(R_1\lambda^2\frac{k}{2} + \lambda k'\sinh(\lambda)\sin(\lambda)\right)\sinh\left(\frac{\lambda x}{c}\right)\sin\left(\frac{\lambda x}{c}\right)\right) \tag{2.11}$$

where

$$\beta = \sqrt{\frac{8G_a t}{Et_a}} \qquad\qquad \lambda = \frac{\gamma c}{t}$$

$$u_2 = \sqrt{\frac{3(1-v^2)}{2}}\frac{1}{t}\sqrt{\frac{P}{tE}} \qquad\qquad \gamma = \sqrt[4]{\frac{6E_a t}{Et_a}}$$

$$k = \frac{\cosh(u_2 c)}{\cosh(u_2 c) + 2\sqrt{2}\sinh(u_2 c)} \qquad k' = \frac{kc}{t}\sqrt{3(1-v^2)\frac{P}{tE}} \tag{2.12}$$

$$R_1 = \cosh(\lambda)\sin(\lambda) + \sinh(\lambda)\cos(\lambda)$$
$$R_2 = \sin(\lambda)\cos(\lambda) - \cosh(\lambda)\sin(\lambda)$$
$$\Delta = \frac{1}{2}(\sin(2\lambda) + \sinh(2\lambda))$$

In the comparison presented in Fig. 2.16, it can be seen that, for the same SLJ, Goland and Reissner (1944) and Volkersen (1938) give similar τ distributions although the Goland and Reissner solution predicts higher τ values at the ends of the overlap.

2.2.1.4 *Limitations of the Classical Analyses*

There is no doubt that the earlier work done mainly by Volkersen (1938) and Goland and Reissner (1944) was a big step forward in the stress analysis of adhesively bonded joints. Nevertheless, their work had several limitations.

1. They do not account for variations of the adhesive stresses through the thickness direction, especially the interface stresses which are important when failure occurs close to the interface.
2. The peak τ occurs at the ends of the overlap, which violates the stress-free condition, as shown in Fig. 2.5. Analyses that ignore the stress free condition overestimate the stress at the ends of the overlap and tend to give conservative failure load predictions.
3. Finally, the adherends were considered as thin beams, ignoring the through-thickness shear and normal deformations. Adherend shear is particularly important in shear-soft adherends such as composites.

These limitations were investigated by Ojalvo and Eidinoff (1978) and Tsai et al. (1998). Ojalvo and Eidinoff investigated the effect of t_a while Tsai et al. investigated the shear and normal deformations in the adherends.

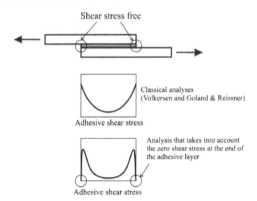

FIG. 2.5 Adhesive τ stress distribution when the stress free condition at the ends of the overlap is verified.

In the analysis of Ojalvo and Eidinoff (1978), the early work by Goland and Reissner (1944) was extended using a more complete shear strain/displacement equation for the adhesive layer to investigate the influence of t_a on the stress distribution. The adhesive τ stress was allowed to vary across the thickness, no matter how thin the adhesive may be, but the adhesive σ stress was maintained constant across the thickness. The adhesive longitudinal normal stress was neglected when compared with the adherend longitudinal normal stress. They concluded that the main difference between the theories that include and those that ignore t_a effects occurs at the ends of the overlap: the maximum τ stress increases and the σ stress decreases with the inclusion of this effect. The effect t_a is more significant with short overlaps, thick adherends and stiff adhesives. This analysis predicts that the peak τ stress occurs at the ends of the overlap, and the adhesive τ stress free condition at the free edge is not satisfied. As a comparison, Fig. 2.6 shows the results using both analytical methods for the ends of a sample overlap.

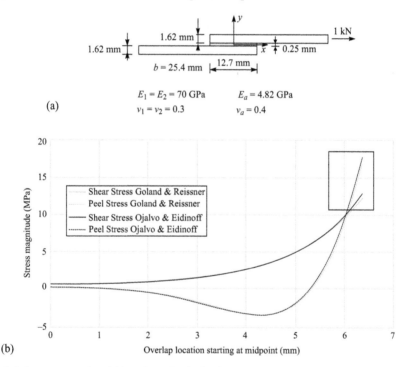

FIG. 2.6 Geometry analysed (a) and results for both Goland & Reissner and Ojalvo & Eidinoff methods (b).

Higher order theories, such as those of Allman (1977) and Chen and Cheng (1983), predict that peak τ stresses occur near the joint ends. Allman (1977) showed that the distance from the joint ends at which the peak adhesive shear stress occurs will depend on the relative flexibility between the adherends and the adhesive. Using a joint with a relatively flexible adhesive layer, Chen and Cheng (1983) concluded that the peak shear stress in the adhesive is not at the end of the overlap but at a distance of 20% the adherend thickness.

Normally, through-thickness (or transverse) shear and normal deformations in the adherends are neglected when compared to the higher value of the adherend longitudinal deformations, as in the case of the classical analyses. However, large adhesive shear and normal transverse stresses will be present at the adhesive/ adherend interfaces. For adherends with relatively low transverse shear and normal moduli, as is the case with laminated composite adherends, these stresses will cause large transverse shear and normal deformations in the adherends, close to the interfaces. Therefore, the adherend transverse shear and normal deformations should be included when laminated composite adherends are used.

Tsai et al. (1998) improved the classical works of Volkersen (1938) and Goland and Reissner (1944) for SLJ, and de Bruyne (1944) for DLJ, to account for the adherend shear deformation, which was assumed to vary linearly across the thickness. They showed that these improved solutions compared better with experimental results than the classical solutions, especially when laminated composite adherends are used. In spite of their limitations, the classical solutions are still a reference for new solutions and also for design purposes, which demonstrates the reason why they are still used. Moreover, the solutions are relatively simple and there is no need for great computer power.

2.2.2 Other Linear Analyses

After the so-called classical works, some authors tried to obtain more general closed-form solutions such as by including non-identical adherends (thickness and material properties) or composite adherends (Adams and Mallick 1992, Allman 1977, Cheng et al. 1991, Lee et al. 2005, Pirvics 1974, Renton and Vinson 1975a, 1975b, Sawa et al. 2000, Srinivas 1975, Wah 1973, Yang and Pang 1996, Yang et al. 1996, Zou et al. 2004). However, as the model gets more general, the governing equations become increasingly complicated and require the use of a computer for solution. There are two classes of solution on a computer. In one, a closed-form function is used which gives stress values, such as in the Goland and Reissner (1944) or Volkersen (1938) models. In the second, the differential equations are solved numerically. Table 2.1 indicates, for each model, how the solution is obtained. For example, the analysis of Pirvics (1974) is one of the most general analyses but requires a numerical solution. This analytical technique is based on the minimization of the internal energy in the longitudinal and transverse directions of an elastic body in the absence of body forces and thermal effects. With this minimization and with the boundary conditions, a set of two independent partial differential equations with two unknowns was obtained, for which a closed-form solution was not found. Therefore, a numerical analysis approach based on the finite difference method is needed. Also, with Pirvics' analysis (Pirvics 1974), plane stress/strain or axisymmetric problems can be analysed, and the longitudinal and transverse normal stresses and the shear stresses in the adherends and in the adhesive layer can be obtained.

2.2.2.1 Through-Thickness Shear and Normal Deformations

As previously mentioned, through-thickness (or transverse) shear and normal deformations in the adherends should be considered, especially when laminated composite adherends are present. The most important of the earlier analyses to account for these deformations were those of Renton and Vinson (1975a), Srinivas (1975) and Allman (1977). Renton and Vinson (1975a) and Srinivas (1975) performed similar analyses, where the adhesive stresses are constant across the thickness and the adhesive longitudinal normal stress is neglected. However, only the Renton and Vinson (1975a) model satisfies the adhesive shear stress free condition at the ends of the overlap. On the other hand, in Allman's analysis (Allman 1977), the adhesive peel stresses vary across the thickness, and it also satisfies the shear stress free condition at the ends of the overlap. As in the previous two analyses, the adhesive shear stress is assumed to be constant through the adhesive thickness and the adhesive longitudinal normal stress is also neglected.

Due to the increase in use of composite materials at that time, Renton and Vinson (1975a, 1975b) suggested that the analysis should take into account not only the anisotropic properties of composites, but also the laminated construction (anisotropic properties of each lamina and lamina fibre orientation). Using a composite laminated plate theory, they developed a linear elastic analysis contemplating two similar or dissimilar adherends for a SLJ, laminated or isotropic. The adherends were symmetric about their own midsurface and each lamina was orthotropic. In addition, thermal effects were also considered. This analysis resulted in two coupled, linear, fourth-order differential equations where twenty-six boundary conditions had to be satisfied to obtain the adhesive τ and σ stresses from a closed-form equation.

Srinivas (1975) noted that considering thin adherends with no transverse shear and normal stresses caused errors in the Goland and Reissner (1944) and Hart-Smith (1973b) results. He developed a refined elastic analysis in which these stress components were considered. Srinivas also considered both SLJ and DLJ. The thickness of the adherends was constant, tapered or stepped. The tapered adherends were idealized as stepped joints. The effect of large deflections in the joint was taken into account, considering a nonlinear geometrical problem. The lap joint was divided into the overlap region and the outer region. The overlap could be divided into more regions depending on the existence of steps or debonds. The governing equations for each region were solved separately and satisfied the boundary conditions at the ends of each region.

For the derivation of the governing equations, the joint was assumed to be in a state of either plane stress or plane strain and a unit width was chosen. The adherends were modelled using two-dimensional linear elasticity and could be either isotropic or a composite. This analysis can be used in balanced or unbalanced joints (different adherend thickness and/or material properties). The adhesive layer was represented by shear springs and tension/compression springs. The adhesive longitudinal normal stress was neglected. The modelling of the adhesive layer by linear springs did not allow through-thickness variation of the adhesive stresses. The governing equations were obtained by integrating the equations of equilibrium of a two-dimensional linear elastic problem and solved with the boundary conditions. This gave τ and σ stresses in both adhesive and adherends. The adherend longitudinal normal stress was obtained and was allowed to vary through the thickness.

Srinivas (1975) did an extensive parametric study of the effect of the transverse and shear deformation in the adherends, and methods of reducing the maximum

τ and σ stresses in the adhesive were investigated. He showed that neglecting the transverse and shear deformation in the adherends gave a good estimate of the maximum adhesive τ and σ stresses for long overlaps or flexible bonds in both longitudinal and thickness directions. To reduce the adhesive τ and σ stresses, he concluded that, for SLJ and DLJ, this can be done decreasing the adhesive modulus, tapering the adherends (although this affects more the adhesive peel stress than the adhesive shear stress) and using a MAJ (see Section 2.2.2.5). The MAJ would have a flexible adhesive at the ends of the overlap, which is a region of high stresses, and a stiff adhesive in the middle of the overlap, which is a region of low stresses. The model of Srinivas (1975) has the disadvantage of needing a numerical solution and ignoring important features such as the variation of the adhesive stresses through the thickness and the stress-free end condition.

Allman's elastic theory (Allman 1977) simultaneously included the effects of bending, stretching and shearing in the adherends, and shearing and tearing (σ stress) actions in the adhesive. He considered symmetric SLJ but he indicated that this method can be applied to other types of joints such as the DLJ. Isotropic and composite adherends were considered. However, for the case of composite adherends, laminated construction was not considered as it was by, for example, Renton and Vinson (1975a). Allman's bending moments and shear forces at the joint ends were calculated as in Goland and Reissner (1944). The adherends and adhesive stress distributions were expressed by stress functions that satisfy all the equations of equilibrium and the stress boundary conditions including the stress-free surface condition in the adhesive at the ends of the overlap. In the adherends, a state of plane strain was used, considering the longitudinal and transverse normal stresses and the shear stress. All these stresses vary across the adherend thickness. In the adhesive layer, only τ and σ stresses were considered. The τ stress was assumed to be constant through the adhesive thickness whereas the σ stress was allowed to vary linearly. Integrating the equations of equilibrium, the entire stress distribution in the joint was determined. The stress functions were obtained by minimizing the strain energy calculated from the equilibrium stress distribution in the joint. Allman also indicated that the incorporation of nonlinear material behaviour can be accomplished if appropriate modifications are made to the strain energy expression.

Allman's approach was also used by Chen and Cheng (1983), Cheng et al. (1991) and Adams and Mallick (1992). Chen and Cheng (1983) considered the geometric nonlinear characteristics of SLJ, as in Allman (1977), using the edge loads determined in the same way as Goland and Reissner (1944). If the adherends are similar, the loads at the two joint edges are equal, but as Cheng et al. (1991) concluded, for dissimilar adherends, the stress distribution is not symmetric and the τ and σ stress peaks at the two joint edges are different. Thus, Cheng et al. (1991) extended the determination of the edge loads from Goland and Reissner (1944) to account for dissimilar adherends. Instead of getting one bending moment factor (k), they obtained two (k_1 and k_2) giving different edge loads at the two overlap ends. For similar adherends, these two factors are equal to k.

Adams and Mallick's analysis (Adams and Mallick 1992) considered both SLJ and DLJ. The formulation for these two types of joint is similar. The adherends could be either isotropic or a composite. The composite adherends were limited to unidirectional composites. In addition, the adherends could be dissimilar with different material properties and/or different thicknesses. The adherends were elastic and the adhesive could deform plastically.

Comparing the analysis by Adams and Mallick (1992) with that by Cheng et al. (1991), the method used to determine the expressions for the stress distribution is similar, being described by two independent stress functions in the longitudinal coordinate and by the classical beam-plate theory of bending in the thickness direction. The longitudinal, peel, and shear stress distributions for each adherend and for the adhesive layer were determined in both works. One of the main differences between these two analyses is that Adams and Mallick (1992) allowed the stresses in the adhesive layer to vary through the thickness direction. The adhesive and adherend longitudinal normal stresses were assumed to vary linearly across the thickness to account for hygrothermal deformation in both adherends and adhesive.

The determination of the stress functions used by Adams and Mallick (1992) was achieved by defining the complementary energy in terms of these functions and then minimizing the energy function. They were not able to get a closed-form solution, so they used a numerical solution based on an equilibrium FEA approach, similar to the numerical solution suggested by Allman (1977). They concluded that, although it is necessary to use a computer implementation, this method is not disadvantageous compared to other closed-form solutions because many of these also require some form of computing power. This method has the advantage of considering adhesive plasticity and allows the variation of the adhesive stresses through the thickness. Adams and Mallick (1992) state that this type of approach has advantages over the displacement methods such as those presented by Renton and Vinson (1975a) and Delale et al. (1981) since their boundary conditions are easier to satisfy and thermal stresses are also included. In addition, the computational power required to get a solution is smaller than for these other methods.

Recently, Zhao and Lu (2009) and Zhao et al. (2014) published two analytical approaches for bonded SLJ. In the first model (Zhao and Lu 2009) the authors have managed to provide an explicit closed-form solution for elastic stresses by basing the analysis on a two-dimensional elasticity theory, which simultaneously includes the complete strain-displacement and the complete stress-strain relationships, and the strict compatibility conditions for the adherends and the adhesive layer. This method also fulfils the boundary stress conditions, including the zero shear stress conditions at the ends of the overlap. They studied the improvements of the formulation when compared with other similar approaches, and found an increase in accuracy from 4 to 13% when compared with the Goland and Reissner's (1944) model for a given geometry. Furthermore, good agreement between the model and FEA results was found, with the disadvantage that the deduced equations are more complex than the ones provided by Goland and Reissner's (1944). The analytical study by Zhao et al. (2014) employs the variational theorem of complementary energy, which allows the determination of σ and τ stresses by minimizing the energy functional. The same approach has been used in other analysis such as Allman (1977) and Chen and Cheng (1983), because it has several advantages, such as avoiding the use of integrals in the displacement method (leading to a simpler formulation) and allowing for thermal stresses to be evaluated (through complementary functions), which is an important aspect when designing bonded joints. The resulting plots of this second study (Zhao et al. 2014) are similar to those obtained in the first study (Zhao and Lu 2009), but with the advantage of allowing further customization due to the mentioned advantages of the variational method.

2.2.2.2 Evaluation of Laminated Composites

Wah (1973) was the first to consider laminated composite adherends. The laminated adherends were symmetrical about their midsurface. The adhesive τ stress was constant through the thickness whereas the adhesive σ stress was allowed to vary. This analysis can also be used for balanced and unbalanced joints. For the two cases, a set of two second-order differential equations is obtained. For unbalanced joints, it is necessary to solve an auxiliary problem to satisfy all the boundary conditions. However, the numerical results show that the corrections introduced by the auxiliary problem are quite negligible. Therefore, in most cases, the solution obtained by the main problem is entirely adequate. Neglecting the auxiliary problem, the computational effort is smaller and the extension of this analysis to the plastic regime is simpler.

The previous analyses were limited to composite adherends that are symmetrical about their midsurface. The case of asymmetric composite adherends in balanced or unbalanced joints was considered more recently by Yang and Pang (1996) and Mortensen and Thomsen (2002) where the coupling effect of the external tensile loading and the induced bending moment due to the asymmetry of the composite laminates is considered. Yang and Pang (1996) considered a SLJ divided into three zones, two of which were outside the overlap and the other being the overlap itself. The adherend behaviour was described by a first-order laminated anisotropic plate theory. The adhesive σ and τ stresses were assumed to be constant through the thickness and the stress free condition at the overlap end was not satisfied. The adhesive longitudinal stress was neglected. A system of six, coupled, second-order ordinary differential equations was obtained for the governing equations in the overlap. This system of equations was solved using a Fourier series. A similar and simpler system of equations was obtained for the two zones outside the overlap. The results were found to be in very close agreement with a FEA model. Yang and Pang (1996) concluded that the use of asymmetric laminates can provide more flexibility in design, although it is more difficult to manufacture asymmetric composite laminates.

In the analysis of Mortensen and Thomsen (2002), the consideration of asymmetric composite adherends in balanced or unbalanced adhesive joints was done by modelling the adherends using classical laminate theory, assuming beams or wide plates in cylindrical bending, and obeying the linear elastic constitutive laws presented by Whitney (1987). This analysis can be used in most types of adhesive joint, particularly SLJ and DLJ, where load and boundary conditions can be chosen arbitrarily. The adhesive layer was modelled in two ways, with linear and nonlinear behaviour (see Section 3). In the first case, the adhesive was assumed to be a homogeneous, isotropic and linear elastic material, modelled as continuously distributed linear tension/compression and shear springs. The thickness of the adhesive layer was assumed to be small compared with the thickness of the adherends, so the adhesive peel and shear stresses do not vary across the adhesive thickness. The use of this spring model has the consequence of not satisfying the condition of zero stress at the ends of the overlap, as in Goland and Reissner's analysis (1944). The authors minimized this limitation by saying that it can be seen as an approximation of the spew fillets formed in real joints which have the capability of transferring τ stress. The analysis results in a set of first order ordinary differential equations that are solved numerically using the 'multi-segment method of integration' since no general closed-form solution is obtainable with the boundary conditions. The 'multi-segment method of integration' consists of dividing the original problem into a finite number of segments, where the solution for each segment can be obtained by direct integration.

More recently, Luo and Tong (2009) studied composite SLJ and determined new equilibrium equations based on a nonlinear geometrical analysis that are in excellent agreement with results from commercial FEA software MSC/NASTRAN. Furthermore, they also modified the effective E_a to account for the lower transverse moduli present in composite adherends, and the developed simplified equations provide solutions that are adequately accurate when compared to the full solutions. Their solutions are of higher accuracy (but also take longer to compute) than the analysis of Sayman (2012), who did a simplified study for a composite SLJ bonded with a ductile adhesive, where both the bending moment and the shear stress variation across the thickness of the adhesive were neglected. This naturally leads to a simplified analysis (similar to Volkersen (1938)) but does account for both the composite adherends' properties and an elasto-plastic analysis besides the elastic formulation. The disadvantages is that σ stresses cannot be obtained, due to neglecting the bending moment, which is a very important factor when studying composite materials because the adherends are usually very weak when solicited in peel due to the inexistence of fibres in that direction.

2.2.2.3 *Simple Analyses to Carry Out Preliminary Estimates of Adhesive Stresses*

Making a review of previous analyses (Allman 1977, Goland and Reissner 1944, Hart-Smith 1973b, Ojalvo and Eidinoff 1978, Renton and Vinson 1975a, Volkersen 1938), Bigwood and Crocombe (1989) concluded that most of them considered only one type of joint configuration, i.e. SLJ or DLJ. They therefore attempted to create a general elastic analysis that permitted the analysis of various configurations of adhesive joints under complex loading, consisting of tensile and shearing forces and a bending moment at the ends of the adherends. They modelled the overlap region as an adherend-adhesive sandwich (Fig. 2.7) that permits the analysis of any configuration that can be simplified to this form.

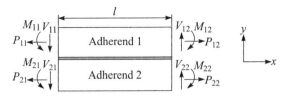

FIG. 2.7 Bigwood and Crocombe's diagram of adherend-adhesive sandwich under general loading.

In the derivation of this general analysis, the adherends were considered as cylindrically bent flat plates connected by an adhesive layer that transferred the load from the adherends through both peel and shear. Only isotropic adherends with constant thickness were considered, although the adherends can be of different thickness and material properties. Transverse shear and normal deformation in the adherends were not considered. The variations of σ and τ stresses in the adhesive through the thickness were neglected to facilitate the introduction of material non-linearity. Therefore, the adhesive τ stress free condition at the ends of the overlap was not satisfied. In addition, the longitudinal direct stress in the adhesive was neglected. The full elastic analysis calculated the distribution of the adhesive σ and τ stresses in the overlap region by using two uncoupled seventh and sixth order differential equations.

In addition to this general analysis, Bigwood and Crocombe (1989) produced two simplified, two-parameter design formulae that accurately determined the adhesive σ and τ stress peaks at the ends of the overlap. For similar adherends, the results yield exact values in relationship to the general analyses but there are limitations for dissimilar adherends. In any case, the formulae provide a simple initial estimate of joint strength. To facilitate the analysis to produce design formulae for determining the peak peel and shear stresses, the adhesive stresses (peel and shear) were uncoupled. The resulting simplified equations for determining the peel and shear stresses are

$$\sigma = A_1 \cos(K_1 x)\cosh(K_1 x) + A_2 \cos(K_1 x)\sinh(K_1 x) + A_3 \sin(K_1 x)\cosh(K_1 x)$$
$$+ A_4 \sin(K_1 x)\sinh(K_1 x) \tag{2.13}$$

$$\tau = C_1 \cosh(K_2 x) + C_2 \sinh(K_2) + C_3$$

The necessary variables are defined as

$$D_1 = \frac{E_1 t_1^3}{12(1-v_1^2)} \quad D_2 = \frac{E_2 t_2^3}{12(1-v_2^2)}$$

$$K_1 = \sqrt[4]{\frac{E_a}{4t_a}\left(\frac{1}{D_1} + \frac{1}{D_2}\right)} \quad K_2 = \sqrt{\frac{4G_a}{t_a}\left(\frac{1-v_1^2}{E_1 t_1} + \frac{1-v_2^2}{E_2 t_2}\right)}$$

$$A_1 = \frac{b_3 R_3 - 2b_2 \sinh(K_1 L)\sin(K_1 L) + b_1 R_6 + b_4 R_1}{R_5}$$

$$A_2 = \frac{b_2 R_2 - b_3 \sinh^2(K_1 L) - b_1 R_4 - b_4 \sinh(K_1 L)\sin(K_1 L)}{R_5}$$

$$A_3 = \frac{b_2 R_2 - b_3 \sin^2(K_1 L) - b_1 R_4 - b_4 \sinh(K_1 L)\sin(K_1 L)}{R_5}$$

$$A_4 = b_1$$

$$B_1 = \frac{b_2 - b_1 \cosh(K_2 L)}{K_2 \sinh(K_2 L)}$$

(2.14)

$$B_2 = \frac{b_1}{K_2}$$

$$B_3 = \frac{b_3}{L} - \frac{b_2 - b_1}{K_2^2 L}$$

$$b_1 = \frac{E_a}{2K_1^2 t}\left(\frac{M_{21}}{D_2} - \frac{M_{11}}{D_1}\right) \qquad b_2 = \frac{E_a}{2K_1^2 t}\left(\frac{M_{22}}{D_2} - \frac{M_{12}}{D_1}\right)$$

$$b_3 = \frac{E_a}{2K_1^3 t}\left(\frac{V_{21}}{D_2} - \frac{V_{11}}{D_1}\right) \qquad b_4 = \frac{E_a}{2K_1^3 t}\left(\frac{V_{22}}{D_2} - \frac{V_{12}}{D_1}\right)$$

$$R_1 = \cosh(K_1 L)\sin(K_1 L) - \sinh(K_1 L)\cos(K_1 L)$$

$$R_2 = \cosh(K_1 L)\sin(K_1 L) + \sinh(K_1 L)\cos(K_1 L)$$

$$R_3 = \cosh(K_1 L)\sinh(K_1 L) - \cos(K_1 L)\sin(K_1 L)$$

$$R_4 = \cosh(K_1 L)\sinh(K_1 L) + \cos(K_1 L)\sin(K_1 L)$$

$$R_5 = \sinh^2(K_1 L) - \sin^2(K_1 L)$$

$$R_5 = \cosh^2(K_1 L) - \cos^2(K_1 L)$$

E is the Young's modulus, t is the thickness, v is Poisson's ratio, D is the flexural rigidity, G is the shear modulus, L is the length of the overlap, M is the bending moment, V is the shear stress, and the subscripts 1, 2 and a represent adherend 1, adherend 2 and the adhesive, respectively.

2.2.2.4 Interface Stresses

Although some of the previous works, such as Adams and Mallick (1992) for example, obtained the stresses at the adherend/adhesive interface, Sawa and his co-workers have done an extensive analysis of these stresses. Sawa et al. (1997) analysed a SLJ with similar isotropic adherends subjected to tensile loads, as in a three-body contact problem. The two-dimensional theory of elasticity was used to evaluate the contact stress distribution at the adherend/adhesive interface. The numerical results showed that by decreasing both t and E of the adherends, the interface shear stress near the overlap ends increased. Sawa et al. (2000) extended the previous work to account for dissimilar isotropic adherends. The Airy stress function was used to express the stress and displacement components, and the boundary conditions permitted the analysis of the three finite strips (top adherend, adhesive, bottom adherend). For each strip, they considered the longitudinal normal stress, the transverse normal stress, and the longitudinal shear stress. The adhesive τ stress free condition at the ends of the overlap was satisfied.

The effects of some important joint parameters were investigated. They found that at the ends of the overlap, the interface stresses increased sharply, indicating that a stress singularity is present. They concluded that when the adhesive thickness is small enough, this singular stress increases as t_a decreases. On the other hand, when t_a is large enough, this singular stress also increases as t_a increases. Therefore, there is a t_a value for which the strength of the singularity is minimum, implying that there is an optimum t_a in terms of joint strength.

Most analytical methods show an improved strength (the stresses decrease as t_a increases) with thicker bondlines. For instance, in the study by Srinivas (1975), the stresses decrease with increase of t_a. However, in practice, the adhesive lap joint strength decreases as the glue line gets thicker. Adams and Peppiatt (1974) tried to explain the discrepancy by saying that thicker bondlines contain more defects such as voids and micro cracks. Another possible reason is the fact that thicker bondlines cause more adherend bending because of load misalignment. The effect of t_a on the lap shear strength can also be explained by interface adhesive-adherend stresses (da Silva et al. 2006, Gleich et al. 2001). Therefore, for realistic failure load predictions, an analytical model should include the adhesive stress variations through the thickness, including the interface stresses. For example, in the case of concrete beams strengthened with a fibre-reinforced polymer (FRP) plate, the most common failure modes are debonding of the FRP plate or tensile fracture of the concrete. Both of these premature failure modes are caused by interfacial stress concentrations in the adhesive layer. Tounsi and Benyoucef (2007) give an analytical solution of the interface stresses in concrete beams which are externally plated with FRP.

2.2.2.5 Mixed Adhesive Joints

The first to introduce the idea of using MAJ to increase joint strength by minimizing the maximum adhesive stresses was Raphael (1966). High modulus adhesives develop high stress concentrations at the ends of the overlap. These stress concentrations can be reduced with the use of a flexible adhesive at the overlap ends. Few closed-form analyses have considered the case of MAJ, where more than one adhesive is used in the overlap. Only two analyses have been found in the literature: Raphael (1966) and Srinivas (1975). Hart-Smith (1973b) also refers this type of joint but only in qualitative terms.

According to Raphael (1966), an ideal joint is one in which the adhesive flexibility and strength properties vary along the overlap. He assumed that failure in the adhesive will occur when the adhesive shear strain exceeds a certain value. Because of greater adhesive shear strains at the edges of the overlap, a ductile and flexible adhesive should be used at the overlap ends (adhesives 1 and 3 in Fig. 2.8), while in the middle a stiff and less ductile adhesive should be used (adhesive 2 in Fig. 2.8). A flexible adhesive is usually less strong than a stiffer adhesive but tolerates a higher strain to failure. Raphael's philosophy was therefore to select the strongest adhesive, having in mind the deformation that it will have to sustain. As a result, the joint will have maximum strength and stiffness in the middle, and maximum flexibility and ductility at the ends. In terms of stress distribution, the maximum τ stress in the stiff adhesive (middle) will be lower than the maximum τ stress in a joint with a rigid adhesive through the entire overlap. Raphael's idea of an optimum joint is a joint where both adhesives are at their ultimate strength when the joint fails. In this way, the joint has the maximum possible strength since the adhesives are at their maximum load bearing capacity. An alternative concept to the MAJ is to profile the adherends or to vary the glue line thickness (Adams et al. 1973, Cherry and Harrison 1970, Erdogan and Ratwani 1971, Helms et al. 1997, Oterkus et al. 2006, Reddy and Sinha 1975, Thamm 1976). The analytical model of Raphael is rather simplistic since it only looks at the τ stress of the adhesive and ignores several important aspects such as the stress free condition and the variation of the adhesive stress through the thickness. A more complete model is necessary in order to make more realistic parametric studies.

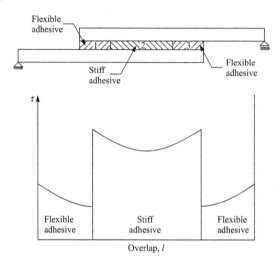

FIG. 2.8 MAJ and schematic adhesive τ stress distribution.

Srinivas (1975) considered MAJ (SLJ and DLJ) with a stiff adhesive in the middle of the overlap and a flexible adhesive at the ends of the overlap as an alternative to joints with a flexible or a stiff adhesive alone. Srinivas (1975) studied the effect of the bond length corresponding to the flexible and stiff adhesives, while keeping l constant, on the maximum adhesive τ and σ stresses. He concluded that by using a MAJ with a long overlap for the stiff adhesive and, consequently, a small overlap for the flexible adhesive, the maximum τ and σ stresses in both adhesives can be reduced in comparison with joints with a stiff or a flexible adhesive alone. On the other hand, if the overlap of the stiff adhesive is small, the stresses in the flexible adhesive are no longer lower than those in a joint with the flexible adhesive alone. The overlap of the stiff adhesive has to be longer than that for the flexible adhesive.

As the overlap of the stiff adhesive increases, the maximum stresses in the stiff adhesive increase and the stresses in the flexible adhesive decrease. There is an optimum ratio between the stiff adhesive overlap and the flexible adhesive overlap corresponding to the minimum stresses in the adhesives.

2.2.2.6 Graded Adhesive Joints

An adhesive joint made with a graded adhesive can be considered a more refined version of a MAJ. Instead of using two adhesives with different properties, a single adhesive, whose properties vary continuously along l, can be used. This way, in order to improve the joint strength, the adhesive must be ductile and flexible at the ends of the overlap and rigid and strong at the middle of the overlap. As some adhesives change their mechanical properties depending on the cure temperature, if the cure temperature is allowed to vary along l, it will create a graded adhesive joint (Carbas et al. 2014). Another way to create a graded adhesive joint is to add particles to the adhesive. If the concentration of particles varies along l, the mechanical properties will also vary, thus creating a graded adhesive joint (Stapleton et al. 2012).

Carbas et al. (2014) developed a simple analytical model that is capable of predicting the strength of functionally graded adhesive joints. In this model, which is based on the Volkersen model, it is assumed that G_a changes with l. The modified differential equation is

$$\frac{d^2\sigma_2(x)}{dx^2} - \frac{G_a(x)}{t_a} \cdot \left(\frac{1}{E_1 t_1} + \frac{1}{E_2 t_2} \right) \cdot \sigma_2(x) + \frac{P}{t_a b \cdot E_2 t_2 t_1} G_a(x) = 0 \qquad (2.15)$$

Because this analysis is based on the Volkersen model, both the adherends and the adhesive are elastic. The model may be solved using a power series approach and assuming a simple elasto-plastic model, making it possible to predict the strength of functionally graded adhesive joints.

2.2.2.7 Spew Fillet

The two-dimensional linear elastic analysis of Frostig et al. (1999) is an extension of their previous work on the analysis of sandwich panels with a transversely flexible or stiff core, with a closed-form high-order (CFHO) theory (Frostig et al. 1992). The principle of virtual displacements, a variational principle, was used to derive the governing equations, the boundary conditions, and the continuity requirements. The CFHO also has the advantage of modelling the stress free condition at the ends of the overlap. The adhesive τ stress was considered constant through the adhesive thickness and the σ stress was allowed to vary through the thickness. The adherends were modelled as linearly elastic thin beams or panels (wide beams) obeying the Bernoulli–Euler assumptions. The stress and deformation fields were uniform across the width. The adherends could be either metal or laminated composites. The shear and transverse normal (through thickness) deformations in the adherends were neglected.

Frostig et al.'s analysis (1999) was the only analytical model that considers the spew fillet. The spew fillet is a surplus of adhesive that results from the manufacturing process that is 'squeezed out' at the ends of the overlap. Real joints are always associated with this surplus material. Adams and Peppiatt (1974) showed with a FEA that the spew fillet is beneficial because it reduces the adhesive τ and σ stresses at the ends of the overlap. The spew fillet was modelled by Frostig et al. (1999) using two approaches, having in mind that when there is a spew fillet the adhesive stress free condition is no longer valid or needed. In the first approach, the adhesive transverse displacement (through its thickness) was equated to the relative transverse displacements of both upper and lower adherends. In the second, the spew fillet was modelled as an inclined equivalent elastic bar with in-plane longitudinal stiffness only, as shown in Fig. 2.9.

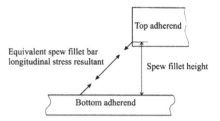

FIG. 2.9 Equivalent spew fillet bar.

2.2.2.8 Bolted-Bonded Joints

There is interest in the study of hybrid bolted-bonded joints that combine the advantages of adhesively bonded joints with the ones of bolted joints. The analysis of such configurations is usually left to FEA because of the complex interactions between the different materials and also the inherent three dimensional analysis requirement (due to the bolted area also having important influence in the width direction). But because of the considerable computing time such an analysis would require, attempts have been made to develop analytical methods for these type of joints (Barut and Madenci 2009, Bois et al. 2013, Paroissien et al. 2007).

Paroissien et al. (2007) developed a semi-analytical two-dimensional model based on the matrix displacement method, the same basis as the one used in FEA. They proposed an elastic formulation that was able to simulate joints made with isotropic adherends with up to three fasteners, and obtained a good agreement between their results and a FEA model. The approach was to formulate a bonded element that took into consideration the boundary conditions in each element due to the fasteners, and then incorporated the element formulation in a MATLAB code together with the rest of the structure. Barut and Madenci (2009) followed a similar approach, but with higher complexity and for composite adherends. By considering the kinematics of the composite laminates, the adhesive and the bolts, the contacts between the bolt/hole and bolt head/laminate, and appropriate boundary conditions, they were able to define equations that provide τ and σ solutions very similar to those obtainable by the FEA analysis. Furthermore, the proposed model allows the combined consideration of vertical distributed loadings in the adherends besides the usual uniaxial solicitations present in SLJ, which has impact in the contact between the different materials of the joint.

Bois et al. (2013) took a fully analytical, although one dimensional, approach which departs from the matrix based method of both the previous models. They divide the overlap in $n + 1$ sections, where n is the number of bolts present in the overlap, define the equilibrium conditions for each section, and using the material constitutive equations of the adhesive, adherends and bolts they formulate a system of $2n + 2$ equations with the same number of unknowns, which can then be solved to obtain the adhesive τ stress. The adhesive τ stress equations has the simple form of

$$\tau_i(x) = \frac{\eta}{w}(-A_i e^{-\eta x} + B_i e^{\eta x}) \tag{2.16}$$

where τ_i is the shear stress in section i, η and w are material and geometric parameters, and A_i and B_i are determined by solving the system of $2n + 2$ equations. Besides the assumption of linear behaviour for both the adhesive and bolts, the model also enables a linear elastoplastic behaviour for the adhesive to be incorporated in the form of a bilinear law that takes into account a plastic shear modulus, the shear yield stress and shear yield strain of the adhesive. The model resolution of the elastoplastic behaviour is based on an iterative scheme to determine the abscissa of the overlap where the shear stress is equal to the adhesive shear yield stress, and has the advantage of also allowing nonlinear bolt stiffness to be incorporated into the model.

2.2.2.9 *Joints Under Impact Loads*

The prediction of the mechanical behaviour of adhesive joints under impact loads has been increasingly gaining importance, mainly due to the demand of impact resistance by the automotive industry, where passenger safety in crash situations is mandatory. For the prediction of impact response, the usual approach is to consider the strain rate dependence and apply it to the results, for example assuming that impact loads correspond to a strain rate of 1000 mm/s.

Zgoul and Crocombe (2004) modelled the rate-dependency of adhesively bonded structures, and found that a rate-dependent von Mises model presented good accuracy on the prediction of the response of SLJ. For the purpose of predicting strain rate dependency, they assumed a base stress-strain curve (for static conditions), and then modelled the new higher strain rate curve using the following relationship for strain

$$\varepsilon_{new} = \varepsilon_{base} + \left[\frac{A}{E}\ln\left(\frac{\dot{\varepsilon}_{p\,new}}{\dot{\varepsilon}_{p\,base}}\right)\right] \tag{2.17}$$

and the stress at any particular point using:

$$\sigma = A\ln(\dot{\varepsilon}_p) + B \tag{2.18}$$

where A and B are constants found from experimental data, E is the Young's modulus, $\dot{\varepsilon}_p$ the plastic strain rate and ε the strain. Very good agreement between this proposed model and experimental tests for 1 mm/min, 10 mm/min and 100 mm/min was obtained, suggesting that the model can be used, even if for only roughly predicting, the behaviour of the SLJ at 1000 mm/min.

Sato performed two distinct studies on the effect of impact loads on SLJ. The first (Sato and Ikegami 2000) studied the dynamic stress response of the joints when subjected to a step loading of a specified intensity. The shear stress values and oscillations throughout the joint as a function of time were modelled, and it was found that immediately after applying the step loading there were stress waves in the overlap region while the ends recorded the highest stress values. Sometime after the applied step load occurred the stresses in the overlap region reached zero while the ends maintained the stresses present in the beginning. This is expected, because applying a step load creates some instability in the beginning that is eliminated when the step load is continuously applied. The second study (Sato 2009) is based on a modified Volkersen model, where the governing equations were changed by adding the inertia force to the static formulation. This allowed for the visualization of the propagation of the stress waves through the overlap as a function of time, and also that in the moment of impact the registered stresses where higher than the value of the applied impact stress, signalling that extra care should be taken when designing joints for impact as the stresses may be higher than expected. A MATLAB code was developed for the semi-analytical formulation and published in another work (Sato 2008), which can be implemented by anyone and used to model the dynamic response of any SLJ configuration, including adherends with dissimilar materials and thicknesses.

2.3 TWO-DIMENSIONAL ELASTO-PLASTIC ANALYSES

In the previous Section, the analyses assumed elastic behaviour for the adhesive layer as well as for the adherends. These analyses are accurate enough for brittle adhesives because they have little or no plastic strain. When adhesives having a large plastic strain to failure are used, such as rubber-modified epoxies, the adhesive plasticity must be included in order to correctly simulate the stress and strain distributions when the adhesive yields. Adherends can yield too, and the analysis needs to account for this behaviour if realistic failure loads are to be predicted.

Material nonlinearity due to plastic behaviour is not often included because of the increased complexity in the mathematical formulation. Normally, nonlinear material behaviour is only implemented in FEA. Most of the analyses presented next assume plastic behaviour only for the adhesive layer.

2.3.1 Hart-Smith's Analysis

One of the most important works considering adhesive plasticity was done by Hart-Smith for SLJ (Hart-Smith 1973b) and DLJ (Hart-Smith 1973a). Hart-Smith developed the analyses of Volkersen (1938) and de Bruyne (1944) for DLJ (Hart-Smith 1973a) and the work done by Goland and Reissner (1944) for SLJ (Hart-Smith 1973b), for which he presented an alternative expression for the bending moment factor.

Hart-Smith's solutions accounted for adhesive plasticity, using an elastic–plastic τ stress model. He also included adherend stiffness imbalance and thermal mismatch. If adhesive plasticity is considered, the joint strength prediction is more accurate than with an elastic analysis. The maximum lap-joint strength was calculated by using the maximum shear strain as the failure criterion. Any stiffness differences between the adherends result in a decrease of the joint strength. To characterize the adhesive behaviour, Hart-Smith chose an elasto-plastic model (see Fig. 2.10) such that the ultimate shear stress and strain in the model are equal to the ultimate shear stress and strain of the adhesive, the two curves having the same strain energy. He showed that any adhesive model defined by two straight lines that have the same failure stress and strain and the same strain energy predicts the same maximum joint strength developed between uniform adherends.

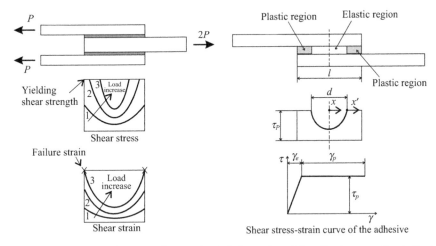

FIG. 2.10 Schematic explanation of shear plastic deformation of the adhesive according to Hart-Smith (1973a).

Any thermal mismatch between adherends decreases the joint strength and this reduction is more significant with increase of adherend thickness and stiffness. The equations require an iterative solution and describe the τ stress and strain distributions for the elastic and plastic regions in the overlap. The formulation is divided in two parts: an elastic analysis and a plastic analysis for the τ stress of the adhesive. For the elastic formulation, the equations are straightforward

$$\tau = A_2 \cosh(2\lambda'x) + C_2$$
$$\sigma = A \cosh(\alpha x) \cos(\alpha x) + B \sinh(\alpha x) \sin(\alpha x) \qquad (2.19)$$

where the constants are defined as:

$$\lambda' = \sqrt{\frac{1+3(1-v^2)/k_b}{4}}\,\lambda \qquad \xi = \sqrt{\overline{P}/D} \qquad D = \frac{Et^3}{12(1-v^2)}$$

$$M = P\left(\frac{t+t_a}{2}\right)\frac{1}{1+\xi c + (\xi^2 x^2/6)} \qquad \alpha^4 = \frac{E_a}{2Dt_a}$$

$$k_b = \frac{D}{Et^3/[12(1-v^2)]} \qquad \lambda = \frac{2G_a}{t_a Et}$$

$$A_2 = \frac{G_a}{t_a Et}\left(\bar{P} + \frac{6(1-v^2)M}{t}\right)\frac{1}{2\lambda' \sinh(2\lambda' c)} \tag{2.20}$$

$$C_2 = \frac{1}{2c}\left(\bar{P} - \frac{A_2}{\lambda'}\sinh(2\lambda' c)\right)$$

$$A = -\frac{E_a M(\sin(\alpha c) - \cos(\alpha c))}{t_a D\alpha^2 e^{\alpha c}}$$

$$B = \frac{E_a M(\sin(\alpha c) + \cos(\alpha c))}{t_a D\alpha^2 e^{\alpha c}}$$

But Hart-Smith also considered adhesive shear stress plasticity, keeping the peel stress elastic. The shear stress is modelled using a bi-linear elastic-perfectly plastic approximation. The overlap is divided into three regions: a central elastic region of length d and two outer plastic regions. The overlap length is l and, for a balanced lap joint, both nonlinear regions have length $(l-d)/2$. Co-ordinates x and x' are defined in these regions, as shown in Fig. 2.10. The problem is solved in the elastic region in terms of the τ stress according to

$$\tau = A_2 \cosh(2\lambda' x) + \tau_p(1-K) \tag{2.21}$$

and in terms of the shear strain, γ, in the plastic region according to

$$\gamma = \gamma_e(1 + 2K[(\lambda' x')^2 + \lambda' x' \tanh(\lambda' d)]) \tag{2.22}$$

A_2, used in τ, is defined as

$$A_2 = \frac{K\tau_p}{\cosh(\lambda' d)} \tag{2.23}$$

where τ_p is the plastic adhesive shear stress. K and d are solved by an iterative approach using the following equations

$$\frac{P}{l\tau_p}\lambda' l = 2\lambda'\frac{l-d}{2} + (1-K)\lambda' d + K\tanh(\lambda' d)$$

$$\frac{P}{\tau_p}\lambda^2\frac{l-d}{2}\left[1 + \frac{3k(1-v^2)}{k_b}\left(1 + \frac{t_a}{t}\right)\right] = 2\frac{\gamma_p}{\gamma_e} + K\left[2\lambda'\frac{l-d}{2}\right]^2 \tag{2.24}$$

$$2\frac{\gamma_p}{\gamma_e} = K\left(\left[2\lambda'\frac{l-d}{2} + \tanh(\lambda' d)\right]^2 - \tanh(\lambda' d)\right)$$

γ_e is the elastic adhesive shear strain and γ_p the plastic adhesive shear strain (see Fig. 2.10). An initial value of the bending moment factor k is given and the system solved for P, K, and d. This process is repeated until there is convergence of k.

For SLJ, the effects of the σ stresses are more pronounced than in DLJ due to the eccentric load path, this being a particular problem for composites which have a low interlaminar tensile strength. This problem gets more serious as the adherend thickness increases, since the total in-plane load carried can increase with thickness, but the through-thickness tensile stresses due to the load transfer mechanism are limited by the transverse tensile strength of the composite. Even for DLJ, the adhesive peel stress can induce composite failure. For sufficiently thin adherends, the peeling stresses are not important. Hart-Smith (1973a) took that into consideration and combined elastic σ stress with plastic τ stresses. He obtained an equation for DLJ that gives the peak σ stress as a function of the peak τ stress. The adhesive σ stresses were confined to the elastic range because the interlaminar tensile strength of the laminate is generally smaller than the peel strength of typical adhesives.

A well-designed joint is one that fails outside the overlap i.e. the adhesive should never be the weak part of the joint. Therefore, for Hart-Smith (1973a), if there is a risk of high σ stresses occurring in the adhesive, this should be minimized by tapering the adherends or by locally thickening the adhesive layer. Several authors proposed analytical solutions for such cases. Thamm (1976) studied scarf joints completely and partially sharpened. Analytical solutions for tapered joints can be found in the literature (Erdogan and Ratwani 1971, Helms et al. 1997, Oterkus et al. 2006, Reddy and Sinha 1975). Cherry and Harrison (1970) developed a simple equation to make the shear stress uniform while Adams et al. (1973) removed the stress concentration by using a quadratic profile for the adhesive layer.

2.3.2 Other Analyses Considering Only Adhesive Nonlinear Behaviour

Bigwood and Crocombe (1990) extended their elastic analysis (Bigwood and Crocombe 1989) to account for adhesive nonlinear behaviour. The model can accommodate the nonlinear stress response of the adhesive and can also be subjected to several forms of loading. A hyperbolic tangent approximation was used for the nonlinear behaviour of the adhesive, which was assumed to behave as a series of nonlinear shear and tensile springs. The adhesive yielding was modelled using the von Mises criterion and a modified von Mises criterion (Raghava et al. 1973). Numerical results were obtained by using a finite-difference method to resolve a set of six nonlinear first-order differential equations.

The Adams and Mallick analysis (1992) referred to in Section 2.2.2.1 also considered elastic–plastic adhesive behaviour. The authors took into account the influence of the adhesive plasticity by using an iterative procedure. Successive load increments are applied until the maximum stress or strain reaches some failure condition or until the full load has been applied. On the other hand, they observed that while the adhesive τ stresses were limited by adhesive yield, the adhesive shear strains followed a similar form of shear strain distribution to that predicted by Volkersen's theory. This brought them to introduce a linear 'effective modulus' solution (Adams and Mallick 1993), equating the energy under the stress–strain curve for the two cases and using the same strain to failure (see Fig. 2.11), avoiding the bilinear adhesive stress–strain curves used by Hart-Smith and allowing a single, linear, analysis to be used.

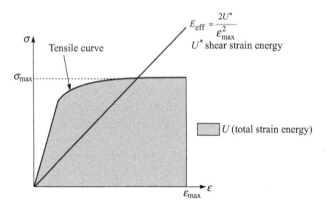

FIG. 2.11 'Effective modulus' solution proposed by Adams and Mallick.

An improvement of Hart-Smith's work was made by Tong (1996), where the adhesive has nonlinear behaviour both in τ and σ. Tong (1996) used the adhesive-adherend sandwich model for the SLJ of Bigwood and Crocombe (1989). The adhesive shear and peel stresses were represented as arbitrary functions of shear strain and tensile strain, respectively. These functions can be the shear stress–strain and the peel stress–strain curves measured by appropriate test specimens. In this analysis, Tong assumed that the shear and peel strains were constant through the adhesive thickness.

For an unbalanced joint, Tong's analysis resulted in two coupled second-order differential equations of the shear and peel strains, which does not permit a closed-form solution if the adhesive is nonlinear. If balanced joints are considered, the two differential equations become uncoupled. One of the differential equations relates the τ stress with the shear strain in the adhesive, and the other relates the σ stress with the peel strain in the adhesive. However, even these uncoupled equations do not permit a closed-form solution. These last equations can be integrated to give relatively simple equations for the failure load in the adhesive layer, in terms of the maximum strain energy density of the adhesive. Tong (1996) devised formulae for the cases of balanced joints in pure shear, pure peel and mixed failure of the adhesive. For the cases of pure shear and pure peel failure, he assumed that failure occurs when the maximum shear/peel stress or strain in the adhesive reaches a limiting value. For a mixed failure, which is usually the case in adhesive joints, two criteria can be used. In the first, called the 'limit criterion', failure in the adhesive occurs when the adhesive shear or peel strength is attained. In the second, called the 'interactive criterion', the failure in the adhesive occurs when the maximum strain energy density in the adhesive attains its allowable value for a combination of shear and peel strain.

In the Mortensen and Thomsen's analysis described in Section 2.2.2 (Mortensen and Thomsen 2002), the adhesive layer was modelled as having both linear and nonlinear material behaviour. The authors took into account the nonlinearity of the adhesive using a secant modulus approach for the non-linear stress/strain curve in conjunction with a modified von Mises yield criterion (Gali et al. 1981). They compared the results of linear adhesive behaviour with those of nonlinear adhesive behaviour and concluded that, even at low loads, the nonlinear adhesive behaviour influences the stress in the adhesive, since the non-linear behaviour of the adhesive tends to reduce the severe stress concentrations at the ends of the overlap as predicted by Hart-smith and others.

As in the Mortensen and Thomsen's analysis, Smeltzer and Klang (2003) considered composite adherends bonded by an adhesive layer with nonlinear behaviour. The analysis considered laminated symmetric composites and the effects of the transverse shear deformation in the adherends using the first-order shear-deformable laminated plate theory. The inelastic adhesive behaviour was modelled using the deformation theory of plasticity with a modified version of the von Mises yield criterion. σ and τ stresses were assumed to be constant through the adhesive thickness, and the τ stress free condition at the ends of the overlap was not considered. A similar analysis was done recently by Yang et al. (2004).

2.3.3 Adherend and Adhesive Nonlinear Behaviour

All the plastic analyses presented so far considered the adherends to have linear elastic behaviour, some considering only isotropic adherends and others considering isotropic or composite adherends. Only three analyses were found in the literature that considered both adherend and adhesive nonlinear behaviour (Crocombe and Bigwood 1992, Grimes and Greimann 1975, Wang et al. 2003).

Grimes and Greimann's analysis (1975) is nonlinear and uses a differential equation approach. They studied three types of joints: SLJ, DLJ and step-lap joints. The joints were considered to be sufficiently wide to assume a state of plane strain. The adherends could be either dissimilar in terms of material (isotropic or orthotropic) and in terms of thickness. If the adherends were orthotropic, the laminates were assumed to be symmetrical about their middle surface. The adherends were modelled as flat plates in bending. τ and σ stresses in the adherends were neglected, considering only the longitudinal normal stress. The adherend nonlinear material behaviour was modelled with the deformation theory of plasticity using the Ramberg and Osgood approximation to the stress–strain curve for both isotropic and orthotropic adherends. This type of modelling for the nonlinear material behaviour was also used by Delale et al. (1981) when they reformulated their elastic analysis to account for adhesive plasticity. In the work of Grimes and Greimann (1975), the τ and σ stresses were assumed constant through the thickness of the adhesive. The adhesive longitudinal normal stress was neglected.

Bigwood and Crocombe (1990) extended their elastic analysis (Bigwood and Crocombe 1989) to account for adhesive nonlinear behaviour. They further extended this analysis (Bigwood and Crocombe 1990) to account for nonlinear behaviour of both adhesive and adherends (Crocombe and Bigwood 1992). The Crocombe and Bigwood (1992) model can accommodate the nonlinear stress response of both the adhesive and the adherends and can also be subjected to several forms of loading. Numerical results were obtained by using a finite-difference method to resolve a set of six nonlinear first-order differential equations. However, this analysis does not account for adherend shear deformation, an important aspect when adherends with relatively low transverse shear modulus are present, as for the case of laminated composite adherends. Wang et al. (2003) extended the work of Crocombe and Bigwood (1992) to account for adherend shear deformation to predict adhesive failure in arbitrary joints subjected to large scale adherend yielding. They concluded that the shear deformation of the adherends increases the stresses near the end of the overlap compared to results from Crocombe and Bigwood's analysis (1992). The results obtained were in close agreement with FEA, except for the localized stress and strain concentrations near the free edge.

Adams et al. (1997) proposed a simple design methodology based on adherend yielding and supported this with experimental data. The simple predictive model is described next. In an SLJ with elastic adherends, the load corresponding to the total plastic deformation of the adhesive (i.e. everywhere in yield) is

$$P_a = \tau_y bl \tag{2.25}$$

where P_a is the failure load of the adhesive joint and τ_y is the yield strength of the adhesive. The direct tensile stress (σ_t) acting in the adherend due to the applied load P is

$$\sigma_t = P/bt \tag{2.26}$$

If there is bending (as per Goland and Reissner) the stress at the inner adherend surface (σ_s) due to the bending moment M is

$$\sigma_s = 6M/bt^2 \tag{2.27}$$

where $M = kPt/2$ (Goland and Reissner 1944). The variable k is the bending moment factor which reduces (from unity) as the lap rotates under load. The stress acting in the adherend is the sum of the direct stress and the bending stress. Thus, the maximum load which can be carried which just creates adherend yield (P_s) is

$$P_s = \sigma_y bt/(1+3k) \tag{2.28}$$

where σ_y is the yield strength of the adherend. For low loads and short overlaps, k is approximately 1. Therefore, for such a case

$$P_s = \sigma_y bt/4 \tag{2.29}$$

However, for joints which are long compared to the adherend thickness, such that $l/t \geqslant 20$, the value of k decreases and tends to zero. In this case, the whole of the cross-section yields in tension and

$$P_s = \sigma_y bt \tag{2.30}$$

For high-strength, non-yielding adherends, the shear strength increases almost linearly with l if the adhesive is sufficiently ductile (10% or more shear strain to failure): in this case, Eq. (2.25) gives the joint strength. For substrates that yield, any type of adhesive (ductile or brittle) may be used and a plateau is reached for a certain value of overlap corresponding to the yielding of the adherend and the joint strength can be predicted using Eqs. (2.29) and (2.30). The design methodology is represented graphically in Fig. 2.12. For intermediate or brittle adhesives and non-yielding adherends, the analysis is less robust and the authors suggest using FEA or a more complete analytical solution.

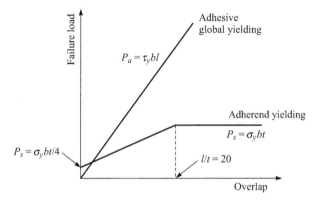

FIG. 2.12 Simple design methodology of SLJ based on the adherend yielding according to Adams et al. (1997).

2.4 THREE-DIMENSIONAL ANALYSES

The analyses considered so far were two-dimensional, assuming that the adhesive lap joints were in plane stress or plane strain in the plane perpendicular to the width direction, but neglecting the stresses across the width direction caused by Poisson's ratio strains in the adherends (Fig. 2.13) and the anticlastic bending of the adherends (Fig. 2.14).

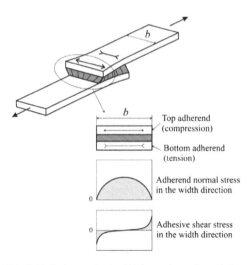

FIG. 2.13 Deformation in SLJ due to lateral straining.

FIG. 2.14 Schematic representation of the anticlastic bending.

The first to consider these effects were Adams and Peppiatt (1973) and more recently Oterkus et al. (2004). The adhesive τ stress and the adherend normal stress acting in the width and longitudinal direction (Fig. 2.13) were considered by Adams and Peppiatt (1973). In the adhesive layer, only the τ stress was considered and the effect of the bending moment was neglected. A set of two second-order partial differential equations was obtained. Two numerical methods were presented to solve this set of equations, an approximate analytical solution (assuming that the longitudinal normal stress in the adherends is constant across the joint width) and a finite-difference solution. Both methods were in close agreement. The authors found that the adherend normal stress in the width direction is maximum in the middle of the joint and zero at the edges. The τ stress in the adhesive is maximum at the edges and zero in the middle (Fig. 2.13). Adams and Peppiatt (1973) showed that, without considering the bending of the joint, adhesive and adherend stresses exist in the width direction. However, these stresses are smaller than the stresses in the longitudinal direction and a two-dimensional analysis is sufficient for most cases. The limiting case for 'wide' joints is Poisson's ratio multiplied by the longitudinal shear stress. There is a maximum transverse tensile stress (across the width) at the mid-section. This explains why when testing in tension unidirectional composites, longitudinal cracks can occur in the specimen.

Similar results were obtained by Oterkus et al. (2004). In addition to the adhesive τ stress, the adhesive σ stress was also considered. The SLJ model was three-dimensional and included nonlinear geometry. The analysis used the Von Karman nonlinear plate theory to model tapered or parallel composite adherends, and the shear-lag theory to model the adhesive layer that could have linear or bilinear behaviour. The adhesive layer was considered to be thin in comparison with the adherend thickness, and did not allow the adhesive stresses to vary across the thickness. The analysis resulted in a system of nonlinear governing equations that did not have a closed-form solution and needed to be solved numerically.

2.5 STRESS DISTRIBUTION COMPARISON

All the previously shown analytical methods and corresponding equations represent the state-of-the-art of analytical analysis for SLJ. The mathematical equations shown are important to fully implement the models, but the results of those equations are equally important, specifically for assessing the accuracy of the solutions. In this chapter, some analytical methods will be used for a specified joint conditions and compared between each other.

Some of the presented models, such as Volkersen (1938), Goland and Reissner (1944), and others, are implemented in JointDesigner (Costa and da Silva 2013), available at www.jointdesigner.com, a web application for the design of adhesive joints. JointDesigner calculates the stress distribution in the SLJ as a function of the chosen analytical model and the joint characteristics, and it also allows to export the results, which is a useful function for the purpose of comparing values between models. More complex methods, such as Bigwood and Crocombe (non-linear analysis) are not implemented at the present time in JointDesigner, and thus Maple® was used to implement the methods and export the resulting stress distributions for comparison purposes. The joint configuration shown in Fig. 2.15 with 1.62 mm thick aluminium adherends will be evaluated.

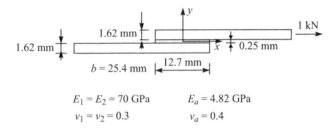

$$E_1 = E_2 = 70 \text{ GPa} \qquad E_a = 4.82 \text{ GPa}$$
$$v_1 = v_2 = 0.3 \qquad v_a = 0.4$$

FIG. 2.15 Sample SLJ configuration considered in all the comparisons through this chapter.

The objective is to visualize the difference between simple and complex formulations, and thus the chosen methods were Volkersen (1938), Goland and Reissner (1944), Ojalvo and Eidinoff (1978), Bigwood and Crocombe Linear (Bigwood and Crocombe 1989), and Bigwood and Crocombe Non-Linear Adhesive (Bigwood and Crocombe 1990). Figure 2.16 presents the distribution for both τ and σ stresses as a function of the overlap, where the x-axis origin ($x = 0$) represents the middle of the overlap.

It can be seen in Fig. 2.16 that Volkersen overestimates the τ stress at the middle of the overlap, and underestimates it at the ends of the overlap, which is the most critical point. This is understandable because it was the first model ever developed, which makes a series of assumptions and simplifications to reach a simple τ stress equation. It is also worth noting that, due to disregarding the bending moment and assuming that the adhesive only deforms in shear, the peel plots do not show results for Volkersen. Also, all method with the exception of Volkersen present very similar results, and one can argue that from an engineering point of view the plots are roughly the same. This proves that the evolution from Volkersen's model to Goland and Reissner's was very significant, and that more complex formulations lead to very marginal gains in accuracy and stress prediction, while increasing the computational time.

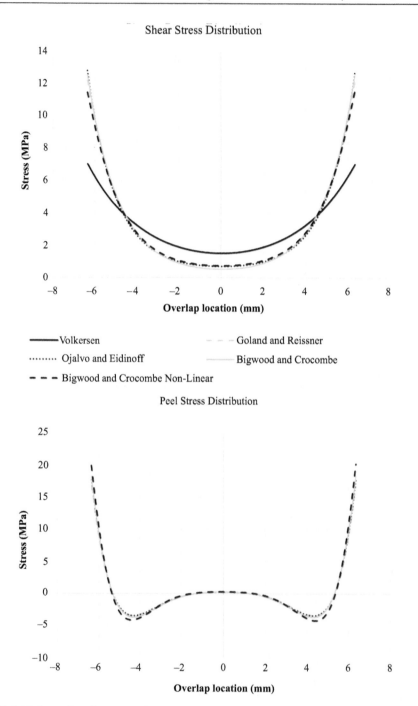

FIG. 2.16 Stress distribution results using various analytical methods for the same geometry.

2.6 CONCLUSIONS

Most of the analytical models for adhesively bonded joints are two-dimensional. In these analyses, it is assumed that the adhesive joints are in a state of plane stress or plane strain in the plane perpendicular to the width direction, neglecting the stresses across the width direction caused by Poisson's ratio strains in the adherends and the anticlastic bending of the adherends. However, there are some exceptions (Adams and Peppiatt 1973, Oterkus et al. 2004) that take into account three-dimensional effects. Nonlinear material behaviour is difficult to incorporate because the analysis becomes very complex, so most of the analyses are linear elastic for both adherends and adhesive. Table 2.1 gives a detailed summary of the available analytical models, indicating the conditions of applicability and the stresses considered. For example, if the joint bending is not severe and the adhesive is brittle, Volkersen's analysis is sufficient. However, if there is yielding of the adhesive and/or the adherends and substantial peeling is present, a more complex model is necessary. The more complete is an analysis, the more complicated it becomes and the more difficult it is to obtain a simple and effective solution.

2.7 REFERENCES

Adams, R. D. and Peppiatt, N. A. 1973. Effect of Poisson's ratio strains in adherends on stresses of an idealized lap joint. The Journal of Strain Analysis for Engineering Design 8(2): 134-139.

Adams, R. D., Chambers, S. H., Del Strother, P. J. A. and Peppiatt, N. A. 1973. Rubber model for adhesive lap joints. The Journal of Strain Analysis for Engineering Design 8(1): 52-57.

Adams, R. D. and Peppiatt, N. A. 1974. Stress analysis of adhesive-bonded lap joints. The Journal of Strain Analysis for Engineering Design 9(3): 185-196.

Adams, R. D. and Peppiatt, N. A. 1977. Stress analysis of adhesive bonded tubular lap joints. Journal of Adhesion 9(1): 1-18.

Adams, R. D. and Mallick, V. 1992. A method for the stress analysis of lap joints. The Journal of Adhesion 38(3-4): 199-217.

Adams, R. D. and Mallick, V. 1993. The effect of temperature on the strength of adhesively-bonded composite-aluminium joints. The Journal of Adhesion 43(1-2): 17-33.

Adams, R. D., Comyn, J. and Wake, W. C. 1997. Structural Adhesive Joints in Engineering, 2nd ed. Chapman & Hall, New York, London.

Allman, D. J. 1977. A theory for elastic stresses in adhesive bonded lap joints. Quarterly Journal of Mechanics and Applied Mathematics 30(4): 415-436.

Barut, A. and Madenci, E. 2009. Analysis of bolted–bonded composite single-lap joints under combined in-plane and transverse loading. Composite Structures 88(4): 579-594.

Bigwood, D. A. and Crocombe, A. D. 1989. Elastic analysis and engineering design formulae for bonded joints. International Journal of Adhesion and Adhesives 9(4): 229-242.

Bigwood, D. A. and Crocombe, A. D. 1990. Non-linear adhesive bonded joint design analyses. International Journal of Adhesion and Adhesives 10(1): 31-41.

Bois, C., Wargnier, H., Wahl, J.-C. and Le Goff, E. 2013. An analytical model for the strength prediction of hybrid (bolted/bonded) composite joints. Composite Structures 97(0): 252-260.

Carbas, R., da Silva, L., Madureira, M. and Critchlow, G. 2014. Modelling of functionally graded adhesive joints. The Journal of Adhesion 90(8): 698-716.

Chen, D. and Cheng, S. 1983. Analysis of adhesive-bonded single-lap joints. Journal of Applied Mechanics, Transactions ASME 50(1): 109-115.

Cheng, S., Chen, D. and Shi, Y. 1991. Analysis of adhesive-bonded joints with nonidentical adherends. Journal of Engineering Mechanics 117(3): 605-623.

Cherry, B. W. and Harrison, N. L. 1970. Note: The Optimum Profile for a Lap Joint. The Journal of Adhesion 2(2): 125-128.

Costa, M. and da Silva, L. F. M. 2013. JointDesigner, www.jointdesigner.com.

Crocombe, A. D. and Bigwood, D. A. 1992. Development of a full elasto-plastic adhesive joint design analysis. The Journal of Strain Analysis for Engineering Design 27(4): 211-218.

da Silva, L. F. M., Rodrigues, T. N. S. S., Figueiredo, M. A. V., de Moura, M. F. S. F. and Chousal, J. A. G. 2006. Effect of adhesive type and thickness on the lap shear strength. Journal of Adhesion 82(11): 1091-1115.

de Bruyne, N. A. 1944. The strength of glued joints. Aircraft Engineering and Aerospace Technology 16(4): 115-118.

Delale, F., Erdogan, F. and Aydinoglu, M. N. 1981. Stresses in adhesively bonded joints: A closed-form solution. Journal of Composite Materials 15(3): 249-271.

Erdogan, F. and Ratwani, M. 1971. Stress distribution in bonded joints. Journal of Composite Materials 5(3): 378-393.

Fernlund, G. and Spelt, J. K. 1994. Mixed-mode fracture characterization of adhesive joints. Composites Science and Technology 50(4): 441-449.

Fernlund, G., Papini, M., Mccammond, D. and Spelt, J. K. 1994. Fracture load predictions for adhesive joints. Composites Science and Technology 51(4): 587-600.

Frostig, Y., Baruch, M., Vilnay, O. and Sheinman, I. 1992. High-order theory for sandwich-beam behavior with transversely flexible core. Journal of Engineering Mechanics 118(5): 1026-1043.

Frostig, Y., Thomsen, O. T. and Mortensen, F. 1999. Analysis of adhesive-bonded joints, square-end, and spew-fillet-high-order theory approach. Journal of Engineering Mechanics 125(11): 1298-1307.

Gali, S., Dolev, G. and Ishai, O. 1981. An effective stress/strain concept in the mechanical characterization of structural adhesive bonding. International Journal of Adhesion and Adhesives 1(3): 135-140.

Gleich, D. M., Van Tooren, M. L. and Beukers, A. 2001. Analysis and evaluation of bondline thickness effects on failure load in adhesively bonded structures. Journal of Adhesion Science and Technology 15(9): 1091-1101.

Goland, M. A. and Reissner, E. 1944. The stresses in cemented joints. Journal of Applied Mechanics 11(1): A17-A27.

Grimes, G. C. and Greimann, L. F. 1975. Analysis of discontinuities, edge effects and joints. pp. 135-230. In: C. C. Chamis (ed.). Composite Materials, Vol. 8. Academic Press, New York.

Hart-Smith, L. J. 1973a. Adhesive-bonded double-lap joints. In NASA Technical Report CR-112235.

Hart-Smith, L. J. 1973b. Adhesive-bonded single-lap joints. In NASA Technical Report CR-112236.

Helms, J. E., Yang, C. and Pang, S. S. 1997. A laminated plate model of an adhesive-bonded taper-taper joint under tension. Journal of Engineering Materials and Technology, Transactions of the ASME 119(4): 408-414.

Lee, J. M., Han, J. K., Kim, S. H., Lee, J. Y., Shin, K. S. and Choi, B. I. 2005. An ex-vivo experimental study on optimization of bipolar radiofrequency liver ablation using per-fusion-cooled electrodes. Acta Radiologica 46(5): 443-451.

Lubkin, J. L. and Reissner, E. 1955. Stress distributions and design data for adhesive lap joints between circular tubes. American Society of Mechanical Engineers 78: 1213-1221.

Luo, Q. and Tong, L. 2009. Analytical solutions for nonlinear analysis of composite single-lap adhesive joints. International Journal of Adhesion and Adhesives 29(2): 144-154.

Mortensen, F. and Thomsen, O. T. 2002. Analysis of adhesive bonded joints: A unified approach. Composites Science and Technology 62(7-8): 1011-1031.

Nemeş, O., Lachaud, F. and Mojtabi, A. 2006. Contribution to the study of cylindrical adhesive joining. International Journal of Adhesion and Adhesives 26(6): 474-480.

Ojalvo, I. U. and Eidinoff, H. L. 1978. Bond thickness effects upon stresses in single-lap adhesive joints. AIAA Journal 16(3): 204-211.

Oplinger, D. W. 1994. Effects of adherend deflections in single lap joints. International Journal of Solids and Structures 31(18): 2565-2587.

Oterkus, E., Barut, A., Madenci, E., Smeltzer III, S. S. and Ambur, D. R. 2004. Nonlinear analysis of bonded composite single-lap joints. Paper read at Proceedings of the AIAA/ASME/ASCE/AHS/ASC 45th Structures, Structural Dynamics, and Materials Conference.

Oterkus, E., Barut, A., Madenci, E., Smeltzer III, S. S. and Ambur, D. R. 2006. Bonded lap joints of composite laminates with tapered edges. International Journal of Solids and Structures 43(6): 1459-1489.

Papini, M., Fernlund, G. and Spelt, J. K. 1994. The effect of geometry on the fracture of adhesive joints. International Journal of Adhesion and Adhesives 14(1): 5-13.

Paroissien, E., Sartor, M., Huet, J. and Lachaud, F. 2007. Analytical two-dimensional model of a hybrid (bolted/bonded) single-lap joint. Journal of Aircraft 44(2): 573-582.

Pirvics, J. 1974. Two dimensional displacement-stress distributions in adhesive bonded composite structures. The Journal of Adhesion 6(3): 207-228.

Raghava, R., Caddell, R. M. and Yeh, G. S. Y. 1973. The macroscopic yield behaviour of polymers. Journal of Materials Science 8(2): 225-232.

Raphael, C. 1966. Variable-adhesive bonded joints. Paper read at Applied Polymer Symposia 3: 99-108.

Reddy, M. N. and Sinha, P. K. 1975. Stresses in adhesive-bonded joints for composites. Fibre Science and Technology 8(1): 33-47.

Renton, J. W. and Vinson, J. R. 1975a. The efficient design of adhesive bonded joints. The Journal of Adhesion 7(3): 175-193.

Renton, J. W. and Vinson, J. R. 1975b. On the behavior of bonded joints in composite material structures. Engineering Fracture Mechanics 7(1): 41-52, IN1-IN6, 53-60.

Ripling, E. J., Mostovoy, S. and Corten, H. T. 1971. Fracture mechanics: A tool for evaluating structural adhesives. The Journal of Adhesion 3(2): 107-123.

Sato, C. and Ikegami, K. 2000. Dynamic deformation of lap joints and scarf joints under impact loads. International Journal of Adhesion and Adhesives 20(1): 17-25.

Sato, C. 2008. Impact. pp. 279-303. In: L. da Silva and A. Öchsner (eds.). Modeling of Adhesively Bonded Joints. Springer, Berlin Heidelberg.

Sato, C. 2009. Dynamic stress responses at the edges of adhesive layers in lap strap joints of half-infinite length subjected to impact loads. International Journal of Adhesion and Adhesives 29(6): 670-677.

Sawa, T., Nakano, K. and Toratani, H. 1997. A two-dimensional stress analysis of single-lap adhesive joints subjected to tensile loads. Journal of Adhesion Science and Technology 11(8): 1039-1062.

Sawa, T., Liu, J., Nakano, K. and Tanaka, J. 2000. A two-dimensional stress analysis of single-lap adhesive joints of dissimilar adherends subjected to tensile loads. Journal of Adhesion Science and Technology 14(1): 43-66.

Sayman, O. 2012. Elasto-plastic stress analysis in an adhesively bonded single-lap joint. Composites Part B: Engineering 43(2): 204-209.

Smeltzer, S. S. and Klang, E. C. 2003. Analysis of elastic-plastic adhesively bonded joints with anisotropic adherends. Paper read at Proceedings of the American Society for Composites 18th Annual Technical Conference.

Srinivas, S. 1975. Analysis of bonded joints. NASA Technical Note TN D-7855.

Stapleton, S. E., Waas, A. M. and Arnold, S. M. 2012. Functionally graded adhesives for composite joints. International Journal of Adhesion and Adhesives 35: 36-49.

Suo, Z. 1990. Delamination specimens for orthotropic materials. Journal of Applied Mechanics, Transactions ASME 57(3): 627-634.

Thamm, F. 1976. Stress distribution in lap joints with partially thinned adherends. Journal of Adhesion 7(4): 301-309.

Tong, L. 1996. Bond strength for adhesive-bonded single-lap joints. Acta Mechanica 117(1): 101-113.

Tounsi, A. and Benyoucef, S. 2007. Interfacial stresses in externally FRP-plated concrete beams. International Journal of Adhesion and Adhesives 27(3): 207-215.

Tsai, M. Y., Oplinger, D. W. and Morton, J. 1998. Improved theoretical solutions for adhesive lap joints. International Journal of Solids and Structures 35(12): 1163-1185.

Volkersen, O. 1938. Die Nietkraftverteilung in zugbeanspruchten Nietverbindungen mit konstanten Laschenquerschnitten. Luftfahrtforschung 15(1/2): 41-47.

Wah, T. 1973. Stress distribution in a bonded anisotropic lap joint. Journal of Engineering Materials and Technology 95(3): 174-181.

Wang, R. X., Cui, J., Sinclair, A. N. and Spelt, J. K. 2003. Strength of adhesive joints with adherend yielding: I. analytical model. The Journal of Adhesion 79(1): 23-48.

Whitney, J. M. 1987. Structural Analysis of Laminated Anisotropic Plates. CRC Press.

Yang, C. and Pang, S. S. 1996. Stress-strain analysis of single-lap composite joints under tension. Journal of Engineering Materials and Technology, Transactions of the ASME 118(2): 247-255.

Yang, C., Huang, H., Tomblin, J. S. and Sun, W. 2004. Elastic-plastic model of adhesive-bonded single-lap composite joints. Journal of Composite Materials 38(4): 293-309.

Yang, C.-h., Kawai, S., Kawakami, Y., Sinozaki, K. and Machiyama, T. 1996. Experimental study on solenoid valves controlled pneumatic diaphragm motor. Paper read at Proceedings of the JFPS International Symposium on Fluid Power, pp. 247-252.

Zgoul, M. and Crocombe, A. 2004. Numerical modelling of lap joints bonded with a rate-dependent adhesive. International Journal of Adhesion and Adhesives 24(4): 355-366.

Zhao, B. and Lu, Z. H. 2009. A two-dimensional approach of single-lap adhesive bonded joints. Mechanics of Advanced Materials and Structures 16(2): 130-159.

Zhao, B., Lu, Z. H. and Lu, Y. N. 2014. Two-dimensional analytical solution of elastic stresses for balanced single-lap joints—Variational method. International Journal of Adhesion and Adhesives 49(0): 115-126.

Zhao, X., Adams, R. D. and da Silva, L. F. M. 2010. A new method for the determination of bending moments in single lap joints. International Journal of Adhesion and Adhesives 30(2): 63-71.

Zou, G. P., Shahin, K. and Taheri, F. 2004. An analytical solution for the analysis of symmetric composite adhesively bonded joints. Composite Structures 65(3-4): 499-510.

3

CHAPTER

Analytical Modeling for Composite Structures

Hyonny Kim

NOMENCLATURE

x, y, z	Coordinates; composite adherend in x-y plane, z is the transverse direction
c	Adhesive joint overlap half length
t_i, t_o	Thickness of the inner, outer adherend
t_a	Thickness of the adhesive layer
E_i, E_o	Effective Young's modulus E_x of the inner, outer adherend in the x-direction
G_i, G_o	Effective transverse shear modulus G_{xz} of the inner, outer adherend in the x-z plane
G_a	Shear modulus of the adhesive
N_i, N_o	Axial x-direction stress resultant in the inner, outer adherend
N_x	Applied x-direction stress resultant quantity acting at the ends of joint (external loading)
u_i, u_o	Axial displacement component in the x-direction in the inner, outer adherend
τ_i, τ_o	Transverse shear stress in the inner, outer adherend acting in the x-z plane

Department of Structural Engineering, University of California San Diego, 9500 Gilman Drive, La Jolla, CA 92093-0085 USA.
E-mail: hyonny@ucsd.edu

γ_i, γ_o	Transverse shear strain in the inner, outer adherend acting in the x-z plane
τ_a	Adhesive shear stress acting in the x-z plane
γ_a	Adhesive shear strain acting in the x-z plane
λ	Geometric-material parameter describing the joint
μ	Transverse shear flexibility parameter
k	Stepped-lap joint step index
m	Total number of steps in the stepped-lap joint

3.1 INTRODUCTION: COMPOSITE vs. METAL ADHERENDS

The analysis of bonded polymer matrix composite joints requires additional considerations due to the complexity of the composite material adherends. In particular, laminated composites may exhibit an additional failure mode, as indicated in Fig. 3.1, which is not found in the bonding of metal adherends. While cohesive failure, interfacial (also known as adhesive or adhesion) failure and net section failure of the adherend are failure modes that also occur for metal adherends, interlaminar failure of the laminated composites can develop as well, particularly for high-strength toughened adhesives. This failure mode is shown in Fig. 3.2 (Park and Kim 2010), where the crack started in the film adhesive of this carbon/epoxy joint and transitioned into the ply nearest to the adhesive-adherend interface due to the composite having lower interlaminar strength than the adhesive. The interlaminar failure mode is sometimes referred to as "first ply failure" of the composite. Thus the interface-adjacent plies are often configured to be aligned to be in the main direction of loading since the interlaminar shear strength is higher along the fiber direction.

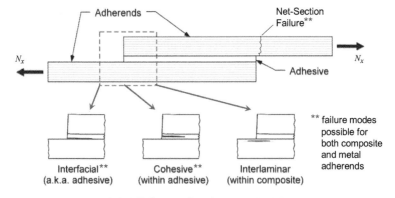

FIG. 3.1 Failure modes of a composite joint.

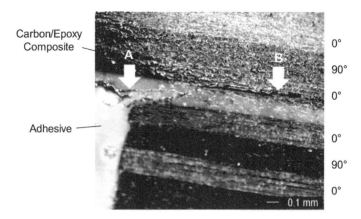

Carbon/Epoxy Composite

Adhesive

0°
90°
0°
0°
90°
0°

0.1 mm

FIG. 3.2 Interlaminar failure mode showing crack starting in the adhesive (A) and transitioning into the 0° carbon/epoxy ply (B) (Park and Kim 2010).

Related to the interlaminar failure mode, composite adherends exhibit an additional complexity due to their heterogeneous microstructure. As shown in Fig. 3.3, a microphotograph of the end region of an adhesive lap joint shows the composite to exhibit complex material and geometric morphology. In this case, the tows (fiber bundles) of a woven glass/epoxy composite show the adherend to be very heterogeneous at length scales that are just larger than the adhesive bondline thickness. Lap joints under in-plane develop complex stress states at these end-regions of the adherends and, thus, when employing analysis models that assume homogeneous and continuous material distribution within the adherends (or at least at length scales small enough to assume homogeneity), one must keep in mind what the actual microstructure looks like during the interpretation of the model results and the prediction of failure modes.

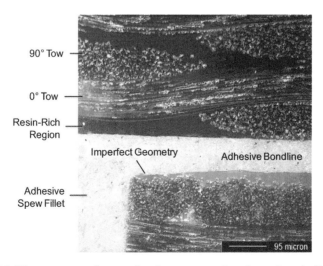

90° Tow

0° Tow

Resin-Rich Region

Imperfect Geometry

Adhesive Bondline

Adhesive Spew Fillet

95 micron

FIG. 3.3 Microstructure of woven glass/epoxy composite adherends and adhesive.

Additional factors that one must consider for the bonding of composite adherends are: proper surface preparation and pre-bond moisture effects. These aspects can significantly weaken the interface adhesion strength, resulting in an interfacial failure mode. As these are materials and processing aspects, the details of these phenomena are not discussed within this chapter, which is primarily focused on the stress analysis of composite bonded joints.

The stress analysis of in-plane loaded bonded joints typically seeks to predict the shear stress within the adhesive. While models developed for metal adherends are capable and in most cases can be applied to composite joints, under certain circumstances the modeling of composite joints should also account aspects that exist when the adherends are made with polymer matrix composite materials. Specifically, when considering joints under multiaxial in-plane loading (Kim and Kedward 2001, 2002, Lee and Kim 2013) or considering bending effects which produce peel stress (Lee and Kim 2005, 2007, 2013), models must account for the fact that laminated composite adherends have independent in-plane Young's modulus E_x and E_y, in-plane shear modulus G_{xy}, transverse shear modulus G_{xz} and G_{yz} (laminate in-plane directions are x and y; z is out-of-plane), and adherend bending stiffness D_{11}, D_{22}, D_{12}, and D_{66} (as defined in Classical Laminated Plate Theory). Unlike metals and isotropic materials, for which the Young's modulus (E) and shear modulus (G) are related via the material's Poisson's ratio (v), and the adherend bending stiffness is related primarily to E by the thickness (cubic power relationship), such properties for composite laminates are dependent on the disposition of fiber orientations and the location of these orientations within the thickness-stacking direction. Thus, since the modulus and adherend bending stiffness parameters are independent of each other, models analyzing composite bonded joints must allow these parameters to be defined independently when considering multiaxial loading and adherend bending effects (Lee and Kim 2013).

Another phenomenon that one should consider when analyzing composite bonded joints is the transverse shear deformation of the adherends (Tsai et al. 1998). This effect is more important for cases of thinner adhesive bondline thickness (Kim 2003) and for laminates with a stronger anisotropy, specifically higher ratios of in-plane Young's modulus to transverse shear modulus (e.g., E_x/G_{xz}) as found in polymer matrix composites. These dependencies and the range at which anisotropy strength and bondline thickness play a strong role in the stress analysis are examined within this chapter.

Closed form analytical models for bonded joint shear stress calculation are presented herein which address two important aspects specific to the use of composite adherends in bonded joints: (i) transverse shear deformation within the composite adherends, and (ii) analysis of composite joints employing ply drop-offs to form stepped-lap joints. A stepped-lap joint configuration allows the design of highly-loaded bonded joints (Hart-Smith 1974) and is commonly used in the bonded repair of aircraft structures.

3.2 COMPOSITE LAP JOINT SHEAR STRESS

The analysis of composite bonded lap joints is presented herein in two stages: (i) adhesive shear stress of lap joints based on shear-lag and (ii) accounting for transverse shear deformation in the adherends for adhesive shear stress calculation.

3.2.1 Volkersen Shear-Lag Model

The baseline model for predicting the shear stresses in adhesively bonded lap joints, often referred to as the Volkersen shear-lag model (Volkersen 1938, Hart-Smith 1973a, Hart-Smith 1973b), is first presented and then improved upon in subsequent sections. The analysis mainly seeks to determine the adhesive shear stress and is applicable to the single-lap and symmetric double-lap joint configurations shown in Fig. 3.4. Note that, for the double-lap joint, the outer adherends must have the same stiffness (function of both material and geometry), and the loading applied to the inner adherend (or lower adherend for single-lap joint) is twice that applied to the outer adherend, as depicted in Fig. 3.4. Based on symmetry, since the inner adherend thickness for the double-lap is twice the thickness of the single-lap, then the models are identical to each other, and the equations developed are directly applicable to both configurations if the aforementioned definitions and conventions are followed. Thus, for the single-lap joint, the adherends will be referred to as inner and outer for the lower and upper adherends, respectively.

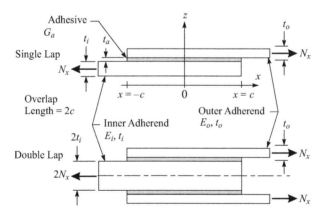

FIG. 3.4 Joint geometry for single-lap or symmetric double-lap.

The external loading N_x is the axial stress resultant acting at the cross sections located at the ends of the joint, i.e., the integration of the axial stress σ_x through the thickness, and can be considered as the force per unit width (width is in y-direction acting in the page direction) applied to the joint. Units of N_x are force per length. The following key assumptions are made in the formulation of this model:

1. Uniform bondline thickness t_a and adherend thickness t_i and t_o.
2. Uniform distribution of shear strain through the adhesive thickness.

3. The adhesive carries only out-of-plane stresses while the adherends carry only in-plane stresses.
4. Linear elastic material behavior in both the adherends and adhesive.
5. The deformation of the adherends in the out-of-plane direction is negligible.

The overlap dimension in the x-direction is $2c$, with $x = 0$ located at the center of the joint, as shown in Fig. 3.4. Thus, the domain of this model is $-c \leq x \leq c$. At any location x, global force equilibrium in the x-direction on a section cut through both adherends (see free body diagram in upper portion of Fig. 3.5) yields the relationship

$$N_i + N_o = N_x \tag{3.1}$$

where N_i and N_o are the axial stress resultants acting within the inner and outer adherends, respectively. For a thin slice dx (see lower portion of Fig. 3.5), x-direction force equilibrium of the outer adherend shows that the adhesive shear stress is related to the x-direction gradient in adherend axial stress resultant

$$\tau_a = \frac{dN_o}{dx} \tag{3.2}$$

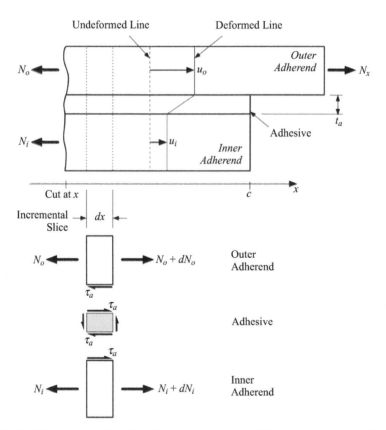

FIG. 3.5 Volkersen model free body diagrams on section of joint and on elemental slice dx.

Similarly, force equilibrium on an elemental slice of the inner adherend results in the following relationship

$$\tau_a = -\frac{dN_i}{dx} \tag{3.3}$$

Comparison of Eqs. 3.2 and 3.3 shows that the gradient in axial stress resultant in the outer adherend is equal and opposite to the gradient in the inner adherend, which is a reflection of the transfer of axial load (between the adherends) via adhesive shear stress, i.e., shear-lag load transfer mechanism. The strength of these gradients in load transfer (via shear) is strongly affected by the relative adherend and adhesive stiffness.

For the classical Volkersen shear-lag model, the adherends are assumed to exhibit no transverse shear deformation (assumption number 5). Thus, the axial displacement components u_i and u_o in the adherends are uniform through the thickness (not function of z) as depicted by the deformed line in Fig. 3.5. This condition will later be relaxed for the transverse shear deformation model. The adhesive shear strain is determined from the differential axial displacements of the adherends divided by t_a

$$\gamma_a = \frac{(u_o - u_i)}{t_a} \tag{3.4}$$

The adhesive shear stress τ_a is calculated by multiplying Eq. 3.4 with the adhesive shear modulus

$$\tau_a = G_a \frac{(u_o - u_i)}{t_a} \tag{3.5}$$

Taking the first derivative of τ_a with respect to x produces x-direction gradients of u_i and u_o, which can be related to the axial strain in the adherends. Using the linear Hooke's Law relationship between axial strain and axial stress, and describing stress in terms of resultants divided by adherend thickness

$$\frac{du_i}{dx} = \frac{N_i}{E_i t_i} \quad \text{and} \quad \frac{du_o}{dx} = \frac{N_o}{E_o t_o} \tag{3.6a, b}$$

yields the intermediate relationship

$$\frac{d\tau_a}{dx} = \frac{G_a}{t_a}\left(\frac{N_o}{E_o t_o} - \frac{N_i}{E_i t_i}\right) \tag{3.7}$$

Finally, using Eq. 3.1 to express N_i in terms of N_x and N_o, and also Eq. 3.2 for the first derivative of τ_a with respect to x, the governing equation of for calculating the axial stress resultant N_o distribution is determined

$$\frac{d^2 N_o}{dx^2} - \lambda^2 N_o + C_o = 0 \tag{3.8}$$

where

$$\lambda^2 = \frac{G_a}{t_a}\left(\frac{1}{E_o t_o} + \frac{1}{E_i t_i}\right) \quad \text{and} \tag{3.9}$$

$$C_o = \frac{N_x G_a}{t_a t_i E_i} \tag{3.10}$$

Thus, the solution approach for computing the adhesive shear stress is to first calculate the x-dependent axial stress resultant N_o via solving Eq. 3.8, and then using Eq. 3.2 to determine the adhesive shear stress. The general solution to the second order ordinary differential equation (Eq. 3.8) is

$$N_o = A_o \cosh(\lambda x) + B_o \sinh(\lambda x) + C_o / \lambda^2 \tag{3.11}$$

where A_o and B_o are constants solved by enforcing the boundary conditions on N_o at the locations $x = -c$ and $+c$ (see outer adherend in Fig. 3.4)

$$N_o(-c) = 0 \tag{3.12}$$
$$N_o(+c) = N_x \tag{3.13}$$

Applying the boundary conditions, the constants are determined to be

$$A_o = \frac{N_x - (2C_o / \lambda_o)}{2 \cosh(\lambda c)} \quad \text{and} \tag{3.14}$$

$$B_o = \frac{N_x}{2 \sinh(\lambda c)} \tag{3.15}$$

Thus, the profile of internal stress resultant N_o acting within the outer adherend over the domain of the problem, from $x = -c$ to $+c$, is determined. The shear stress in the adhesive is then calculated from the spatial gradient in N_o per Eq. 3.2 as

$$\tau_a^{Volk}(x) = \frac{dN_o}{dx} = \lambda \left[\left(\frac{N_x}{2} - \frac{C_0}{\lambda^2} \right) \frac{\sinh(\lambda x)}{\cosh(\lambda c)} + \frac{N_x}{2} \frac{\cosh(\lambda x)}{\sinh(\lambda c)} \right] \tag{3.16}$$

This equation will be referred to as the Volkersen model and can be considered as the baseline model used for calculating shear stress in the adhesive. Most notably, the model accounts only for the axial deformations of the adherends, in the direction of loading, and particularly does not account for transverse shear deformations effects acting through the thickness of the adherends.

3.2.2 Transverse Shear Deformation Effects

For fiber-reinforced polymer matrix composites, high in-plane Young's modulus can exist due to fibers being aligned within the plane, while the out-of-plane stiffness is relatively much lower mainly due to those properties being matrix-dominated. The transverse shear deformations through the thickness are significant enough to affect the adhesive shear stress and, thus, they should be accounted for.

The classical Volkersen model is improved by allowing the in-plane displacement component to vary through the thickness of the adherends, thereby accounting for transverse shear deformation effects. As shown in Fig. 3.6, the axial displacements u_i and u_o are varying through the thickness (z-direction) due to transverse shear deformation, namely from the strain component γ_{xz}. This profile of varying axial displacement is determined from the transverse shear stress acting within the adherends. The transverse shear stress τ_{xz} in the inner and outer adherends are hereafter referred to as τ_i and τ_o, respectively. They are assumed to vary through be z-direction, with known values: (i) τ_a at the adhesive-to-adherend interface, and (ii) zero at the outer traction-free surfaces (see τ_i and τ_o, profiles in Fig. 3.6). These transverse shear stress distributions are assumed to vary quadratically, represented as

$$\tau_o = \tau_a \left(1 - \frac{2z_o}{t_o} + \frac{z_o^2}{t_o^2} \right) \text{ and} \tag{3.17}$$

$$\tau_i = \tau_a \frac{z_i^2}{t_i^2} \tag{3.18}$$

where z_i and z_o are local coordinate systems defined for the inner and outer adherends, respectively. As depicted in Fig. 3.6, accounting for transverse shear deformation results in lower axial displacement at the adhesive-adherend interface and, thus, for a given applied load, the peak adhesive shear stress is predicted to be lower in comparison to the Volkersen model.

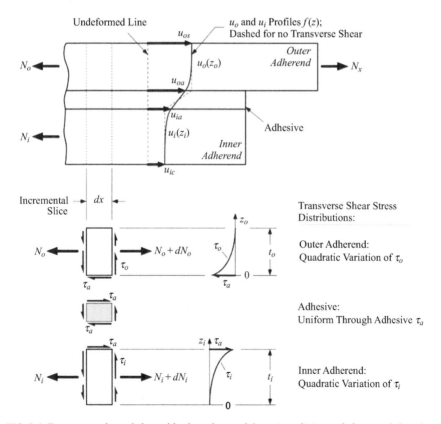

FIG. 3.6 Transverse shear deformable shear-lag model section of joint and elemental slice dx.

Since the z-direction displacement component w is assumed to be small, the transverse shear strain γ_{xz} in the adherends is expressed in terms of u_i and u_o. These are related to the transverse shear stress via Hooke's law as

$$\gamma_o = \frac{du_o}{dz_o} = \frac{\tau_o}{G_o} \quad \text{and} \tag{3.19}$$

$$\gamma_i = \frac{du_i}{dz_i} = \frac{\tau_i}{G_i} \tag{3.20}$$

Seeking expressions for u_i and u_o, Eqs. 3.17 and 3.18 are first substituted into Eqs. 3.19 and 3.20 before integrating with respect to z_o and z_i, which thereby yields the z-dependent adherend axial displacements

$$u_o = \left(\int_0^{z_o} \gamma_o dz_o \right)\Bigg|_{u_o(0)=u_{oa}} = \frac{\tau_a}{G_o}\left(z_o - \frac{z_o^2}{t_o} + \frac{z_o^3}{3t_o^2} \right) + u_{oa} \tag{3.21}$$

$$u_i = \left(\int_0^{z_i} \gamma_i dz_i \right)\Bigg|_{u_i(t_i)=u_{ia}} = \frac{\tau_a}{3G_i}\left(\frac{z_i^3}{t_i^2} - t_i \right) + u_{ia} \tag{3.22}$$

Note that the constants of integration in Eqs. 3.21 and 3.22 were determined by evaluating the displacement u_o and u_i at the adhesive-adherend interface locations, specifically at $z_o = 0$ and $z_i = t_i$ (see Fig. 3.6). The resulting expressions contain information about the adhesive shear stress and the adherend displacement at the adhesive interface locations, with both of these quantities being functions of the x-coordinate. From Eqs. 3.21 and 3.22 the axial strain in both adherends can be computed and related to the axial stress via the adherend's Young's modulus

$$\sigma_o = E_o \varepsilon_o = E_o \frac{\partial u_o}{\partial x} = \frac{1}{G_o}\frac{\partial \tau_a}{\partial x}\left(z_o - \frac{z_o^2}{t_o} + \frac{z_o^3}{3t_o^2} \right) + \frac{\partial u_{oa}}{\partial x} \quad \text{and} \tag{3.23}$$

$$\sigma_i = E_i \varepsilon_i = E_i \frac{\partial u_i}{\partial x} = \frac{1}{3G_i}\frac{\partial \tau_a}{\partial x}\left(\frac{z_i^3}{t_i^2} - t_i \right) + \frac{\partial u_{ia}}{\partial x} \tag{3.24}$$

The axial stress resultants are now computed by integrating Eqs. 3.23 and 3.24 through the thickness of each adherend

$$N_o = \int_0^{t_o} \sigma_o dz_o = E_o t_o \left(\frac{du_{oa}}{dx} + \frac{d\tau_a}{dx}\cdot\frac{t_o}{4G_o} \right) \quad \text{and} \tag{3.25}$$

$$N_i = \int_0^{t_i} \sigma_i dz_i = E_i t_i \left(\frac{du_{ia}}{dx} - \frac{d\tau_a}{dx}\cdot\frac{t_i}{4G_i} \right) \tag{3.26}$$

Note that the partial derivatives were converted into ordinary derivatives since τ_a, u_{oa}, and u_{oi} are only functions of x. The adhesive shear stress is composed, as before in Eq. 3.5, in terms of the adherend axial displacements, with specific care to use the quantities at the adhesive-adherend interface u_{oa} and u_{ia}

$$\tau_a = G_a \gamma_a = G_a \frac{(u_{oa} - u_{ia})}{t_a} \tag{3.27}$$

Taking the first derivative of τ_a with respect to x and recognizing that the x-direction gradients of u_o and u_i can be related to the adherend axial stress resultants using Eqs. 3.25 and 3.26, and furthermore using the force equilibrium relationships Eqs. 3.1 and 3.2, the governing equation accounting for transverse shear deformation of the adherends is derived as

$$\frac{d^2 N_o}{dx^2} - \frac{\lambda^2}{\mu^2} N_o + \frac{C_o}{\mu^2} = 0 \tag{3.28}$$

where μ is the shear flexibility parameter

$$\mu^2 = 1 + \frac{G_a}{t_a}\left(\frac{t_i}{4G_i} + \frac{t_o}{4G_o}\right) \tag{3.29}$$

and λ and C_o are the same grouping of terms defined by the Volkersen model, given by Eqs. 3.9 and 3.10. The solution of Eq. 3.28 follows the same process and boundary conditions as the Volkersen solution in Eqs. 3.11 to 3.16, with the primary difference being the inclusion of transverse shear flexibility parameter

$$N_o = A_o \cosh\left(\frac{\lambda}{\mu}x\right) + B_o \sinh\left(\frac{\lambda}{\mu}x\right) + C_o/\lambda^2 \tag{3.30}$$

where

$$A_o = \frac{N_x - (2C_o/\lambda_o)}{2\cosh\left(\dfrac{\lambda}{\mu}c\right)} \quad \text{and} \tag{3.31}$$

$$B_o = \frac{N_x}{2\sinh\left(\dfrac{\lambda}{\mu}c\right)} \tag{3.32}$$

The shear stress in the adhesive is calculated from the spatial gradient in N_o per Eq. 3.2

$$\tau_a^{TSD}(x) = \frac{dN_o}{dx} = \frac{\lambda}{\mu}\left[\left(\frac{N_x}{2} - \frac{C_o}{\lambda^2}\right)\frac{\sinh\left(\dfrac{\lambda}{\mu}x\right)}{\cosh\left(\dfrac{\lambda}{\mu}c\right)} + \frac{N_x}{2}\frac{\cosh\left(\dfrac{\lambda}{\mu}x\right)}{\sinh\left(\dfrac{\lambda}{\mu}c\right)}\right] \tag{3.33}$$

The result in Eq. 3.33 will be referred to as the transverse shear deformable (TSD) model. Note that the adherend transverse shear moduli G_i and G_o are contained in the denominator of the shear flexibility parameter μ, so in the limit that these parameters tend towards infinity (i.e., no transverse shear flexibility), then $\mu = 1$ and the Volkersen solution is exactly obtained. This shear flexibility parameter μ allows for the effects of the adherend shear deformations to be readily accounted for, which is especially important for polymer matrix composite adherends as they have much higher shear flexibility than isotropic materials.

3.2.3 Example Calculation

The two solutions are compared for the analysis of a woven carbon/epoxy composite lap joint having geometry shown in Fig. 3.4. The joint parameters are given in Table 3.1.

Table 3.1 Joint parameters.

Member	Parameters
Outer adherend	$t_o = 3.0$ mm, $E_o = 68.9$ GPa, $G_o = 3.45$ GPa
Inner adherend	$t_i = 2.0$ mm, $E_i = 68.9$ GPa, $G_i = 3.45$ GPa
Adhesive	$t_a = 0.15$ mm, $G_a = 1.46$ GPa
Total overlap length	$2c = 60$ mm

For an adhesive bondline thickness $t_a = 0.15$ mm, the adhesive shear stress solutions predicted by the Volkersen (Eq. 3.16) and TSD (Eq. 3.33) models are shown in Fig. 3.7 normalized by the average shear stress, calculated as

$$\tau_{ave} = \frac{N_x}{2c} \tag{3.34}$$

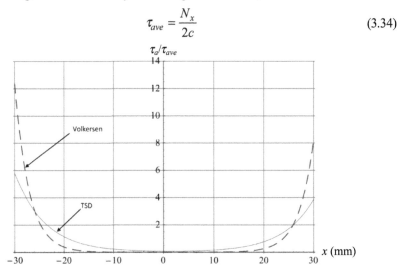

FIG. 3.7 Adhesive shear stress for Volkersen and TSD models, for $t_a = 0.15$ mm.

The maximum adhesive shear stress develops at the ends of the joint, at $x = -c$ or at $x = +c$ (joints with identical adherend stiffness, i.e., $E_o t_o = E_i t_i$, have same peak shear stress at $x = +c$ and $-c$). For the particular set of joint parameters given in Table 3.1 (typical for carbon/epoxy composite of either woven fabric plies or of unidirectional plies in quasi-isotropic layup), it can be seen that the Volkersen model predicts a much higher adhesive shear stress than the TSD model. It has been shown that the TSD model more accurately matches the peak shear stress as predicted by detailed Finite Element Analysis models (Kim 2003).

Specifically, for the joint analyzed in this example, the peak shear stress occurs at $x = -c$ since the inner adherend has lower axial stiffness, and the Volkersen model predicts a peak shear stress that is 2.13 times higher than the TSD model. It is noted that both the Volkersen and TSD shear stress profiles have the same area, reflecting that for a given applied load N_x, the shear stress balances this loading and maintains force equilibrium.

The effects of transverse shear deformation are highly dependent on both the adherend transverse shear modulus (G_i and G_o) and the adhesive bondline thickness (t_a). To explore this dependency, the ratio of peak adhesive shear stress calculated by the TSD model over the Volkersen model is plotted in Fig. 3.8 for a range of bondline thickness values. Separate contours are shown for a range of ratios of inner adherend shear modulus over in-plane Young's modulus (G_i/E_i).

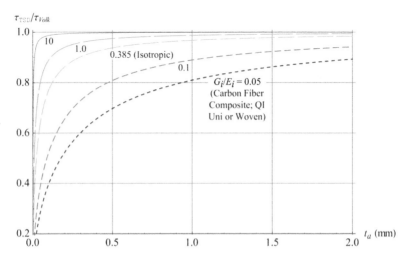

FIG. 3.8 Comparison of peak adhesive shear stress predicted by TSD and Volkersen models at location of highest stress ($x = -c$) for varying bondline thickness and ratios of adherend transverse shear modulus vs. in-plane Young's modulus; $c = 30$ mm.

The results in Fig. 3.8 show that, as the ratio G/E decreases, the transverse shear effect becomes more important to consider. Typical values for polymer matrix composite used in the example calculation yield a ratio of transverse shear modulus to in-plane Young's modulus, $G_i/E_i = 0.05$ (only inner adherend values calculated since the outer adherend has same properties). Unidirectional-dominated laminates will have even higher Young's modulus while the transverse shear modulus will remain roughly the same, and thus G/E will be even lower (0.02-0.03), resulting in even more sensitivity to transverse shear deformation. In contrast, for isotropic materials, the ratio $G/E = 0.385$ (for typical Poisson's ratio of $v = 0.3$). Thus, peak shear stress calculated by the Volkersen model is over-predicted by only ~10% for isotropic adherends having a bondline thickness of 0.3 mm (for joint parameters in this example calculation). Higher ratios of G/E are also shown in Fig. 3.8, with the hypothetical case of $G/E = 10$ (shear modulus higher than Young's modulus) which represents the almost-rigid case with no transverse shear deformation (i.e., the Volkersen model).

Figure 3.8 also shows that the transverse shear effect is more important to consider for joints having thinner adhesive bondline. The thinner bondline provides a more stiff load transfer path, resulting in higher and more spatially localized shear stresses, thereby making transverse shear effects even more important to consider. The range of bondline thickness plotted in Fig. 3.8 spans a realistic range, where typical aerospace joints are 0.12 to 0.3 mm. Joints fabricated with paste-adhesive joints can get to over 2 mm thick. Adherend thickness control and joint tooling during assembly affects bondline thickness considerably, with "pinch-off" of the joint possibly producing (undesirable) bondline thickness less than 0.05 mm (Kim 2003).

3.3 STEPPED-LAP COMPOSITE JOINT

3.3.1 Overview of Stepped-Lap Joint

For bonded joints composed of thin adherends, single-step joints are typically designed to fail in the adherend material. However, when bonding thick and high-strength adherends together, the failure of the adhesive material is usually the limiting factor defining the maximum load carrying capability. Joints configured as a single-step furthermore have the disadvantage that the peak shear stress is high at the ends of the overlaps with often no shear stress within the interior of the joint (see adhesive shear stress profile for single-step illustrated in Fig. 3.9). Increasing the joint overlap length does not lower the peak adhesive shear stress that is localized in the near-end regions of the joint. Thus, the interior region of a long-overlap single-step joint does not contribute to carry the applied load since the non-zero shear stress in the outer regions balance the applied loading N_x, as illustrated in Fig. 3.9.

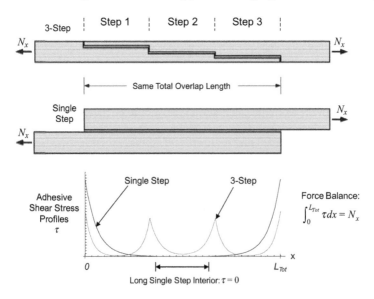

FIG. 3.9 Qualitative comparison of 3-step and single-step bonded joint adhesive shear stress profiles.

In contrast, a stepped-lap joint consisting of multiple discrete overlap segments, or steps, transfers load between the adherends more gradually, i.e., step-by-step, and thus the peak shear stress can potentially be significantly reduced. Furthermore, the interior of the joint can be engaged in carrying some portion of the applied load, as reflected by the development of non-zero shear stress. This is illustrated in Fig. 3.9 for a 3-stepped-lap joint. Comparison of the single-step and 3-step joint adhesive shear stress profiles shows a significant reduction in peak shear stress and more participation of the interior portion of the joint in carrying the load via non-zero shear stress developing at each of the step-to-step transitions. Note that the two shear stress profiles shown in Fig. 3.9 have the same total area, as the adhesive shear stress acting over the lap length balances the applied force N_x. Therefore, joints having stepped-lap configuration can be designed to carry much higher loads than single-step joints due to two mechanisms: (i) gradual load transfer from step to step, and (ii) engaging the interior portion of the joint via development of non-zero shear stress. The design of the joint, e.g., the number of steps and thickness transitions, strongly affects the resulting adhesive shear stress and, thus, a design-oriented adhesive shear stress analysis is presented herein.

It should be noted that, for composite materials, since the adherends are built up via individual plies of discrete thickness, the ply drop-offs are inherently in a stepped-lap geometry, e.g., as depicted in Fig. 3.10 for various configurations. Furthermore, for composites, the stiffness of each ply is strongly affected by the fiber orientation. Dropping off plies having fibers oriented in the main-axis direction of the joint (refer to this as 0° orientation) will have a very large change in stiffness in comparison to plies that are oriented at 90° or ±45° directions, thereby strongly affecting the adherend stiffness from step to step.

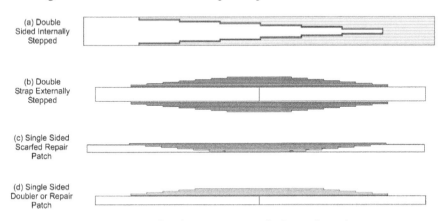

(a) Double Sided Internally Stepped

(b) Double Strap Externally Stepped

(c) Single Sided Scarfed Repair Patch

(d) Single Sided Doubler or Repair Patch

FIG. 3.10 Examples of composite stepped-joint configurations.

3.3.2 Stepped-Lap Joint Model

The general model describing a stepped-lap joint is shown in Fig. 3.11. This model and the accompanying analysis is applicable to the case of a symmetric double-lap multi-stepped joint, or single-lap multi-stepped joint. Furthermore, both internally-stepped (e.g., cases (a) and (c) in Fig. 3.10) and externally-stepped joints (cases (b) and (d) in Fig. 3.10) can be analyzed with this model. The assumptions of the Volkersen model are enforced, and it is furthermore assumed that load transfer between adherends occurs only across the adhesive lap region, with no (negligible) load transfer via the butted end faces at each step transition (i.e., no load transfer through the butted end at the far right side of the joint depicted in Fig. 3.11). This assumption is justified by the fact that butt joints in tension fail at much lower loads than shear lap joints. Thus, while adhesive surely fills the butted ends, any load transfer at these locations cannot be relied upon for high levels of applied load.

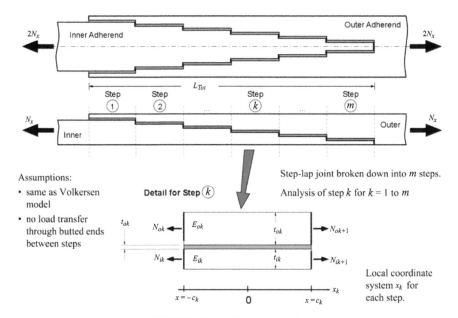

FIG. 3.11 Stepped-lap joint model.

The stress analysis of the stepped-lap joint employs the previously-developed Volkersen model for a single-step by considering each step k separately, with a more general set of boundary conditions acting at each end of the joint, as shown in Fig. 3.11. For a different loading acting on each face, the load is transferred from one adherend to the other via adhesive shear stresses. The governing equation for each step k is similar to Eq. 3.8, but with subscript k indicating values specific to the properties and geometry of each step k

$$\frac{d^2 N_o}{dx_k^2} - \lambda_k^2 N_o + C_{ok} = 0 \tag{3.35}$$

where

$$\lambda_k^2 = \frac{G_a}{t_{ak}} \left(\frac{1}{E_{ok} t_{ok}} + \frac{1}{E_{ik} t_{ik}} \right) \text{ and} \tag{3.36}$$

$$C_{ok} = \frac{N_x G_a}{t_{ak} t_{ik} E_{ik}} \tag{3.37}$$

Note that Eq. 3.35 is valid over the domain of one step, which is defined by each step's local coordinate system x_k, and each step has a different length $2c_k$ as indicated in Fig. 3.11. The general solution to the governing equation is applicable within the domain $-c_k \le x_k \le c_k$ for step k and can be written as

$$N_o = A_{ok} \cosh(\lambda_k x_k) + B_{ok} \sinh(\lambda_k x_k) + C_{ok} / \lambda_k^2 \tag{3.38}$$

The constants A_{ok} and B_{ok} are determined by enforcing the boundary conditions, i.e., stress resultants N_{ok} and N_{ok+1} acting on outer adherend in step k, as shown in Fig. 3.11

$$N_o(-c_k) = N_{ok} \tag{3.39}$$
$$N_o(+c_k) = N_{ok+1} \tag{3.40}$$

The resulting expressions for A_{ok} and B_{ok} are applicable for each step k

$$A_{ok} = \frac{N_{ok} + N_{ok+1} - (2C_{ok}/\lambda_k^2)}{2\cosh(\lambda_k c_k)} \quad \text{and} \tag{3.41}$$

$$B_{ok} = \frac{N_{ok+1} - N_{ok}}{2\sinh(\lambda_k c_k)} \tag{3.42}$$

Having A_{ok} and B_{ok}, the axial stress resultant within each step k can be calculated using Eq. 3.38, and the adhesive shear stress is then calculated by Eq. 3.3, resulting in

$$\tau_a(x_k) = \lambda_k \left[\left(\frac{N_{ok} + N_{ok+1}}{2} - \frac{C_{ok}}{\lambda_k^2} \right) \frac{\sinh(\lambda_k x_k)}{\cosh(\lambda_k c_k)} + \left(\frac{N_{ok+1} - N_{ok}}{2} \right) \frac{\cosh(\lambda_k x_k)}{\sinh(\lambda_k c_k)} \right] \tag{3.43}$$

Local maximum shear stress values for each step are located at the overlap ends, at $x_k = +c_k$ and $-c_k$

$$(\tau_a)_{x_k = \pm c_k} = \lambda_k \left[\pm \left(\frac{N_{ok} + N_{ok+1}}{2} - \frac{C_{ok}}{\lambda_k^2} \right) \tanh(\lambda_k c_k) + \frac{N_{ok+1} - N_{ok}}{2\tanh(\lambda_k c_k)} \right] \tag{3.44}$$

While Eq. 3.43 describes the adhesive shear stress profile in each step, the quantities N_{ok} are yet unknown. These are the values of the stress resultant N_o acting at the faces of each step, i.e., equal and opposite internal forces acting on the faces in between steps. For a total of m steps existing across the total joint overlap length, there are $m + 1$ values of N_o to specify. As an example, a three-step joint ($m = 3$) is shown in Fig. 3.12 with the quantities N_{o2} acting between steps 1 and 2, and N_{o3} acting between steps 2 and 3. The quantities N_{o1} and N_{o4} are the force quantities acting at the outermost ends of the overlap, labeled per the numbering assignment defined in Fig. 3.11. These outermost N_o values are always known, however, being the outer boundary conditions for the model configuration as shown in Fig. 3.11

$$N_{o1} = 0 \tag{3.45}$$
$$N_{om+1} = N_x \tag{3.46}$$

FIG. 3.12 Example of 3-step joint ($m = 3$) with stress resultants acting at ends of each step.

Thus, for the total $m + 1$ values of N_o to solve for, applying the boundary conditions Eqs. 3.45 and 3.46 reduces the number of unknowns to $m - 1$. So in the example in Fig. 3.12 with three steps ($m = 3$), the two unknown quantities remaining are N_{o2} and N_{o3} (i.e., $m - 1 = 2$). These $m - 1$ values are found by enforcing the condition that the adhesive shear stress is continuously-valued between each step.

$$\left(\tau_a\right)_{x_k = +c_k}^{\text{Step } k} = \left(\tau_a\right)_{x_{k+1} = -c_{k+1}}^{\text{Step } k+1}$$

(3.47)

Eq. 3.47 is applied at each step-to-step interface, and thus $m - 1$ equations are established to solve for the $m - 1$ unknown values N_{o2}, N_{o3}, ... N_{om}. Using Eq. 3.44 for steps k and $k + 1$, Eq. 3.47 becomes

$$(\alpha_k - \beta_k)N_{ok} + (\alpha_k + \beta_k + \alpha_{k+1} + \beta_{k+1})N_{ok+1}$$
$$+ (\alpha_{k+1} - \beta_{k+1})N_{ok+2} - 2\alpha_k \frac{C_{ok}}{\lambda_k^2} - 2\alpha_{k+1}\frac{C_{ok+1}}{\lambda_{k+1}^2} = 0 \quad (3.48)$$

for $k = 1$ to $m - 1$, where

$$\alpha_k = \frac{\lambda_k \tanh(\lambda_k c_k)}{2} \text{ and } \beta_k = \frac{\lambda_k}{2 \tanh(\lambda_k c_k)} \quad (3.49, 3.50)$$

and

$$\alpha_{k+1} = \frac{\lambda_{k+1} \tanh(\lambda_{k+1} c_{k+1})}{2} \text{ and } \beta_{k+1} = \frac{\lambda_{k+1}}{2 \tanh(\lambda_{k+1} c_{k+1})} \quad (3.51, 3.52)$$

Eq. 3.48 provides a system of $m - 1$ linear equations in terms of N_{ok} which can be more readily solved by composing these equations in matrix form

$$\begin{bmatrix} \text{Matrix } M \\ \text{Size}\,(m-1) \times (m-1) \end{bmatrix} \begin{Bmatrix} N_{o2} \\ N_{o3} \\ \vdots \\ N_{ok} \\ \vdots \\ N_{om} \end{Bmatrix} = \begin{Bmatrix} 2\alpha_1 C_{o1}/\lambda_1^2 + 2\alpha_2 C_{o2}/\lambda_2^2 \\ 2\alpha_2 C_{o2}/\lambda_2^2 + 2\alpha_3 C_{o3}/\lambda_3^2 \\ \vdots \\ 2\alpha_k C_{ok}/\lambda_k^2 + 2\alpha_{k+1}C_{ok+1}/\lambda_{k+1}^2 \\ \vdots \\ 2\alpha_{m-1}C_{om-1}/\lambda_{m-1}^2 + 2\alpha_m C_{om}/\lambda_m^2 - (\alpha_m - \beta_m)N_x \end{Bmatrix}$$

(3.53)

where the matrix M is composed of terms multiplied to N_{ok} (accounting for $N_{o1} = 0$ and $N_{om+1} = N_x$)

$$\begin{bmatrix} \alpha_1+\beta_1+\alpha_2+\beta_2 & \alpha_2-\beta_2 & 0 & 0 & \cdots & 0 & 0 \\ \alpha_2-\beta_2 & \alpha_2+\beta_2+\alpha_3+\beta_3 & \alpha_3-\beta_3 & 0 & \cdots & 0 & 0 \\ 0 & \alpha_3-\beta_3 & \alpha_3+\beta_3+\alpha_4+\beta_4 & \alpha_4-\beta_4 & \cdots & 0 & 0 \\ \vdots & & & & & & \\ 0 & 0 & 0 & 0 & \cdots & \alpha_{m-1}-\beta_{m-1} & \alpha_{m-1}+\beta_{m-1}+\alpha_m+\beta_m \end{bmatrix}$$

$$(3.54)$$

for $k = 1$ to $m - 1$ (i.e., the 1st row in Eqs. 3.53 and 3.54 corresponds to $k = 1$, 2nd row $k = 2$, ... until $k = m - 1$). Upon solving for the $m - 1$ values of N_{ok}, the axial stress resultant distribution within each step is known via Eq. 3.38, and the adhesive shear stress profiles can be calculated for each step using Eq. 3.43. The solution calculates the outer adherend axial stress resultant N_o profile within each step, with the values determined by Eq. 3.53 acting at each step transition, as illustrated for example in Fig. 3.13. Here, the axial stress resultant N_o can be seen to gradually decrease from N_x at the far right hand side of the joint, to zero at the far left hand side. The shear stress profiles for each step, also shown in Fig. 3.13, have continuous values at each step transition, as enforced by Eq. 3.47, and typically have low (even zero) values of shear stress near the middle of each step due to the low gradients in axial stress resultant N_o at these locations.

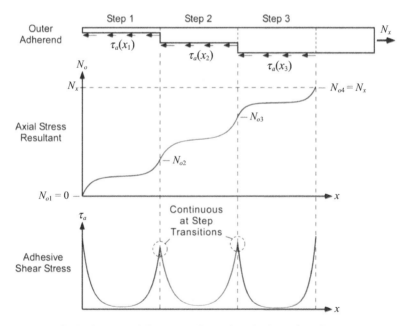

FIG. 3.13 Free-body diagram of the outer adherend with plots of axial stress resultant and adhesive shear stress profiles.

3.3.3 Simplified Equations for Long Step Overlaps

A simplification is possible for joint configurations having a combination of parameter values and sufficient overlap length of each step that results in the term tanh (λc) to effectively equal 1 in the Eqs. 3.49 to 3.52. For the product $\lambda c \geq 3$, the value of tanh (λc) is within 0.5% of 1.00, and thus Eqs. 3.49 to 3.52 all approximately are equivalent to $\lambda/2$ (specializing λ for steps k and $k + 1$). Eq. 3.48 then simplifies to an approximate solution applied to each step k allowing one to directly calculate N_{ok+1}

$$(\lambda_k + \lambda_{k+1})N_{ok+1} - \frac{C_{ok}}{\lambda_k} - \frac{C_{ok+1}}{\lambda_{k+1}} = 0 \quad \text{for} \quad k = 1 \text{ to } m - 1 \qquad (3.55)$$

Eq. 3.55 provides $m - 1$ equations that allow one to directly calculate the $m - 1$ unknown stress resultant values acting between each step. For example, for step 1, Eq. 3.55 calculates the value of N_{o2}. With these values determined, Eqs. 3.38 and 3.43 can be used to calculate the profiles of axial stress resultant and adhesive shear stress within each step.

3.3.4 Example of Calculation

An example calculation is provided for a four-step joint bonding a Ti alloy to a carbon/epoxy composite. All relevant joint parameter information is given in Fig. 3.14, with uniform bondline thickness throughout and the same overlap length $2c = 38.1$ mm for each step. Thus, the total overlap length, L_{Tot}, is 152.4 mm.

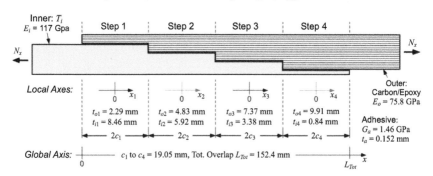

FIG. 3.14 Example of a stepped-lap joint.

The joint parameter information is first used to calculate the important quantities for each step, as summarized in Table 3.2, which will be used in conjunction with Eqs. 3.48 to 3.53 to solve for the values of N_{ok}. Note that the last column in the table is used for assessing whether the simplified solution (Eq. 3.55) can be used. Both the full and approximate solutions are presented here for a general applied load N_x.

Table 3.2 Calculated quantities for each step.

Step no. k	λ_k (1/m)	C_{ok} (N/m³)	$\lambda_k c_k$
1	255	$9670N_x$	4.86
2	200	$13800N_x$	3.81
3	203	$24200N_x$	3.88
4	332	$97600N_x$	6.33

Since there are four steps ($m = 4$), then three unknown N_{ok} values ($m - 1 = 3$) must be determined. The quantities in Table 3.2 are used to write the 3×3 matrix relationship (Eq. 3.53) describing this four-step joint (units for contents within 3×3 matrix are 1/m, while N_{ok} and N_x are in units N/m)

$$\begin{bmatrix} 455.0 & -0.196 & 0 \\ -0.196 & 403.5 & -0.175 \\ 0 & -0.175 & 535.7 \end{bmatrix} \begin{Bmatrix} N_{o2} \\ N_{o3} \\ N_{o4} \end{Bmatrix} = \begin{Bmatrix} 107N_x \\ 188N_x \\ 413N_x \end{Bmatrix} \qquad (3.56)$$

This set of equations is used to solve for the unknown values of N_{ok}

$$N_{o2} = 0.235N_x, \; N_{o3} = 0.466N_x, \text{ and } N_{o4} = 0.771N_x \qquad (3.57\text{a, b, c})$$

with the outer values $N_{o1} = 0$ and $N_{o5} = N_x$ being known boundary conditions. The distribution of axial stress resultant N_o throughout the length of the outer adherend is plotted in Fig. 3.15. Note that the vertical axis is normalized by the applied loading N_x, and the horizontal axis is normalized by the total overall overlap length L_{Tot}. Furthermore, since the N_{ok} profiles are calculated for each step using a local axis x_k that is centered within each step, a simple axis offset must be applied to allow plotting all the profiles together onto one global axis as shown in Fig. 3.15. The value of N_o can be seen to decrease from right to left, as load is transferred into the inner adherend, with the values in Eq. 3.57 of N_{ok} calculated using Eq. 3.56 shown to be the values at the step transitions. Conversely, the stress resultant carried by the inner adherend, N_i, can be calculated using Eq. 1 as $N_x - N_o$, and will be found to increase from the right to left side of the joint due to load transfer between adherends.

The adhesive shear stress profiles within each step, calculated using Eq. 3.43, are shown plotted together in Fig. 3.16 onto a common global axis (normalized by the total overlap length). The vertical shear stress axis is normalized by the average shear stress, calculated as the applied external load N_x divided by the total overlap length, $\tau_{ave} = N_x/L_{Tot}$. Local maximum values of shear stress occur at each step-to-step transition, as shown in Fig. 3.16. These local shear stress maxima correspond to the locations of high gradient in N_o, as shown in Fig. 3.15. Globally, the maximum adhesive shear stress occurs at the far right hand side of the joint, at $x = L_{Tot}$ (or at $x_4 = c_4$), with a value of 5.86 τ_{ave}.

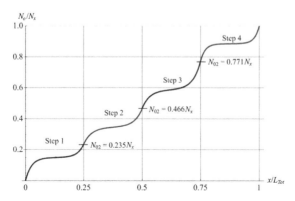

FIG. 3.15 Four-step joint example problem normalized axial stress resultant distribution.

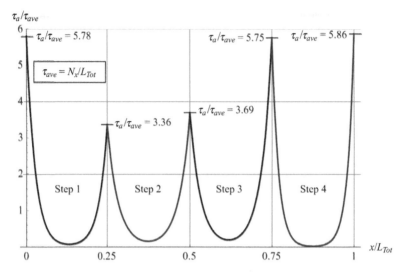

FIG. 3.16 Four-step joint example problem normalized adhesive shear stress distribution.

Failure onset within the adhesive can be determined by comparing the peak shear stress ($\tau/\tau_{ave} = 5.86$ for this example), to the adhesive shear strength. For a typical value of adhesive shear strength $\tau_{ult} = 37.9$ MPa, the load associated with the onset of joint failure can be determined as (recall $\tau_{ave} = N_x/L_{Tot}$)

$$N_{xfail} = (\tau_{ult}L_{Tot})/5.86 = 986 \text{ N/mm} \tag{3.58}$$

For brittle adhesives or those exhibiting little ductility, this load would correspond to final failure of the joint. For joints with ductile adhesives undergoing significant plastic strain to failure, Eq. 3.58 would be a prediction of initial yielding of the adhesive, and nonlinear elasto-plastic type analysis (Hart-Smith 1973b, Lee and Kim 2007) would be necessary to predict ultimate failure of the joint.

As a comprehensive check for other failure modes, the stress within the adherends should also be checked. For this example problem, the average axial stress in each adherend is calculated in a simple manner as N_i/t_i and N_o/t_o, recognizing that these quantities vary with x location. However, for composite adherends having plies oriented at various angles, a more accurate ply-by-ply analysis should be conducted, using the axial stress resultant values N_i and N_o in a laminate analysis calculation. Note that since the inner adherend stress resultant N_i is not directly solved for, it is calculated using Eq. 3.1 as $N_x - N_o$. The resulting average axial stress profiles are plotted in Fig. 3.17 for the loading value $N_x = 986$ N/mm, as calculated in Eq. 3.58. The maximum stress in the inner adherend occurs in the 4th step, while the maximum stress in the outer adherend occurs in the 3rd step, as indicated in Fig. 3.17. These values are well below the typical yield strength of the Ti alloy (~800 MPa) and the failure strength of a multi-directional carbon/epoxy composite laminate (over 500 MPa). Thus, it is expected that the adhesive will be the first material to fail for this example (i.e., cohesive failure mode).

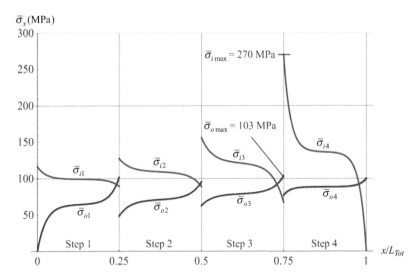

FIG. 3.17 Average axial stresses in the outer and inner adherends for an applied loading of $N_x =$ 986 N/mm.

Finally, the simplified approximate solution is applied to this example problem. Observing in Table 3.2, the product of $\lambda_k c_k$ is tabulated to assess if the simplification can be employed. The lowest value is 3.81 in step 2, so the simplified solution would be applicable since tanh (3.81) = 0.999 is approximately equal to 1. Using Eq. 3.55 to directly calculate N_{ok} yields the values

$$N_{o2} = 0.235N_x, \quad N_{o3} = 0.466N_x, \text{ and } N_{o4} = 0.771N_x \qquad (3.59\text{a,b,c})$$

which are identical to the values listed in Eq. 3.57. The use of Eq. 3.55 is equivalent to setting the off-diagonal terms in Eqs. 3.54 to zero. This simplified solution is accurate for the condition when $\lambda_k c_k > 3$. Otherwise, the full solution composed by Eq. 3.53 should be used.

3.4 CONCLUSIONS

Closed form models predicting the stresses within adhesively bonded joints have been presented. These models account for particular features exhibited by composite adherends: transverse shear effects with transverse shear modulus decoupled from in-plane Young's modulus, and step-tapered adherend thickness transitions, which reflect ply drop offs and stepped-lap scarf joints used in repairs. These models, being in equation form, provide relationships that directly show how various parameters affect the resulting stresses. Since these models are relatively simple to execute and quick to compute, they are ideally suited for design and trade studies considering the effects of various materials or geometry. Finally, the capability of any model is limited by the knowledge of accurate material properties. Poor manufacturing and particularly improper pre-bond surface preparation could lead to low interfacial strengths, thereby causing the joint to fail at much lower loads than would be predicted by these models. Therefore, great attention must be paid to manufacturing process control and surface preparation, in addition to accurate stress analysis.

3.5 REFERENCES

Hart-Smith, L. J. 1973a. Adhesive-bonded single-lap joints, NASA-Langley Contract Report, NASA-CR-112236.

Hart-Smith, L. J. 1973b. Adhesive-bonded double-lap joints, NASA-Langley Contract Report, NASA-CR-112235.

Hart-Smith, L. J. 1974. Analysis and design of advanced composite bonded joints, NASA-Langley Contract Report, NASA-CR-2218.

Kim, H. and Kedward, K. T. 2001. Stress analysis of adhesive bonded joints under in-plane shear loading. The Journal of Adhesion 76: 1-36.

Kim, H. and Kedward, K. T. 2002. The design of in-plane shear and tension loaded bonded composite lap joints. Journal of Composites Technology and Research 24: 297-307.

Kim, H. 2003. The Influence of adhesive bondline thickness imperfections on stresses in composite joints. The Journal of Adhesion 79: 621-642.

Lee, J. and Kim, H. 2005. Analysis of a single lap asymmetric bonded joint under tension and eccentricity moment. The Journal of Adhesion 81: 443-472.

Lee, J. and Kim, H. 2007. Elasto-plastic analysis of adhesively bonded symmetric single lap joints under in-plane tension and edge moments. The Journal of Adhesion 83: 837-870.

Lee, J. and Kim, H. 2013. Stress and strain analysis of symmetric composite single lap joints under combined tension and in-plane shear loading. pp. 1-22. In: S. Kumar and K. L. Mittal (eds.). Advances in Modeling and Design of Adhesively Bonded Systems. Scrivener Publishing LLC, Salem, Massachusetts.

Park, H. and Kim, H. 2010. Damage resistance of single lap adhesive composite joints by transverse ice impact. International Journal of Impact Engineering 37: 177-184.

Tsai, M. Y., Oplinger, D. W. and Morton, J. 1998. Improved theoretical solutions for adhesive lap joints. International Journal of Solids and Structures 35: 1163-1185.

Volkersen, O. 1938. Die Niektraftverteilung in zugbeanspruchten mit konstanten laschenquerschritten. Luftfahrtforschung 15: 41-47.

4

C H A P T E R

Analytical Modelling of Dynamic and Impact Loads

Ramzi Othman

4.1 INTRODUCTION

Adhesive bonding is increasingly used in automobile, aeronautical, electrical engineering, and many other engineering fields. In these applications, adhesively bonded joints can face dynamic loadings. A good design of these joints is based on the analysis of the stress distributions along the joint. This can be undertaken using either numerical or analytical approaches. The second approach is based on a mathematical formulation of the problem, which is not straightforward. Moreover, this approach is based on a simplification of the real mechanical problem through some assumptions on the geometry, deformation and/or stress state. Nevertheless, the analytical approaches can lead to closed-form solutions that are appreciated to have a first and rapid idea about the dominant mechanical phenomenon in the joints. Only few works have dealt with the dynamic analysis of adhesively bonded joints using the analytical approach. They will be briefly reviewed in the next section. The third section will focus on the analytical models predicting the harmonic responses of the adhesively bonded joints. The fourth section emphasizes on the impact responses. The fifth section gives a brief introduction on how the viscoelastic behavior can be taken into consideration in the analytical modelling of the impact and harmonic responses.

4.2 BRIEF STATE-OF-THE-ART

The derivation of analytical solutions is possible through some assumptions. However, this simplification effort can end up by closed-form solutions, which are used to carry out the first analysis or design. Since the work of Volkersen (1938), several analytical solutions have been proposed to model the static response of

Mechanical Engineering Department, Faculty of Engineering, King Abdulaziz University, P.O. Box 80248, Jeddah 21589, Saudi Arabia.
E-mail: Rothman1@kau.edu.sa

adhesive joints. da Silva et al. (2009a, 2009b) have written a comprehensive review on this topic.

The dynamic response of adhesively bonded joints involves longitudinal and/ or transverse deformation. Rao and Crocker (1990) dealt with single-lap joints and assumed that the substrates are beams. He and Rao (1992a, 1992b) derived the vibration equations of single-lap joints by applying the energy method and Hamilton's principle. Dealing also with the vibration of single-lap joints, Ingole and Chatterjee (2010) have used the Euler-Bernoulli free-free beam theory to calculate the natural frequencies and modes. Yuceoglu et al. (1996) were interested in adhesively bonded orthotropic plates. They used the Mindlin plate theory and a Levy-type solution to predict free bending vibration.

Vaziri and co-workers (2001, 2002a, 2002b, 2004, 2006) have focused on the effects of voids on the harmonic responses of several adhesive joints. They considered that the adherends follow an elastic behavior. The adhesive is assumed to be viscoelastic. Vaziri et al. (2001, 2004) dealt with the response of single-lap joints which are subjected to a peeling force. They considered the substrates as Euler-Bernoulli beams. Vaziri and Nayeb-Hashemi (2006) used the same assumptions to predict the dynamic response of a repaired composite beam with an adhesively bonded patch. Vaziri and Nayeb-Hashemi (2002a) were also interested in tubular joints. Mainly, they studied the effects of annular void on their torsional response. They assumed that the adhesive shears in the circumferential direction while the adherends shear in the axial direction. They also investigated the effects of annular voids on tubular joints subjected to axial harmonic loads (Vaziri and Nayeb-Hashemi 2002b). To this aim, they extended the basic shear-lag models of Volkersen (1938) to consider dynamic loadings by including the adhesive and adherends inertia. The solution was expressed in terms of Bessel functions.

Dealing with single-lap composite joints, Pang et al. (1995) studied the impact response of single-lap joints subjected to a transverse load. Firstly, they calculated the impact force using a mass-spring model. Subsequently, they derived the impact response of the composite joint by considering a first-order laminate plate theory for the substrates. Helms et al. (2001) calculated an equivalent stiffness for a hybrid laminate-metal joint. In terms of the response to a longitudinal load, Sato (2008, 2009) derived the stress at the edge of the adhesive layer. He extended the simple shear-lag model (Volkersen 1938) to take into account the inertia of adherends. However, the adhesive inertia was neglected.

Tsai et al. (1998) proposed a modified shear-lag analytical model which takes into account the shear deformation in the substrates. Actually, the simple shear-lag model neglects any shear deformation in the adherends as it assumes that these adherends are much stiffer than the adhesive. Thus, Volkerson's model is not valid when the shear stiffness of the adhesive is of the same magnitude as the shear stiffness of the substrates. The Tsai et al. (1998) model was validated under static loads applied on simple- and double-lap joints. Based on this model, Challita and Othman (2012) have obtained the harmonic response of a double-lap joint. Moreover, Hazimeh et al. (2015) calculated the impact stress response at the edge of the adhesive layer of a double-lap joint of semi-infinite length. Challita and Othman (2012) and Hazimeh et al. (2015) have considered that both the adhesive and the substrates have an elastic behavior. Following Sato (2008, 2009), they have neglected the inertia effects on the adhesive layer. Compared to the simple shear-lag solutions, the harmonic and impact solutions obtained by the modified shear-lag model are closer to the stress response obtained by the finite element method. This is mainly if the adhesive shear stiffness cannot be neglected when compared to the adherends' shear stiffness.

This chapter focuses on the dynamic shear response of the adhesively bonded joints subjected to axial loads. With regard to the above literature review, three main models are presented: Sato's model (Sato 2008, 2009), the Vaziri and Nayeb-Hashemi model (Vaziri and Nayeb-Hashemi 2002b) and the Challita et al. model (Challita and Othman 2012, Hazimeh et al. 2015). The basic assumptions of these models are synthesized in Table 4.1. The coming section will discuss the analytical models predicting the harmonic response of adhesive joints. Subsequently, Section 4.4 will discuss the equations giving the impact response at the edge of the adhesive layer. Next, Section 4.5 briefly extends the harmonic and impact solutions to take into account the viscoelastic behavior of the adhesive and substrates. This chapter will be closed by a conclusion that sum up the main ideas.

Table 4.1 Analytical models.

	Sato's model	Vaziri and Nayeb-Hashemi's model	Challita et al. model
Adhesive inertia	No	Yes	No
Substrates' inertia	Yes	Yes	Yes
Shear deformation of the substrates	No	No	Yes
Joint type	Single-lap	Tubular joint	Double-lap
Impact/Harmonic loads	Impact	Harmonic	Impact and Harmonic
Elastic/ Viscoelastic	Elastic substrates and elastic adhesive	Elastic substrates and viscoelastic adhesive	Elastic substrates and elastic adhesive

4.3 HARMONIC RESPONSE

4.3.1 Sato's Model

4.3.1.1 Basic Equations

The shear-lag model was first introduced by Volkersen (1938) to predict the shear stress distribution in double-lap adhesive joints submitted to static loads. The adhesive layer and substrates are considered to be elastic. A pure shear deformation is assumed in the adhesive, whereas only one-dimensional axial deformation is allowed for the substrates. Both shear deformation in the adhesive layer and axial deformation in the substrates are assumed as constant through-the-thickness. This model assumes that the shear stiffness of the adherends is much higher than the shear stiffness of the adhesive layer. Put in other words, the shear stiffness of the adherends is supposed to be high enough to oppose and block any rotation of substrates' cross-sections, which then remain perpendicular to the applied load.

In this section, we are dealing with single-lap and double-lap joints. Their geometrical characteristics are shown in Fig. 4.1. In terms of the single-lap joint, we consider that a harmonic load $P_1(t) = Re(\tilde{P}_1 e^{i\omega t})$ is applied at the right end of the lower substrate (Fig. 4.2(a)), where $\tilde{P}_1$ is the magnitude of the harmonic load, $Re(z)$ denotes the real part of the complex number z, ω is the angular frequency of the harmonic load and t is time. In the case of the double-lap joint, the central substrate is submitted to $2P_1(t)$ at its right edge (Fig. 4.2(b)). The left end of the single-lap joint's upper substrate and the left ends of the double-lap joint's outer substrates are clamped (Fig. 4.2). Within this framework, the double-lap joint's problem is symmetric with respect to the x axis. Thus, it is possible to study the problem for half of the geometry (for example: $y \geq 0$) and extend the solution to the second part using symmetry. Considering the symmetry, solving the problem of the loaded double-lap joint represented in Fig. 4.2(b) on the half of geometry corresponds to solving the problem of the loaded single-lap joint schematized in Fig. 4.2(a). Based on the shear-lag assumptions and considering symmetry, the shear stress distribution in the adhesive layer of the single-lap joint and the shear stress distribution in the adhesive layers of the double-lap joints should be alike. Actually, there should be some differences which are due to the peeling stresses; however, they are ignored by the shear-lag model. That's why this model results in similar stress distribution in single-lap and double-lap joints. Henceforth, the analysis will be focused on the single-lap joints. The reader can derive the solutions of double-lap joints using symmetry.

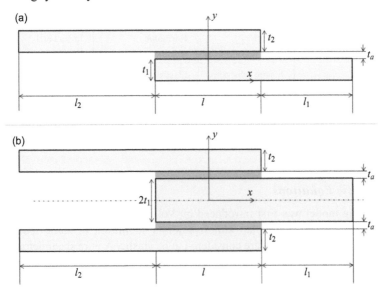

FIG. 4.1 Geometrical characteristics of (a) single-lap joint, and (b) double-lap joint.

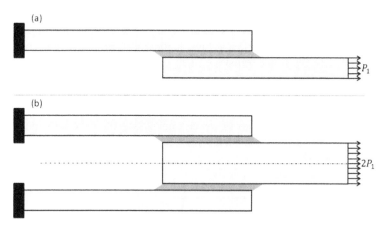

FIG. 4.2 Loaded (a) single-lap joint, and (b) double-lap joint.

Considering the hypothesis of the shear-lag model, the substrates' displacements, axial stresses, and axial strains are uniform through-the-thickness and then only depend on the horizontal coordinate x and time t. Similarly, the adhesive shear stress and shear strain are constant through-the-thickness. Let $u_1(x, t)$, $\varepsilon_1(x, t)$ and $\sigma_1(x, t)$ be the displacement, strain and stress, respectively, of the lower substrate. Likewise, let $u_2(x, t)$, $\varepsilon_2(x, t)$ and $\sigma_2(x, t)$ be the displacement, strain and stress, respectively, of the upper substrate. Moreover, let $\gamma_a(x, t)$ and $\tau_a(x, t)$ be the shear strain and shear stress, respectively, in the adhesive layer. As we are interested in harmonic response in this section, any function $f(x, t)$ is written as the real number of a complex number of angular frequency ω, i.e., $f(x, t) = Re(\tilde{f}e^{i\omega t})$, where $\tilde{f}$ is the complex amplitude.

In this section, the adhesive and adherends are assumed as elastic. Let E_1 and E_2 be the Young's moduli of the lower and upper adherends and G_a be the shear modulus of the adhesive. Using Hooke's law, the axial stresses in the adherends read:

$$\sigma_j(x, t) = E_j \varepsilon_j(x, t) = E_j \frac{\partial u_j}{\partial x}(x, t) \tag{4.1}$$

where $j = 1$ or 2, whereas the shear stress in the adhesive reads:

$$\tau_a(x, t) = G_a \gamma_a(x, t) \tag{4.2}$$

The shear deformation in the adhesive is induced by the axial displacement of the substrates. Recall that the adhesive shear strain and the adherends' displacements are independent of the vertical coordinate y. Thus, the adhesive shear strain is given by:

$$\gamma_a(x, t) = \frac{u_2(x, t) - u_1(x, t)}{t_a} \tag{4.3}$$

where t_a is the adhesive layer thickness. The main assumption of the shear-lag model is that the substrates only carry axial loads whereas the adhesive layer only carries transverse loads. Isolating an overlap region of unit width and length dx, the substrates are submitted to axial forces at x and $x + dx$, and also a transverse loads at the adherend-adhesive interfaces (Figs. 4.3(a) and 4.3(c)). The adhesive layer is only submitted to the latter transverse load (Fig. 4.3(b)). The free-body diagram of Fig. 4.3 shows that assuming constant through-the-thickness shear stress/strain signifies that the adhesive layer inertia is insignificant and *vice versa*. Actually, the last hypothesis is fair as adhesive layers are mostly made up from materials of lower densities and are also of lower thicknesses than the adherends. If the adhesive layer inertia was considered then the adhesive shear stress would vary through-the-thickness.

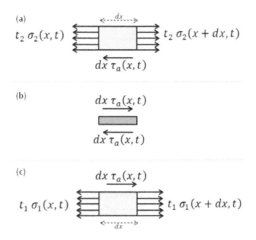

FIG. 4.3 Free-body diagram of a dx-long single-lap joint of unit width: (a) upper substrate, (b) adhesive, and (c) lower substrate.

Considering Figs. 4.3(a) and 4.3(c), and applying the second Newton's law to the lower adherend yields (Sato 2008):

$$t_1 \frac{\partial \sigma_1}{\partial x} + \tau_a(x, t) = \rho_1 t_1 \frac{\partial^2 u_1}{\partial t^2} \tag{4.4}$$

where t_1 and ρ_1 are the thickness and density, respectively, of the lower substrate. Similarly, applying the second Newton's law to the upper adherend yields:

$$t_2 \frac{\partial \sigma_2}{\partial x} - \tau_a(x, t) = \rho_2 t_2 \frac{\partial^2 u_2}{\partial x^2} \tag{4.5}$$

where t_2 and ρ_2 are the thickness and density, respectively, of the upper substrate. Substituting, in Eqs. (4.4) and (4.5), the expression of the adherends' axial stresses of Eq. (4.1) and the adhesive's shear stress of Eqs. (4.2) and (4.3), gives two coupled partial differential equations of the axial displacements u_1 and u_2. More precisely, Eqs. (4.4) and (4.5) are equivalently written (Sato 2008, 2009):

$$t_1 E_1 \frac{\partial^2 u_1}{\partial x^2} + G_a \frac{u_2(x, t) - u_1(x, t)}{t_a} = \rho_1 t_1 \frac{\partial^2 u_1}{\partial t^2} \tag{4.6}$$

and

$$t_2 E_2 \frac{\partial^2 u_2}{\partial x^2} - G_a \frac{u_2(x,t) - u_1(x,t)}{t_a} = \rho_2 t_2 \frac{\partial^2 u_2}{\partial t^2} \tag{4.7}$$

Respectively, where E_1 is the lower adherend Young's modulus, E_2 is the upper adherend Young's modulus, and G_a is the adhesive shear modulus. These equations can be simplified if the adherends have the same thickness $t_1 = t_2 = t_s$ and the same material properties, i.e., $E_1 = E_2 = E_s$ and $\rho_1 = \rho_2 = \rho_s$. Assuming similar adherends yields:

$$t_s E_s \frac{\partial^2 u_1}{\partial x^2} + G_a \frac{u_2(x,t) - u_1(x,t)}{t_a} = \rho_s t_s \frac{\partial^2 u_1}{\partial t^2} \tag{4.8}$$

and

$$t_s E_s \frac{\partial^2 u_2}{\partial x^2} - G_a \frac{u_2(x,t) - u_1(x,t)}{t_a} = \rho_s t_s \frac{\partial^2 u_2}{\partial t^2} \tag{4.9}$$

Eqs. (4.8) and (4.9) are coupled equations in terms of u_1 and u_2. In order to get rid of the coupling, they will be solved in terms of $(u_1 + u_2)$ and $(u_1 - u_2)$. More precisely, adding Eqs. (4.8) and (4.9) gives:

$$\frac{\partial^2(u_1 + u_2)}{\partial x^2} = \frac{\rho_s}{E_s} \frac{\partial^2(u_1 + u_2)}{\partial t^2} \tag{4.10}$$

while subtracting Eqs. (4.8) and (4.9) leads to:

$$\frac{\partial^2(u_2 - u_1)}{\partial x^2} + \frac{2G_a}{t_a t_s E_s}(u_2 - u_1) = \frac{\rho_s}{E_s} \frac{\partial^2(u_2 - u_1)}{\partial t^2} \tag{4.11}$$

Let $c_s = \sqrt{\dfrac{E_s}{\rho_s}}$ be the velocity of the longitudinal waves in substrates and

$k_s = \sqrt{\dfrac{2G_a}{t_a t_s E_s}}$ be a constant that depends on the adhesive and adherends thicknesses, the adhesive shear modulus, and the adherends axial modulus. Actually, k_s gives the square root of twice the adhesive shear stiffness-to-substrates axial stiffness ratio. As we are interested in the harmonic response, the partial differential equations (4.10) and (4.11) give the following ordinary differential equations:

$$\frac{\partial^2(\tilde{u}_1 + \tilde{u}_2)}{\partial x^2} = -\xi^2(\tilde{u}_1 + \tilde{u}_2) \tag{4.12}$$

and

$$\frac{\partial^2(\tilde{u}_2 - \tilde{u}_1)}{\partial x^2} = \zeta^2(\tilde{u}_2 - \tilde{u}_1) \tag{4.13}$$

where $u_1(x,t) = Re(\tilde{u}_1(x)e^{i\omega t})$, $u_2(x,t) = Re(\tilde{u}_2(x)e^{i\omega t})$, $\xi^2 = \dfrac{\omega^2}{c_s^2}$ and $\zeta^2 = k_s^2 - \xi^2$. The wave number ξ is real number. The constant ζ is a real number up to a critical angular frequency $\omega_c = k_s c_s$. Let's focus on the low frequency range where $\omega < \omega_c$. In this framework, the solutions of the ordinary equations (4.12) and (4.13) read:

$$\tilde{u}_1(x) + \tilde{u}_2(x) = A(\omega)e^{-i\xi x} + B(\omega)e^{i\xi x} \qquad (4.14)$$

and

$$\tilde{u}_1(x) - \tilde{u}_2(x) = \alpha(\omega)e^{-\zeta x} + \beta(\omega)e^{\zeta x} \qquad (4.15)$$

respectively, where A, B, α and β are complex constants that depend on the angular frequency ω. Eqs. (4.14) and (4.15) gives the solutions for the sum and difference of the adherend's magnitudes of displacement. The difference between the two displacements is proportional to the shear stress in the adhesive layer. Therefore, the complex magnitude of the adhesive shear stress reads:

$$\tilde{\tau}_a(x, \omega) = \frac{G_a}{t_a}(\alpha(\omega)e^{-\zeta x} + \beta(\omega)e^{\zeta x}) \qquad (4.16)$$

Moreover, summing Eq. (4.14) and Eq. (4.15) gives the complex magnitude of the lower adherend displacement:

$$\tilde{u}_1(x, \omega) = \frac{1}{2}(A(\omega)e^{-i\xi x} + B(\omega)e^{i\xi x} + \alpha(\omega)e^{-\zeta x} + \beta(\omega)e^{\zeta x}) \qquad (4.17)$$

whereas subtracting the Eqs. (4.15) from (4.14) gives the complex magnitude of the upper adherend displacement:

$$\tilde{u}_2(x, \omega) = \frac{1}{2}(A(\omega)e^{-i\xi x} + B(\omega)e^{i\xi x} - \alpha(\omega)e^{-\zeta x} - \beta(\omega)e^{\zeta x}) \qquad (4.18)$$

The complex constants A, B, α, and β should be determined considering the boundary conditions.

4.3.1.2 Boundary Conditions

In order to determine the ω-dependent constants A, B, α, and β, we will concentrate on the overlap region. We assume the virtual boundary conditions represented in Fig. 4.4. This type of virtual boundary conditions has two advantages. First, it works with any real boundary, as R_1 and R_2 are the forces/reactions transmitted by the substrates from the real boundary conditions to the edge of the overlap region. Second, this type of boundary condition is homogeneous and the derivation of solutions is simpler than in the case of non-homogeneous boundary conditions, i.e., one force condition on a substrate and a displacement condition on the other adherend.

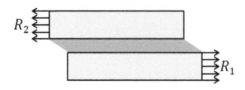

FIG. 4.4 Virtual boundary conditions in the overlap region.

The boundary conditions are force boundary conditions. More precisely, the right end of the upper substrate and the left end of the lower substrate are free. Let $\tilde{N}_1(x, \omega)$ be the amplitude of the axial force applied on a cross-section x of the lower substrate. Similarly, let $\tilde{N}_2(x, \omega)$ be the amplitude of the axial force applied on a cross-section x of the upper substrate. The boundary conditions lead to $\tilde{N}_1(0, \omega) = \tilde{N}_2(l, \omega) = 0$. Moreover, the right end of the lower adherend and the left end of the upper adherend receive the reactions $\tilde{R}_1(\omega)$ and $\tilde{R}_2(\omega)$. Hence, $\tilde{N}_1(l, \omega) = \tilde{R}_1(\omega)$ and $\tilde{N}_2(0, \omega) = \tilde{R}_2(\omega)$. We need now to express the axial forces in the overlap region. Using Eqs. (4.17) and (4.18), it is possible to obtain the axial strain by differentiating these equations with respect to the coordinate x. Subsequently, it is possible to derive the axial forces by multiplying the axial strain by the product $E_s t_s$. More precisely,

$$\tilde{N}_1(x) = \frac{iE_s t_s \xi}{2}(-Ae^{-i\xi x} + Be^{i\xi x}) + \frac{E_s t_s \zeta}{2}(-\alpha e^{-\zeta x} + \beta e^{\zeta x}) \qquad (4.19)$$

whereas summing the Eqs. (4.14) and (4.15) gives the complex magnitude of the upper adherend displacement:

$$\tilde{N}_2(x) = \frac{iE_s t_s \xi}{2}(-Ae^{-i\xi x} + Be^{i\xi x}) - \frac{E_s t_s \zeta}{2}(-\alpha e^{-\zeta x} + \beta e^{\zeta x}) \qquad (4.20)$$

Here we have assumed that substrates are of width 1 m. Substituting the boundary conditions in Eqs. (4.19) and (4.20) gives a system of four ordinary linear equations with four unknowns. Solving these equations yields after simplifications:

$$A(\omega) = \frac{\tilde{R}_2(\omega)e^{i\xi l} - \tilde{R}_1(\omega)}{2t_s E_s \xi \sin(\xi l)} \qquad (4.21)$$

$$B(\omega) = \frac{\tilde{R}_2(\omega)e^{-i\xi l} - \tilde{R}_1(\omega)}{2t_s E_s \xi \sin(\xi l)} \qquad (4.22)$$

$$\alpha(\omega) = \frac{\tilde{R}_2(\omega)e^{\zeta l} + \tilde{R}_1(\omega)}{2t_s E_s \zeta \sinh(\zeta l)} \qquad (4.23)$$

and

$$\beta(\omega) = \frac{\tilde{R}_2(\omega)e^{-\zeta l} + \tilde{R}_1(\omega)}{2t_s E_s \zeta \sinh(\zeta l)} \qquad (4.24)$$

In order to simplify the calculations, we will limit the study to the overlap region. Consequently, we assume that the load $\tilde{P}_1$ is directly applied at the right edge of the lower substrate (limited to the overlap region) as shown in Fig. 4.5, i.e., $\tilde{P}_1 = \tilde{R}_1$. Moreover, the left edge of the upper substrate (limited to the overlap region) is cantilevered, i.e., $\tilde{u}_2(0, \omega) = 0$.

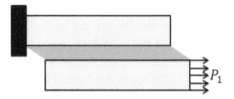

FIG. 4.5 Simplified boundary conditions in the overlap region.

We would like now to consider the real boundary condition $\tilde{u}_2(0, \omega) = 0$ (Fig. 4.5). More precisely, Eq. (4.18) is considered, where we assume $x = 0$. Later we substitute $\alpha(\omega)$, $\beta(\omega)$, $A(\omega)$ and $B(\omega)$ by their expressions in the Eqs. (4.21) to (4.24). Considering that this displacement is null and solving gives the reaction at the support, which reads:

$$\tilde{R}_2(\omega) = \frac{\dfrac{1}{\xi \sin(\xi l)} + \dfrac{1}{\zeta \sinh(\zeta l)}}{\dfrac{1}{\xi \tan(\xi l)} - \dfrac{1}{\zeta \tanh(\zeta l)}} \tilde{P}_1(\omega) \tag{4.25}$$

Eq. (4.25) closes the solution as it gives the reaction at the support $\tilde{R}(\omega)$ in terms of the applied load $\tilde{P}_1(\omega)$. Once $\tilde{R}_2(\omega)$ is calculated, it is possible to calculate $\alpha(\omega)$, $\beta(\omega)$, $A(\omega)$ and $B(\omega)$ using Eqs. (4.21) to (4.24). Subsequently, it is possible to compute the axial forces (Eqs. (4.19) and (4.20)), the displacements (Eqs. (4.17) and (4.18)) and the shear stress in the adhesive (Eq. (4.16)).

It is also possible to derive the transfer function of the system as the ratio of the displacement at the right end of the lower substrate to the applied load. More precisely,

$$\tilde{H}(\omega) = \frac{\tilde{u}_1(l, \omega)}{\tilde{P}_1(\omega)} \tag{4.26}$$

4.3.2 Challita et al. Model

4.3.2.1 Basic Equations

In this section, we are interested in predicting the harmonic response of single-lap and double-lap adhesively bonded joints starting from the assumptions of Tsai et al. (1998). The geometries of the single-lap and double-lap joints are those used in Section 4.3.1.1 and depicted in Fig. 4.1. The boundary conditions represented in Fig. 4.2 are also considered here. Similar materials are used for the two substrates. As for Section 4.3.1.1, analytical solutions are here derived only for the single-lap joint since the solutions for the double-lap joints can be deduced by using symmetry.

Tsai et al. (1998) has made the following assumptions: (i) the adherends and adhesive move in the axial (x-)direction so there is no motion in the transverse direction; (ii) the shear stress in the adherends is linearly dependent on the through-the-thickness coordinate y; and (iii) the shear stress in the adhesive is independent of y; thus, the shear stress is proportional to the difference between the displacement at the adhesive-adherend interfaces. Assuming an elastic behavior for the adhesive and substrates, the axial stress in the adherends and the shear stress in the substrates and adhesive write:

$$\tilde{\sigma}_i(x, y, \omega) = E_s \tilde{\varepsilon}_i(x, y, \omega), \, i = 1, 2 \tag{4.27}$$

$$\tilde{\tau}_i(x, y, \omega) = G_s \tilde{\gamma}_i(x, y, \omega), \, i = 1, 2 \tag{4.28}$$

and

$$\tilde{\tau}_a(x, \omega) = G_a \tilde{\gamma}_a(x, \omega) \tag{4.29}$$

The shear stresses in the adherends are proportional of the through-the-thickness coordinate y. The lower surface of the lower adherend and the upper surface of the upper adherend are free. Thus, $\tilde{\tau}_1(x, y = 0, \omega) = 0$ and $\tilde{\tau}_2(x, y = t_a + 2t_s, \omega) = 0$. Consequently, the substrates shear stresses are given by:

$$\tilde{\tau}_1(x, y, \omega) = \frac{y}{t_s}\tilde{\tau}_a(x, \omega) \tag{4.30}$$

and

$$\tilde{\tau}_2(x, y, \omega) = \left(1 - \frac{y - (t_a + t_s)}{t_s}\right)\tilde{\tau}_a(x, \omega) \tag{4.31}$$

No transverse motion is allowed for the substrates, which only move along the x-direction. Therefore, the shear strain can either be calculated as $\tilde{\gamma}_i(x, y, \omega) = \frac{\partial}{\partial y}\tilde{u}_i(x, y, \omega)$ or also using Eqs. (4.28) and (4.29). Subsequently, the substrates' displacements are obtained by integrating the shear strain along the coordinate y. After simplifications, the axial displacements of the lower and the upper substrates read:

$$\tilde{u}_1(x, y, \omega) = \tilde{u}_{i1}(x, \omega) + \frac{y^2 - t_s^2}{2t_s G_s}\tilde{\tau}_a(x, \omega) \tag{4.32}$$

and

$$\tilde{u}_2(x, y, \omega) = \tilde{u}_{i2}(x, \omega) - \frac{(y - 2t_s - t_a)^2 - t_s^2}{2t_s G_s}\tilde{\tau}_a(x, \omega) \tag{4.33}$$

respectively, where $\tilde{u}_{i1}(x, \omega) = \tilde{u}_1(x, y = t_s, \omega)$ and $\tilde{u}_{i2}(x, \omega) = \tilde{u}_2(x, y = t_s + t_a, \omega)$. The axial strain is then obtained using $\tilde{\varepsilon}_i(x, y, \omega) = \frac{\partial}{\partial x}\tilde{u}_i(x, y, \omega)$. Later, Eq. (4.27) is applied to derive the axial stresses:

$$\tilde{\sigma}_1(x, y, \omega) = E_s\left(\frac{\partial \tilde{u}_{i1}}{\partial x} + \frac{y^2 - t_s^2}{2t_s G_s}\frac{\partial \tilde{\tau}_a}{\partial x}\right) \tag{4.34}$$

and

$$\tilde{\sigma}_2(x, y, \omega) = E_s\left(\frac{\partial \tilde{u}_{i2}}{\partial x} - \frac{(y - 2t_s - t_a)^2 - t_s^2}{2t_s G_s}\frac{\partial \tilde{\tau}_a}{\partial x}\right) \tag{4.35}$$

Subsequently, the axial forces are derived while integrating the axial stresses along the coordinate y. After simplifications, the axial forces are written as:

$$\tilde{N}_1(x, \omega) = E_s t_s \frac{\partial \tilde{u}_{i1}}{\partial x} - \frac{E_s t_s^2}{3G_s} \frac{\partial \tilde{\tau}_a}{\partial x} \tag{4.36}$$

and

$$\tilde{N}_2(x, \omega) = E_s t_s \frac{\partial \tilde{u}_{i2}}{\partial x} + \frac{E_s t_s^2}{3G_s} \frac{\partial \tilde{\tau}_a}{\partial x} \tag{4.37}$$

Considering the free-body diagrams depicted in Fig. 4.3 and applying the second Newton's law to parts of substrates of length dx yields after simplifications:

$$\frac{d\tilde{N}_1}{dx} = -\rho_s t_s \omega^2 \tilde{u}_{i1}(x, \omega) - \left(1 - \frac{\rho_s t_s^2 \omega^2}{3G_s}\right) \tilde{\tau}_a(x, \omega) \tag{4.38}$$

and

$$\frac{d\tilde{N}_2}{dx} = -\rho_s t_s \omega^2 \tilde{u}_{i2}(x, \omega) + \left(1 - \frac{\rho_s t_s^2 \omega^2}{3G_s}\right) \tilde{\tau}_a(x, \omega) \tag{4.39}$$

The shear stress in the adhesive is independent of the transverse coordinate y. Therefore, it is proportional to the difference between the displacements of the adhesive-adherends interfaces, i.e.,

$$\tilde{\tau}_a(x, \omega) = \frac{G_s}{t_a}(\tilde{u}_{i2}(x, \omega) - \tilde{u}_{i1}(x, \omega)) \tag{4.40}$$

In order to obtain a differential equation for $\tilde{\tau}_a$, we first subtract Eq. (4.36) from Eq. (4.37) and then we differentiate with respect to x. Second, we subtract Eq. (4.38) from Eq. (4.39). Equating the two results yields:

$$\frac{\partial^2 \tilde{\tau}_a}{\partial x^2} - \zeta'^2 \tilde{\tau}_a = 0 \tag{4.41}$$

which is equivalently written:

$$\frac{\partial^2(\tilde{u}_{i2} - \tilde{u}_{i1})}{\partial x^2} - \zeta'^2(\tilde{u}_{i2} - \tilde{u}_{i1}) = 0 \tag{4.42}$$

where $\zeta'^2 = \frac{k_s^2}{\lambda} - \xi^2$, and $\lambda = 1 + \frac{2t_s G_a}{3t_a G_s}$. Recall that $k_s^2 = \frac{2G_a}{t_s t_a E_s}$ and $\xi^2 = \frac{\rho_s \omega^2}{E_s}$.

Likewise, we add Eq. (4.36) to Eq. (4.37) and then we differentiate with respect to x. Besides, we add Eq. (4.38) to Eq. (4.37). Equating the two results gives:

$$\frac{\partial^2(\tilde{u}_{i2} + \tilde{u}_{i1})}{\partial x^2} + \xi^2(\tilde{u}_{i2} + \tilde{u}_{i1}) = 0 \tag{4.43}$$

The ordinary differential equation (4.43) is exactly the ordinary differential equation (4.12). Moreover, the ordinary differential equation (4.42) is similar the Eq. (4.13) except for ζ being replaced by ζ'. Consequently, the solutions of Eqs. (4.42) and (4.43) are written as:

$$\tilde{u}_{i2}(x, \omega) - \tilde{u}_{i1}(x, \omega) = \alpha'(\omega)e^{-\zeta'(\omega)} + \beta'(\omega)e^{\zeta'(\omega)} \tag{4.44}$$

and

$$\tilde{u}_{i2}(x, \omega) + \tilde{u}_{i1}(x, \omega) = A'(\omega)e^{-i\xi(\omega)} + B'(\omega)e^{i\xi(\omega)} \tag{4.45}$$

respectively. $\alpha'(\omega)$ and $\beta'(\omega)$ are non-propagating damped waves. Besides, $A'(\omega)$ and $B'(\omega)$ are propagating non-damped waves. Eqs. (4.44) and (4.45) can be used to obtain separately $\tilde{u}_{i1}$ and $\tilde{u}_{i2}$. More precisely, subtracting Eq. (4.45) from Eq. (4.44) gives the displacement at the lower adherend-adhesive interface:

$$\tilde{u}_{i1}(x) = \frac{1}{2}(A'e^{-i\xi x} + B'e^{i\xi x} - \alpha'e^{-\zeta'x} - \beta'e^{\zeta'x}) \tag{4.46}$$

whereas summing Eqs. (4.14) and (4.15) gives the displacement at the upper adherend-adhesive interface:

$$\tilde{u}_{i2}(x) = \frac{1}{2}(A'e^{-i\xi x} + B'e^{i\xi x} + \alpha'e^{-\zeta'x} + \beta'e^{\zeta'x}) \tag{4.47}$$

The complex constants A', B', α', and β' should be determined considering the boundary conditions.

4.3.2.2 Boundary Conditions

In this section, we consider the same approach as the one considered in Section 4.3.1.2. First, we assume the boundary conditions depicted in Fig. 4.4, i.e., homogeneous force boundary conditions.

First, we need to derive the general expressions for the axial forces and average axial displacements in the substrates. Namely, the results obtained in Eqs. (4.40), (4.46) and (4.47) can be substituted in Eqs. (4.36) and (4.37) to get the axial forces. After simplifications, we obtain:

$$\tilde{N}_1 = \frac{it_s E_s \xi}{2}(-A'e^{-i\xi x} + B'e^{i\xi x}) + \frac{t_s E_s \zeta'}{2}(-\alpha'e^{-\zeta'x} + \beta'e^{\zeta'x}) \tag{4.48}$$

and

$$\tilde{N}_2 = \frac{it_s E_s \xi}{2}(-A'e^{-i\xi x} + B'e^{i\xi x}) - \frac{t_s E_s \zeta'}{2}(-\alpha'e^{-\zeta'x} + \beta'e^{\zeta'x}) \tag{4.49}$$

Similarly, substituting Eqs. (4.40), (4.46) and (4.47) in Eqs. (3.32) and (3.33) and calculating the through-the-thickness average gives:

$$\langle \tilde{u}_1 \rangle_y = \frac{1}{2}(A'e^{-i\xi x} + B'e^{i\xi x}) + \frac{1}{2}(\alpha'e^{-\zeta'x} + \beta'e^{\zeta'x}) \tag{4.50}$$

and

$$\langle \tilde{u}_2 \rangle_y = \frac{1}{2}(A'e^{-i\xi x} + B'e^{i\xi x}) - \frac{1}{2}(\alpha'e^{-\zeta'x} + \beta'e^{\zeta'x}) \tag{4.51}$$

The four waves α', β', A', and B' are determined using the boundary conditions at the extremities of the two substrates (Fig. 4.5). Namely, the right end of the upper substrate and the left end of the lower substrate are free; consequently,

$\tilde{N}_1(0, \omega) = \tilde{N}_2(l, \omega) = 0$. Besides, the right end of the lower adherend receives the external load; thus, $\tilde{N}_1(l, \omega) = \tilde{P}_1(\omega)$. Finally, the left end of the upper adherend is cantilevered; hence, $\tilde{u}_2(0, \omega) = 0$. However, we prefer not writing the boundary conditions in terms of displacements but only writing them in terms of forces. More precisely, we will consider that a force equal to the reaction of the support $\tilde{R}_2$ is rather applied to the left end of upper substrate (Fig. 4.4), i.e., $\tilde{N}_2(0, \omega) = \tilde{R}_2(\omega)$. This greatly helps in calculating a close-form solution. The reaction $\tilde{R}_2$ can later be determined using the boundary condition $\tilde{u}_2(0, \omega) = 0$. Substituting the boundary conditions in Eqs. (4.48) and (4.49) gives a system of four ordinary linear equations with four unknowns; the solution of which writes after simplifications:

$$A'(\omega) = \frac{\tilde{R}_2(\omega)e^{i\xi l} - \tilde{R}_1(\omega)}{2t_s E_s \xi \sin(\xi l)} \tag{4.52}$$

$$B'(\omega) = \frac{\tilde{R}_2(\omega)e^{-i\xi l} - \tilde{R}_1(\omega)}{2t_s E_s \xi \sin(\xi l)} \tag{4.53}$$

$$\alpha'(\omega) = \frac{\tilde{R}_2(\omega)e^{\zeta' l} + \tilde{R}_1(\omega)}{2t_s E_s \zeta' \sinh(\zeta' l)} \tag{4.54}$$

and

$$\beta'(\omega) = \frac{\tilde{R}_2(\omega)e^{-\zeta' l} + \tilde{R}_1(\omega)}{2t_s E_s \zeta' \sinh(\zeta' l)} \tag{4.55}$$

We would like now to go back to the boundary condition $\tilde{u}_2(0, \omega) = 0$. Actually, the improved shear-lag model assumes that the displacement is varying with the thickness. Thus, a boundary condition where the displacement is constant is impossible to write (Challita and Othman 2012). Therefore, we will only consider that the average displacement is null, i.e., $\langle\tilde{u}_2\rangle_y(0, \omega) = 0$. More precisely, Eq. (4.51) is considered where we consider $x = 0$. Later we substitute $\alpha'(\omega)$, $\beta'(\omega)$, $A'(\omega)$ and $B'(\omega)$ by their expressions in Eqs. (4.52) to (4.55). Considering that this average displacement is null and solving gives the reaction at the support, which reads:

$$\tilde{R}_2(\omega) = \frac{\dfrac{1}{\xi\sin(\xi l)} + \dfrac{1}{\zeta'\sinh(\zeta' l)}}{\dfrac{1}{\xi\tan(\xi l)} - \dfrac{1}{\zeta'\tanh(\zeta' l)}} \tilde{P}_1(\omega) \tag{4.56}$$

Eq. (4.56) closes the solution for the Challita et al. model as it gives the reaction at the support $\tilde{R}_2(\omega)$ in terms of the applied load $\tilde{P}_1(\omega)$. Once $\tilde{R}_2(\omega)$ is calculated, it is possible to calculate $\alpha'(\omega)$, $\beta'(\omega)$, $A'(\omega)$ and $B'(\omega)$ using Eqs. (4.52) to (4.55). Subsequently it is possible to compute axial forces (Eqs. (4.48) and (4.49)), average displacements (Eqs. (4.50) and (4.51)) and the shear stress in the adhesive (Eq. (4.40)). It is also possible to determine the transfer function of the system as the ratio of the average displacement at the right end of the lower substrate to the applied load, more precisely:

$$\tilde{H}(\omega) = \frac{\langle \tilde{u}_2 \rangle_y (l, \omega)}{\tilde{P}_1(\omega)} \tag{4.57}$$

4.3.2.3 Comparison with Sato's Model

Chalita et al.'s model has some similarities with Sato's model and their solutions have multiple common points. Thus, Eqs. (4.48) and (4.49) are quite similar to Eqs. (4.19) and (4.20). Eqs. (4.52) to (4.57) are also similar to Eqs. (4.21) to (4.26). However, there are two main differences.

The first difference between the solutions of the two models comes from the constant $\lambda = 1 + \dfrac{2t_s G_a}{3t_a G_s}$, which depends on $\dfrac{G_a}{t_a}$, the shear stiffness of the adhesive layer, and $\dfrac{G_s}{t_s}$, the shear stiffness of one adherend. More precisely, the constant λ depends on the ratio of the two stiffnesses, i.e., $\dfrac{G_a}{t_a} \Big/ \dfrac{G_s}{t_s}$. The constant λ is involved in the solutions given by Challita et al.'s model. However, it is not used to express the solutions of Sato's model. Actually, Sato's model expresses the solutions in terms of the constant ζ, such as $\zeta^2 = k_s^2 - \xi^2$. Instead, Challita et al.'s model uses ζ, which is defined by $\zeta'^2 = \dfrac{k_s^2}{\lambda} - \xi^2$. The main difference comes from the constant λ. If the substrates are much stiffer than the adhesive layer, then $\dfrac{G_a}{t_a} \ll \dfrac{G_s}{t_s}$ and λ converges to 1. In this case ζ' converges to $\left(\zeta' \underset{\lambda \to 1}{\to} \zeta \right)$. This means that if the substrates' shear stiffness is much higher than the adhesive layer shear stiffness, then Sato's model and Challita et al.'s model will predict the same axial forces (See Eqs. (4.19), (4.20), (4.48) and (4.49)) and average displacements (See Eqs. (4.17), (4.18), (4.50) and (4.51)) in the substrates. They will also predict the same shear stress in the adhesive layer.

The second difference between the two solutions is due to the fact that Challita et al.'s model allows the substrates' shear stresses to vary in terms of the through-the-thickness coordinates, whereas Sato's model assumes that the shear stress in constant through-the-thickness. Thus, the substrates' displacements, shear strains and shear stresses calculated by Sato's model can only be compared to the substrates average displacements, average shear strains and average shear stresses predicted by Challita et al.'s model.

Sato's model (Sato 2008, 2009) and Challita et al.'s model (Challita and Othman 2012, Hazimeh et al. 2015) were recently developed. Only one comparison was undertaken by Challita and Othman (2012), reproduced in Figs. 4.6 and 4.7. In this comparison, the substrates are assumed polymers. Thus, the shear stiffness of the adhesive layer is comparable to the shear stiffness of the substrates. In this case, Challita et al.'s solutions are closer to the three-dimensional (3D) finite-element results than the solutions predicted by Sato's model.

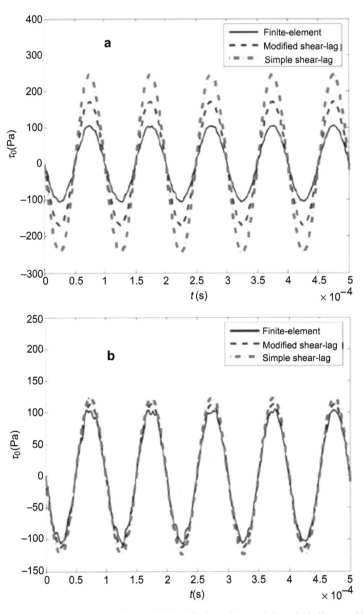

FIG. 4.6 Comparison between Sato's model (simple shear-lag model) and Challita et al.'s model (modified shear-lag model) - Shear stress in the adhesive joint in terms of time: (a) $x = 0.5$ mm and (b) $x = 0.6$ mm. (reproduced from Challita and Othman (2012), with permission from Elsevier Masson SAS).

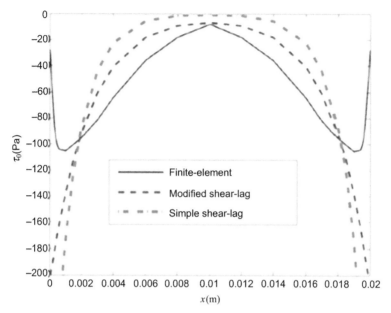

FIG. 4.7 Comparison between Sato's model (simple shear-lag model) and Challita et al.'s model (modified shear-lag model) - Shear stress in the adhesive joint in terms of x at time $t =$ 425 μs. (reproduced from Challita and Othman (2012), with permission from Elsevier Masson SAS).

4.3.3 Vaziri and Nayeb-Hashemi Model

Vaziri and Nayeb-Hashemi (2002b) developed their model to predict the harmonic behavior of a tubular joint. They modified the static shear-lag model (Volkersen 1938) to take into account inertia effects. Sato (2008, 2009) considered the inertia of the substrates and neglected the inertia of the adhesive layer. Vaziri and Nayeb-Hashemi (2002b) included in their equation the inertia of the adhesive layer and the substrates. Moreover, they considered a viscoelastic behavior for the adhesive.

In this section, we will derive the equations assuming that the substrates and the adhesive are elastic. The viscoelastic behavior will be considered in Section 4.5. Moreover, the equations here are derived for the tubular joint shown in Fig. 4.8. As Vaziri and Nayeb-Hashemi (2002b) based their equations on the simple shear-lag model, only normal stresses are involved in the substrates and only shear stresses are involved in the adhesive layer.

Let u_1 and u_1 be the displacements of the inner and outer cylinders, respectively. Focusing on the overlap region, the equilibrium equation for the inner cylinder is written:

$$E_1 \pi (R_2^2 - R_1^2) \frac{\partial^2 u_1}{\partial x^2} + (2\pi R_2) \tau_{1a} - \pi (R_2^2 - R_1^2) \rho_1 \frac{\partial^2 u_1}{\partial t^2} = 0 \qquad (4.58)$$

where E_1 and ρ_1 are the Young's modulus and density of the inner cylinder, and τ_{1a} is the shear stress at the interface between the adhesive and the inner cylinder. Similarly, the equilibrium equation for the outer cylinder can be written:

$$E_2 \pi (R_4^2 - R_3^2) \frac{\partial^2 u_2}{\partial x^2} - (2\pi R_3)\tau_{2a} - \pi (R_4^2 - R_3^2)\rho_1 \frac{\partial^2 u_2}{\partial t^2} = 0 \qquad (4.59)$$

where E_2 and ρ_2 are the Young's modulus and density of the outer cylinder, and τ_{2a} is the shear stress at the interface between the adhesive and the outer cylinder. In the Vaziri and Nayeb-Hashemi's model, the inertia effects in the adhesive are not neglected. Consequently, there is no reason to put τ_{2a} equal to τ_{1a}. On the opposite, if the Sato's model had been considered, then we would have written that τ_{2a} and τ_{1a} are equal. This is not the case here.

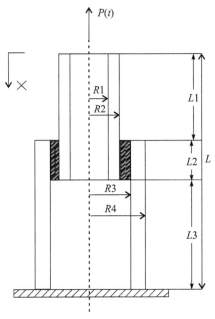

FIG. 4.8 Schematic diagram of a tubular bonded joint under a harmonic axial load. (reproduced from Vaziri and Nayeb-Hashemi (2002b) with permission from Elsevier).

Considering a region of the adhesive layer of length dx, the equilibrium equation is written as:

$$G_a \frac{\partial u_a}{\partial r} + r \frac{\partial}{\partial r}\left(G_a \frac{\partial u_a}{\partial r}\right) - r\rho_a \frac{\partial^2 u_a}{\partial t^2} = 0 \qquad (4.60)$$

where G_a and ρ_a are the shear modulus and density, respectively, of the adhesive layer. In this Eq. (4.60) we have also considered that the shear stress in the adhesive is given by:

$$\tau_a = G_a \frac{\partial u_a}{\partial r} \qquad (4.61)$$

In this section, we are mainly interested in harmonic solutions. Thus, Eqs. (4.58), (4.59) and (4.60) simplify to:

$$E_1 \pi (R_2^2 - R_1^2) \frac{\partial^2 \tilde{u}_1}{\partial x^2} + (2\pi R_2)\tilde{\tau}_{1a} + \pi (R_2^2 - R_1^2)\rho_1 \omega^2 \tilde{u}_1 = 0 \qquad (4.62)$$

$$E_2\pi(R_4^2 - R_3^2)\frac{\partial^2 \tilde{u}_2}{\partial x^2} - (2\pi R_3)\tilde{\tau}_{2a} + \pi(R_4^2 - R_3^2)\rho_1\omega^2\tilde{u}_2 = 0 \qquad (4.63)$$

and

$$G_a\frac{\partial \tilde{u}_a}{\partial r} + r\frac{\partial}{\partial r}\left(G_a\frac{\partial \tilde{u}_a}{\partial r}\right) + r\rho_a\omega^2\tilde{u}_a = 0 \qquad (4.64)$$

Eq. (4.64) can be solved using Bessel functions and boundary conditions at the adhesive interfaces with the inner and outer cylinders. Namely, the adhesive displacement reads:

$$u_a = (Y_0(\delta R_3)J_0(\delta r) - J_0(\delta R_3)Y_0(\delta r))\frac{\tilde{u}_1}{z} + (J_0(\delta R_2)Y_0(\delta r) - Y_0(\delta R_2)J_0(\delta r))\frac{\tilde{u}_2}{z}$$
$$(4.65)$$

where J_0 and Y_0 are the Bessel functions of order zero and the Bessel function of the second kind of order zero, respectively, $\delta = \sqrt{\dfrac{\rho_a\omega^2}{G_a}}$ and $z = J_0(\delta R_2)Y_0(\delta R_3) - Y_0(\delta R_2)J_0(\delta R_3)$. Eq. (4.65) gives the displacement in the adhesive layer in terms of the tubes' displacements. We need now to express the shear stresses at the adherend-adhesive interfaces and then substitute in the differential equations (4.62) and (4.63) in order to cancel u_a.

The shear stress at inner cylinder-adhesive interface $\tilde{\tau}_{1a}$ is obtained by differentiating Eq. (4.65) with respect to the radial coordinate r, then multiplying by the shear modulus G_a and finally substituting R_2 to r. The shear stress $\tilde{\tau}_{1a}$ writes:

$$\tilde{\tau}_{1a} = G_a(\varphi(R_2)\tilde{u}_1 + \psi(R_2)\tilde{u}_2) \qquad (4.66)$$

where, J_1 and Y_1 are the Bessel functions of order one and the Bessel function of the second kind of order one, respectively, and

$$\psi(r) = (J_0(\delta R_3)Y_1(\delta r) - Y_0(\delta R_3)J_1(\delta r))\frac{\delta}{z} \qquad (4.67)$$

and

$$\psi(r) = (J_0(\delta R_2)Y_1(\delta r) - Y_0(\delta R_2)J_1(\delta r))\frac{\delta}{z} \qquad (4.68)$$

Using a similar procedure, the shear stress $\tilde{\tau}_{2a}$ is written:

$$\tilde{\tau}_{2a} = G_a(\varphi(R_3)\tilde{u}_1 + \psi(R_3)\tilde{u}_2) \qquad (4.69)$$

Substituting the new expressions of $\tilde{\tau}_{1a}$ and $\tilde{\tau}_{2a}$ in Eqs. (4.62) and (4.63) gives the two following differential equations:

$$E_1\pi(R_2^2 - R_1^2)\frac{\partial^2 \tilde{u}_1}{\partial x^2} + (2\pi R_2)G_a(\varphi(R_2)\tilde{u}_1 + \psi(R_2)\tilde{u}_2) + \pi(R_2^2 - R_1^2)\rho_1\omega^2\tilde{u}_1 = 0 \quad (4.70)$$

and

$$E_2\pi(R_4^2 - R_3^2)\frac{\partial^2 \tilde{u}_2}{\partial x^2} - (2\pi R_3)G_a(\varphi(R_3)\tilde{u}_1 + \psi(R_3)\tilde{u}_2) + \pi(R_4^2 - R_3^2)\rho_1\omega^2\tilde{u}_2 = 0 \quad (4.71)$$

Eqs. (4.70) and (4.71) make a system of coupled two second-order linear differential equations. Solving these equations gives $\tilde{u}_1$ and $\tilde{u}_2$, and later with Eq. (4.65) it is possible to calculate $\tilde{u}_a$.

Using this model, Vaziri and Nayeb-Hashemi (2002b) have shown, for example, that the first natural frequency increases as the ratios $\dfrac{G_a}{E_1}$ and $\dfrac{E_2}{E_1}$ increase (Fig. 4.9). They also showed that the stress concentration, in the adhesive layer, is more important for higher $\dfrac{G_a}{E_1}$ ratios (Fig. 4.10). This result is similar to single-lap or double-lap joints as the adhesive shear stress concentration increases as the adhesive stiffness tends to increase relatively to the substrates' stiffness.

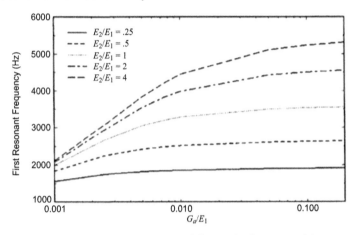

FIG. 4.9 Effect of adhesive/adherents elastic modulus on the first natural frequency of tubular joints. (reproduced from Vaziri and Nayeb-Hashemi (2002b) with permission from Elsevier).

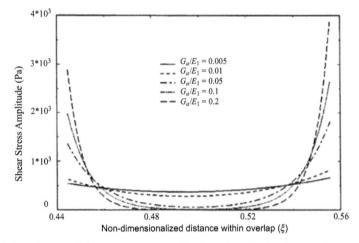

FIG. 4.10 Distribution of the shear stress amplitude at the adhesive/inner cylinder interface. (reproduced from Vaziri and Nayeb-Hashemi (2002b) with permission from Elsevier).

4.4 IMPACT RESPONSE

4.4.1 Sato's Model

The approach to obtain the impact response is quite similar to the one used to obtain the harmonic response. More precisely, the space-time Eqs. (4.1) to (4.11), derived in Section 4.3.1, hold here too. The complex representation used in Section 4.3.1 is fine for deriving the harmonic response of single- or double-lap joints. In order to calculate the impact response, the Laplace transforms will be used here. Let $F(s)$ be the Laplace transform of the function $f(t)$.

Applying the Laplace transform on the partial differential equations (4.10) and (4.11) gives the following the ordinary differential equation:

$$\frac{\partial^2 (U_1 + U_2)}{\partial x^2} = \hat{\xi}^2 (U_1 + U_2) \tag{4.72}$$

and

$$\frac{\partial^2 (U_2 - U_1)}{\partial x^2} = \hat{\zeta}^2 (U_2 - U_1) \tag{4.73}$$

where U_1 and U_2 are the Laplace transforms of u_1 and u_2, and, $\hat{\xi}^2 = \dfrac{s^2}{c_s^2}$ and $\zeta^2 = k_s^2 + \hat{\xi}^2$. Moreover, Eqs. (4.72) and (4.73) assume that the joint is at rest at time $t = 0$. The solutions of the ordinary equations (4.72) and (4.73) read:

$$U_1(x) + U_2(x) = \hat{A}(s)\cosh(\hat{\xi}x) + \hat{B}(s)\sinh(\hat{\xi}x) \tag{4.74}$$

and

$$U_1(x) - U_2(x) = \hat{C}(s)\cosh(\hat{\zeta}x) + \hat{D}(s)\sinh(\hat{\zeta}x) \tag{4.75}$$

respectively, where $\hat{A}, \hat{B}, \hat{C}$ and $\hat{D}$ are complex constants that depend on the complex variable s. Eqs. (4.74) and (4.75) give the solutions for the sum and difference of the Laplace transforms of the adherend's displacement. Therefore, the Laplace transform of the adhesive shear stress reads:

$$T_a(x, s) = \frac{G_a}{t_a}(\hat{C}(s)\cosh(\hat{\zeta}x) + \hat{D}(s)\sinh(\hat{\zeta}x)) \tag{4.76}$$

Moreover, summing Eq. (4.74) from Eq. (4.75) gives the Laplace transform of the lower adherend's displacement:

$$U_1(x, s) = \frac{1}{2}(\hat{A}(s)\cosh(\hat{\xi}x) + \hat{B}(s)\sinh(\hat{\xi}x) + \hat{C}(s)\cosh(\hat{\zeta}x) + \hat{D}(s)\sinh(\hat{\zeta}x)) \tag{4.77}$$

and subtracting Eqs. (4.75) from (4.74) gives the Laplace transform of the upper adherend displacement:

$$U_2(x, s) = \frac{1}{2}(\hat{A}(s)\cosh(\hat{\xi}x) + \hat{B}(s)\sinh(\hat{\xi}x) - \hat{C}(s)\cosh(\hat{\zeta}x) - \hat{D}(s)\sinh(\hat{\zeta}x)) \tag{4.78}$$

The constants $\hat{A}, \hat{B}, \hat{C}$ and $\hat{D}$ could be determined by considering the boundary conditions. More precisely, a stress incident wave $\sigma_i(t)$ is assumed to propagate in the lower substrate and arrives at $t = 0$ at the right edge of the overlap region (Fig. 4.11). Moreover, the upper substrate is assumed to be semi-infinite.

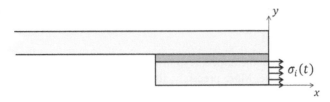

FIG. 4.11 Schematic of the boundary conditions of the impact problem.

Considering and solving for these boundary conditions leads to:

$$\hat{A}(s) = -\hat{B}(s) = -\frac{\Sigma_i(s)}{E_s\hat{\xi}} \qquad (4.79)$$

and

$$\hat{C}(s) = -\hat{D}(s) = \frac{\Sigma_i(s)}{E_s\hat{\zeta}} \qquad (4.80)$$

where $\Sigma_i(s)$ is the Laplace transform of the incident wave.

We would like to focus on the stress that is transmitted at the edge of the adhesive layer, i.e., $\tau_a(x = 0, t)$. Without any loss of generality, the reference point ($x = 0$) is moved to the right edge of the overlap region. Substituting Eq. (4.80) in Eq. (4.76) and considering $x = 0$, the Laplace transform of the shear stress, at the right edge of the adhesive layer, reads:

$$T_a(0, s) = \frac{G_a}{t_a} \frac{\Sigma_i(s)}{E_s\hat{\zeta}} \qquad (4.81)$$

The transfer function between $T_a(0, s)$ and $\Sigma_i(s)$, reads:

$$H(0, s) = \frac{G_a}{t_a E_s\hat{\zeta}} = \frac{G_a c_s}{t_a E_s} \frac{1}{\sqrt{s^2 + c_s^2 k_s^2}} \qquad (4.82)$$

The response of the adhesive layer to an impulse wave is the inverse-Laplace transform of the transfer function. It is then written:

$$\tau_a^{imp}(0, t) = \mathcal{L}^{-1}(H(0, s)) = \frac{G_a c_s}{t_a E_s} J_0(c_s k_s t) u(t) \qquad (4.83)$$

where J_0 is the Bessel function of order zero and $u(t)$ is the Heaviside unit step function. The indicial response, or the response of the joint to Heaviside step wave, is simply the time integration of Eq. (4.83). Namely,

$$\tau_a^{ind}(0, t) = \frac{G_a c_s}{t_a E_s} \int_0^t J_0(c_s k_s t') dt' \qquad (4.84)$$

4.4.2 Challita et al. Model

The same approach, proposed by Sato (2008, 2009) to predict the impulse and indicial shear stress at the edge of the overlap region, can be applied to the Challita and Othman (2012) framework. This was carried out by Hazimeh et al. (2015). More precisely, the complex representation approach is now replaced by the use of Laplace transforms. This leads to the following equations:

$$\tau_a^{imp}(0,t) = \mathcal{L}^{-1}(H(0,s)) = \frac{G_a c_s}{t_a E_s \lambda} J_0\left(\frac{c_s k_s}{\sqrt{\lambda}}t\right)u(t) \qquad (4.85)$$

and

$$\tau_a^{ind}(0,t) = \frac{G_a c_s}{t_a E_s \lambda}\int_0^t J_0\left(\frac{c_s k_s}{\sqrt{\lambda}}t'\right)dt' \qquad (4.86)$$

Challita et al. model gives solutions, Eqs. (4.85) and (4.86), which are very close to Sato's equation, i.e., Eqs. (4.83) and (4.84), except that Eqs. (4.85) and (4.86) involve the parameter λ, which takes into account the relative stiffness between the substrates and the adhesive layer. Hazimeh et al. (2015) made a comparison between the two models (Fig. 4.12). If the substrates are much stiffer than the adhesive layer (Fig. 4.12(a)), the two models give similar results, whereas if the adhesive layer stiffness cannot be neglected then the two results split and the Challita et al. results are closer to the results that are obtained by 3D finite element simulations (Fig. 4.12(b)).

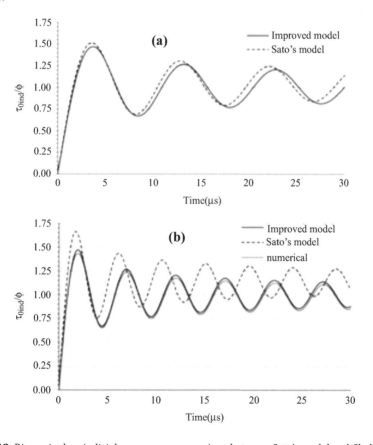

FIG. 4.12 Dimensionless indicial response – comparison between Sato's model and Challita et al. (improved) model: (a) steel substrates and (b) magnesium substrates. (reproduced from Hazimeh et al. (2015) with permissions from Elsevier).

4.5 VISCOELASTIC BEHAVIOR

The adhesives are polymers and their mechanical behavior involves viscoelasticity. Moreover, some substrates can involve viscoelastic behavior too. Vaziri and Nayeb-Hashemi (2002b) have considered complex shear modulus for the adhesive layer. This approach can be extended to consider the viscoelastic behavior of the adherends. Moreover, it can also be used with Sato's model and Challita et al.'s model.

In the above frequency-dependent equations developed in Section 4.3, the substrates' Young's and shear moduli E_s, G_s should be replaced by the complex Young's and shear moduli $E_s^* = E_s(1 + i\eta_s)$ and $G_s^* = G_s(1 + i\eta_s)$, respectively, where $i = \sqrt{-1}$ and η_s is the substrates' loss factor. Similarly, the adhesive shear modulus G_a should be substituted by the complex shear modulus $G_a^* = G_a(1 + i\eta_a)$, where η_a is the adhesive loss factor. For example, the constants c_s, k_s, ξ and ζ should be substituted by c_s^*, k_s^*, ξ^* and ζ^* such as:

$$c_s^* = \sqrt{\frac{E_s^*}{\rho_s}} \tag{4.87}$$

$$k_s^* = \sqrt{\frac{2G_a^*}{t_a t_s E_s^*}} \tag{4.88}$$

$$\xi^{*2} = \frac{\omega^2}{c_s^{*2}} \tag{4.89}$$

and

$$\zeta^{*2} = k_s^{*2} - \xi^{*2} \tag{4.90}$$

In terms of the Challita et al.'s model, the constants λ and ζ' should be replaced by the constants λ^* and ζ'^*, such as:

$$\lambda^* = 1 + \frac{2t_s G_a^*}{3t_a G_s^*} \tag{4.91}$$

and

$$\zeta'^{*2} = \frac{k_s^{*2}}{\lambda^*} - \xi^{*2} \tag{4.92}$$

Vaziri and Nayeb-Hashemi (2002b) showed that the amplitude of the adhesive shear stress decreases as the loss factor increases. However, this factor does not affect the shear stress distribution along the joint. Unfortunately, no equivalent results can be found for Sato's and Challita et al.'s models as they have considered an elastic behavior for the substrates and the adhesive.

4.6 CONCLUSIONS

The chapter focused on the dynamic shear response of the adhesively bonded joints. With regard to the literature review, three main models have been proposed: Sato's model, the Vaziri and Nayeb-Hashemi model and the Challita et al. model. The basic assumptions of these models are synthesized in Table 4.1. The three models are based on the shear-lag model, which assumes axial deformation of the substrates and shear deformation of the adhesive layer. Sato's model extended the shear-lag model by considering the inertia of the adherends. The Vaziri and Nayeb-Hashemi model takes into account the adhesive layer inertia. Challita et al. model improves Sato's model by considering the shear deformation in the adherends, which is important as soon as the stiffness of the adhesive layer cannot be neglected. Challita et al. and Sato' models give solutions that are close to the results obtained by 3D finite element analysis if the substrates are much stiffer than the adhesive. If the adhesive shear stiffness cannot be neglected when compared to the substrates shear stiffness, then only the Challita et al. model continue to give solutions that are close to the 3D finite element results.

4.7 REFERENCES

Challita, G. and Othman, R. 2012. Analytical model of the double-lap bonded joints response to harmonic loads. European Journal of Mechanics A/Solids 34: 149-158.

da Silva, L., das Neves, P., Adams, R. and Spelt, J. 2009a. Analytical models of adhesively bonded joints—part I: Literature survey. International Journal of Adhesion and Adhesives 29: 319-330.

da Silva, L., das Neves, P., Adams, R., Wang, A. and Spelt, J. 2009b. Analytical models of adhesively bonded joints—part II: Comparative study. International Journal of Adhesion and Adhesives 29: 331-341.

Hazimeh, R., Khalil, K., Challita, G. and Othman, R. 2015. Analytical model of double-lap bonded joints subjected to impact loads. International Journal of Adhesion and Adhesives 57: 1-8.

He, S. and Rao, M. D. 1992a. Vibration analysis of adhesively bonded lap joint, Part I: Theory. Journal of Sound and Vibration 152: 405-416.

He, S. and Rao, M. D. 1992b. Longitudinal vibration and damping analysis of adhesively bonded double-strap joints. Journal of Vibration and Acoustics 114: 330-337.

Helms, J. E., Li, G. and Pang, S. S. 2001. Impact response of a composite laminate bonded to a metal substrate. Journal of Composite Materials 35: 237-252.

Ingole, S. B. and Chatterjee, A. 2010. Vibration analysis of single lap adhesive joint: Experimental and analytical investigation. Journal of Vibration and Control 17: 1547-1556.

Pang, S. S., Yang, C. and Zhao, Y. 1995. Impact response of single-lap composite joints. Composites Engineering 5: 1011-1027.

Rao, M. D. and Crocker, M. J. 1990. Analytical and experimental study of the vibration of bonded beams with a lap joint. Journal of Vibration and Acoustics 112: 444-451.

Sato, C. 2008. Impact. pp. 279–304. In: L. F. M. da Silva and A. Öchsner (eds.). Modeling of Adhesively Bonded Joints. Springer, Berlin.

Sato, C. 2009. Dynamic stress responses at the edges of adhesive layers in lap strap joints of half-infinite length subjected to impact loads. International Journal of Adhesion and Adhesives 29: 670-677.

Tsai, M. Y., Oplinger, D. W. and Morton, J. 1998. Improved theoretical solutions for adhesive lap joints. International Journal of Solids and Structure 35: 1163-1185.

Vaziri, A., Hamidzadeh, H. R. and Nayeb-Hashemi, H. 2001. Dynamic response of adhesively bonded single-lap joints with a void subjected to harmonic peeling loads. Proceedings of the Institution of Mechanical Engineers, Part K: Journal of Multi-body Dynamics 215: 199-206.

Vaziri, A. and Nayeb-Hahsemi, H. 2002a. Dynamic response of the tubular joint with an annular void subjected to a harmonic torsional loading. Proceedings of the Institution of Mechanical Engineers, Part K: Journal of Multi-body Dynamics 216: 361-370.

Vaziri, A. and Nayeb-Hashemi, H. 2002b. Dynamic response of tubular joints with an annular void subjected to a harmonic axial load. International Journal of Adhesion and Adhesives 22: 367-373.

Vaziri, A., Nayeb-Hashemi, H. and Hamidzadeh, H. R. 2004. Experimental and analytical investigation of the dynamic response of a lap joint subjected to harmonic peeling loads. Journal of Vibration and Acoustics 126: 84-91.

Vaziri, A. and Nayeb-Hashemi, H. 2006. Dynamic response of a repaired composite beam with an adhesively bonded patch under a harmonic peeling load. International Journal of Adhesion and Adhesives 26: 314-324.

Volkersen, O. 1938. Die nietkraftoerteilung in zubeanspruchten nietverbindungen mit konstanten loschonquerschnitten. Luftfahrtforshung 15: 41-47.

Yuceoglu, U., Toghi, F. and Tekinalp, O. 1996. Free bending vibrations of adhesively bonded orthotropic plates with a single lap joint. Journal of Vibration and Acoustics 118: 122-134.

5

CHAPTER

Continuum Mechanics Modelling by Finite Elements

Luca Goglio

5.1 INTRODUCTION

Historically, the analytical models presented in Chapters 2-4 of this book have been fundamental to understand and describe the behaviour of the bonded joints, giving information on the stress-strain distribution, the role of the geometrical parameters (thicknesses, overlap length) and the stiffness of the assembly. They are also useful nowadays, whenever a quick analysis of a joint is needed, in both the research and industrial context, especially because they can be easily implemented on a spreadsheet or a similar computer tool.

However, as the joint under study exceeds a certain threshold of complication (unfortunately not much high), the analytical description of the problem becomes likely unfeasible. This is due to several concurrent aspects.

- The geometries that the analytical models can deal with are essentially the single-lap, double-lap, or butt joint, with usually flat or cylindrical adherends (da Silva et al. 2009). The scarf joint has been studied with more difficulty (Gleich et al. 2000). The problem must be studied in two dimensions (plane stress, plane strain, axisymmetric). Except for rare exceptions (e.g. Bigwood and Crocombe 1989), the only loading cases considered are the tensile-shear or peel.
- To reduce the number of unknowns, and simplify their dependence on the coordinates, the stress components in the adhesive taken into account must be limited to the most important ones, typically peel and shear, which, moreover, are assumed constant through the thickness or varying in a simplified manner. The consequence is that local equilibrium and boundary conditions are not, or are only partially, satisfied.

Politecnico di Torino, Department of Mechanical and Aerospace Engineering, Corso Duca degli Abruzzi, 24 – 10129 Torino, Italy.

E-mail: luca.goglio@polito.it

- Local details such as spews of adhesive and chamfers of the adherends, which are very influential on joint strength, cannot be described analytically.
- Only the elastic behaviour may be considered to obtain equations that can be solved in closed form.

Therefore, all aspects which conflict with these limitations lead to the need for a numerical solution of the problem. Conceptually, two different cases can be distinguished. In the first case, an analytical model of the joint is developed, not suitable to closed-form solution due to the complication of the equation set, thus the problem is solved by means of a computer program developed ad hoc. The approach is usually based on iterative procedures or finite-difference schemes; examples can be found in both early (Hart-Smith 1973) and recent (Xu and Li 2010) studies. In the second case, which is the subject of this chapter, the joint is modelled and the results are obtained by means of the Finite Element Method (FEM).

In general terms, the Finite Element Method can be defined as a method to solve a problem in physics or engineering by discretization of the continuum domain in zones of finite size, the *finite elements*; in each of these, the solution is approximated by functions (usually low-order polynomials) of the spatial coordinates. The coefficients of these functions depend on the values assumed by the physical quantity involved by the problem (displacement, temperature, etc.) in selected points, called *nodes*. Such nodal values are determined by considering the equations pertaining to the problem (for instance, in a structural case, equilibrium, compatibility, material behaviour) and the boundary conditions. This approach leads to an equation set that is solved numerically.

Early applications of the FEM appeared in mid twentieth century and by the end of the 1960s, the method was already well established, especially for structural applications (Zienkiewicz and Cheung 1967). Taking advantage of the dramatic development of computers in the following decades, it became popular as a routine tool of analysis in the industrial field as well. A great part of the success was also due to the availability, starting from the late 1980s, of graphical computer interfaces that made easier the preparation of the model and the analysis of the results (pre-/post-processing).

As far as adhesive bonding was concerned, FEM modelling appeared soon to be the natural tool to analyse joints, overcoming the difficulties of the analytical study mentioned above. The first reported work dates back to the early 1970s. Wooley and Carver (1971) used a two-dimensional model to assess the stress concentration factors in lap joints, as a function of overlap length to adherend thickness ratio and adhesive thickness to adherend thickness ratio, for different values of adherend elastic modulus to adhesive elastic modulus ratio. Adams and Peppiatt (1974) carried out a two-dimensional analysis to assess the effect of the adhesive fillet on the state of stress and found that the maximum occurs at the adherend corner. Similarly, they carry out the stress analyses of tubular lap joints (Adams and Peppiatt 1977) and axisymmetric butt joints (Adams et al. 1978), in both cases under axial or torsional load, pointing out again the role of the adhesive fillet on stress concentrations.

The FEM was soon used to study cases beyond the elastic regime. Crocombe and Adams (1982) carried out an analysis of the peel test, accounting for the elasto-plastic behaviour of the adherend and adhesive as well as for the large displacements. Later, Adams and Harris (1987) studied the effect of small geometry changes on the local stress distribution close to the adherend end, including plasticity of the adhesive. Crocombe (1989) used elasto-plastic models of single- and double-lap joints to support the failure criterion termed "global yielding".

Another fruitful application of the FEM is modelling fracture mechanics of the joints. In addition to the capability of dealing with cases of complicated geometry (as in stress analysis), FEM models essentially contribute in two tasks: calculation, for the different fracture modes, of the stress intensity factor (SIF) or of the strain energy release rate (SERR). Early and recent examples of these applications can be found in Crocombe and Adams (1981) (SIF), Hamoush and Ahmad (1989) (SERR), Pradhan et al. (1995) (SERR), Abdel Wahab (2000) (SERR) and Noda and Lan (2012) (SIF). Finite elements have been extensively applied also to the study of fatigue crack growth (e.g. Abdel Wahab et al. (2004), Pirondi and Nicoletto (2004), Quaresimin and Ricotta (2006)).

An approach, based on the FEM, that in recent years has been successfully exploited to describe the progressive failure under static or fatigue loading is the Cohesive Zone Model (CZM). It is based on special elements that, when strained, behave elastically up to a given threshold, then soften until complete failure. For instance, lap joints (Campilho et al. 2013), bonded metal laminates (Katnam et al. 2011), double cantilever beam specimens (de Moura et al. 2008) have been modelled in this way.

FEM modelling has been also applied to study environmental effects on adhesives and joints. Mubashar et al. (2009) modelled the diffusion of moisture in the adhesive and Mubashar et al. (2011) developed a method to predict the strength of joints exposed to moisture. Han et al. (2014) simulated the coupled effect of moisture, temperature, and stress on the long-term degradation.

To conclude this brief summary, it can be remarked that the use of FEM modelling in the study of adhesive joints has become widespread in the last two decades and is the commonest approach nowadays. The number of related researches presented in the literature is so large that it is impossible to cite them in brief. The interested reader can find a comprehensive review in He (2011).

The aim of this chapter is to give an introduction to the fundamentals of the use of FEM in modelling adhesive joints. The scope is mainly (although not exclusively) limited to the cases that can be treated and the types of results that can be obtained by considering elastic behaviour of the materials. Moreover, for the sake of conciseness, it is assumed that the reader is already familiar with the fundamental concepts involved in FEM modelling, i.e. nodes, elements, shape functions, stiffness matrices, etc. Should this not be the case, such a background can be found in the large amount of available literature about the FEM. Section 5.2 discusses the results that can be obtained by a "standard" continuum analysis, in particular by the application of classical failure criteria based on stresses, considering brittle or ductile behaviour. This approach can be useful to the design for service in elastic conditions of the joint; moreover, if an elasto-plastic analysis is also performed, then the ultimate strength of the joint can be predicted in some cases. The problem created by the need, on the one hand, for a very fine mesh to model the adhesive and, on the other hand, for an acceptable size of the model, is dealt with in Section 5.3, and some techniques proposed to overcome this difficulty are reported. Section 5.4 concerns the stress singularity due to geometry and bi-material (adhesive/adherend) interface. This peculiar aspect, in addition to making convergence impossible, requires a different perspective to assess the criticality of the stress state. Another peculiar aspect of adhesive joint modelling is that, even when loading is low enough so that the materials remain in elastic regime, geometrical non-linearity may occur. This is due to the flexibility of the joints, which may permit rotations and displacements having non negligible effects on the equilibrium conditions. The problem is considered in Section 5.5. Finally, the conclusions are drawn in Section 5.6.

Further and more sophisticated types of FEM simulations are treated specifically in subsequent chapters of the book. Chapters 6-8 are dedicated to CZM: Chapter 6 introduces this technique, Chapter 7 discusses the identification of the relevant parameters, and Chapter 8 presents the application to fatigue. Chapter 9 deals with Damage Mechanics, for both static and fatigue applications. Finally, Chapter 10 presents the Extended Finite Element Method (XFEM), which is able to model the formation and growth of cracks without the need for a pre-defined path, contrary to the case of CZM. Chapters 11 and 12 deal with modelling applications.

5.2 FAILURE PREDICTION BASED ON "STANDARD" CONTINUUM ANALYSIS

In a typical load-displacement curve (Fig. 5.1) of the commonest bonded joint, the lap joint, three distinct stages can be identified in sequence: (i) nearly linear rise, (ii) markedly non-linear rise, and (iii) decrease.

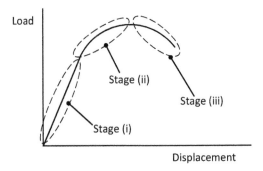

FIG. 5.1 Example of load-displacement curve of a bonded joint.

Stage (i) corresponds to the elastic behaviour where the deviation from linearity (if any) depends on two reasons. The first one is that the adhesives, as do the polymers in general, may exhibit non-linear elasticity. This is particularly true for low-modulus, rubber-like adhesives (e.g. silanes); conversely, high-modulus structural adhesives (e.g. epoxies, which have an E modulus of some GPa) are linear elastic for most of the ascending part of their stress-strain curve. The second reason is significant in case of slender adherends loaded axially: due to the offset between them, a bending moment is generated which makes the sandwich adherends+adhesive rotate. Such a rotation tends to reduce the moment arm and, therefore, the load on the joint (Goland and Reissner 1944). Stage (ii) is mainly due to yielding, both in the adhesive and in the adherends, which reduces the stiffness of the joint, as well as to the rotation effect cited above. The load increases up to a maximum, which represents the ultimate load of the joint. The transition from part (i) to (ii) can be more or less evident, depending on joint properties. Stage (iii) corresponds to the collapse of the joint, involving gross plastic deformation and/or fracture, and it is less interesting from the joint design viewpoint, but can be important to characterize the material.

The fact that the ultimate load occurs at the end of stage (ii) can lead to the conclusion that an elastic analysis of the joint is useless for design purposes. However, if one thinks of a generic structural element made of ductile material, the ultimate load occurs after a non-linear behaviour due to plasticity in that case as well. Nonetheless, typical design calculations are based on the elastic limit. In this perspective, it appears reasonable to apply the approach based on elastic calculations also to the case of bonded joints, taking as limit the load value which ends stage (i).

Another conceptual difficulty in using the results of an elastic stress analysis to predict joint strength is that the actual stress distribution in the adhesive is complicated, as it varies both along the overlap and through-the-thickness. Choosing to adopt the maximum value of the distribution as design value can be misleading, since the singularity that may appear at the interface end makes the result mesh-dependent (this aspect will be discussed in more detail in Section 5.4). Therefore, stress values not affected by the singularity must be utilized. The commonest choice is to consider the stresses, in particular their peak values, at the mid-thickness of the adhesive layer (e.g. Goglio et al. 2008). Alternatively, if the situation on the adhesive/adherend interface is considered, the stresses at a chosen distance from the singularity are taken into account (for instance, Sancaktar and Narayan 1999, and Gleich et al. 2001, considered as representative the stresses in the third node of the mesh from the end of the adhesive layer).

A peculiar aspect of modelling adhesive joints with finite elements is related to the characteristic dimensions of the parts. In most cases, the orders of magnitude of the thicknesses are some mm for the adherends and below one mm for the adhesive layer. To achieve a generally satisfactory description of the stress-strain field, but still neglecting the singularities, the minimum number of elements in the adhesive thickness must be in the range 5-10 approximately; the number of elements along the overlap (and the width, in a three-dimensional case) is dictated by the aspect ratio, which must not exceed some units. Thus the typical element size is of the order of tenths of mm. The number of elements to model the adherends is consequently influenced, since nodal positions in the adherend/adhesive interfaces must match. It is revealed that, for a typical overlap length of some tens of mm, the number of elements required to model the bond is not negligible, however it can be processed with the modern software and hardware. The situation becomes critical when a complete structure, and not just the joint zone, must be modelled, since adopting such an element size for the overall structure would be unsustainable. So, a dramatic variation in mesh size – very refined in the bonds and coarse far from them – must be adopted. Figure 5.2 schematizes two examples for a bi-dimensional case, using triangles (Fig. 5.2b) or quadrilaterals (Fig. 5.2c). The former scheme is simpler, especially for automatic meshing, but the latter can be preferred due to the higher order shape functions of the quadrilaterals. In any case, care must be taken to avoid excessive distortion of the elements (quantitatively, internal angles close to 0° or 180° and aspect ratio larger than a few units). However, it is not always possible to achieve a satisfactory trade-off between fine mesh size of the joint and acceptable total number of nodes/elements of a large structure. To overcome this difficulty, special techniques for efficient joint modelling have been proposed, and they are reported in Section 5.3.

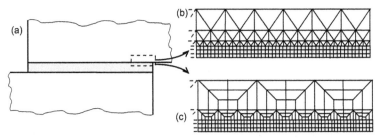

FIG. 5.2 Differences in thickness of adherends and adhesive layer (a) and details of mesh transition obtained with (b) triangles or (c) quadrilaterals.

Once the significant stresses have been identified, the problem of adopting a suitable failure criterion arises. For design purposes, without investigating the post-elastic behaviour, a simple approach suggested by Goglio et al. (2008) is based on a limit envelope in the peel stress - shear stress plane and obtained empirically by testing specimens in different stress combinations. This approach was proposed using stresses calculated analytically, but it can also be applied using the stresses obtained from a FEM analysis in the mid-thickness of the adhesive. A noteworthy application of the elastic analysis presented in the literature is the study of improved geometries at the ends of the overlap (e.g. chamfering, tapering or rounding the adherend, shaping the adhesive spew), designed to reduce the stresses. Examples can be found in Lang and Mallick (1998, 1999) and Rispler et al. (2000).

Regarding the case of a relatively brittle adhesive, the classical Galileo-Rankine failure criterion based on maximum stress can be adopted (e.g. Adams and Peppiatt (1974)). According to it, the ideal stress coincides with the maximum principal stress. However, fortunately for the joint safety, this case is rather uncommon nowadays, since the producers tend to introduce in the adhesives toughening agents such as rubber particles. Therefore, this approach is of limited application. Moreover, it is not obvious what stress should be taken as representative, as the peak values related to the singularity are mesh-dependent.

If the FEM analysis also intends to include inelastic behaviour, then the relevant conceptual tools must be used. The interested reader can refer to relevant references – for instance, Bathe (1996) or de Souza Neto et al. (2008) – only the essential aspects are reported here in a brief and simple form. To deal with elasto-plastic behaviour, three main points are involved in addition to the stress-strain relationship: (i) yield criterion; (ii) flow rule, and (iii) hardening rule.

Point (i) defines the stress level at which the material ceases to behave elastically and plastic flow starts; formally it can be stated as

$$f(\sigma, k) = 0 \tag{5.1}$$

where σ is the stress tensor, k is a set of state variables related to material properties and accumulated plastic strain, and f is the function which defines the adopted criterion (e.g. Mises).

Point (ii) is related to the evolution of the plastic flow; the commonest choice is to use the same function f adopted in point (i), which is called an associated flow rule

$$d\varepsilon_p = d\lambda \frac{\partial f}{\partial \sigma} \tag{5.2}$$

where $d\varepsilon_p$ is the plastic strain increment and $d\lambda$ is the plastic multiplier.

Point (iii) refers to the way the yield condition changes due to plastic flow; typical cases are the isotropic hardening, in which the limit condition "grows" in the stress space keeping its centre, and the kinematic hardening, in which the limit condition "moves" keeping its size. Assuming that the total strain increment $d\varepsilon$ is the sum of an elastic and a plastic ($d\varepsilon_p$) part, the stress-strain relationship is defined in incremental way as follows

$$d\sigma = \mathbf{C}\,(d\varepsilon - d\varepsilon_p) \tag{5.3}$$

in which $\mathbf{C}$ defines the elasto-plastic constitutive law, related to material properties and accumulated plastic strain; therefore, $\mathbf{C}$ is not constant through the analysis as in elasticity and must be progressively recomputed to account for the evolution. The equation set which is obtained via the typical FEM procedure (definition of the stiffness matrices of the elements, assembly, application of the boundary conditions) is solved by applying a suitable method for non-linear problems, for example the Newton-Raphson scheme.

Another important aspect which still has not been mentioned here is that appropriate measures of stress and strain must be adopted, since the possibility of large displacements and strains makes the usual engineering definitions insufficient. The underlying problem is that, when stating the equilibrium, it is no longer possible to neglect the change from undeformed to deformed state (as usually done in elastic problems); moreover, the equations describing the behaviour of the material must not depend on rigid body rotations. Several definitions of stress and strain have been proposed to fulfil these requirements; they must be chosen so that the adopted stress and strain measures are energetically conjugated, that is, their product must give an energy (or a power, considering the time rate) per unit volume. A typical chosen pair is formed by the first Piola-Kirchhoff stress tensor with the rate of deformation tensor. Defining the deformation gradient

$$F_{ij} = \frac{\partial x_i}{\partial X_j} \tag{5.4}$$

where x_i and $X_j (i, j = 1, 2, 3)$ are, respectively, the coordinates in the deformed and undeformed configurations, the first Piola-Kirchhoff stress reads

$$P_{ij} = J\sigma_{ik} \frac{\partial X_j}{\partial x_k} \tag{5.5}$$

where J is the determinant of the deformation gradient, σ_{ik} is the Cauchy stress and the tensor notation implies summation of the repeated indices over their variation range. The rate of deformation is defined as

$$D_{ij} = \frac{1}{2}\left(\frac{\partial \dot{x}_i}{\partial x_j} + \frac{\partial \dot{x}_j}{\partial x_i} \right) \tag{5.6}$$

where the dot stands for the time derivative. Other common choices use the Jaumann rate of stress, or the Truesdell rate of stress, as stress measure. An interested reader should refer to textbooks on continuum mechanics for details.

Modern commercial FEM codes for non-linear analysis implement these concepts and offer the capability of solving the related structural problems; the main difficulty for the user is often in knowing the values of the various coefficients which define material properties. With regard to the adhesives, in the case of a ductile one, in the first instance the same hypotheses used for ductile metals can be adopted, in particular those based on the second deviatoric invariant J_2' (Hencky, Mises). For example, Sheppard et al. (1998) presented a damage model based on the accumulated plastic strain in the adhesive, with the equivalent plastic strain defined according to the Mises criterion. This allowed for identifying local zones of damage, contrary to the criterion of global yielding.

However, since the yield of the adhesives depends, as for polymers in general, also on the hydrostatic pressure, most of the analyses have been carried out using criteria which account for this aspect, such as the one proposed by Raghava et al. (1973). Crocombe and Adams (1982) used a yield surface (in the stress space) function of J_2' and of the first (hydrostatic) invariant J_1, the latter dependence can be expressed also as a function of tensile yield stress to compressive yield stress ratio. This approach has been used until present times. For example, in Hassanipour and Oechsner (2011), a pressure sensitive criterion is implemented and the integration of the equations is obtained via an implicit Euler algorithm. Karachalios et al. (2013) studied the failure of single lap joints to identify the failure modes and damage loci corresponding to different values of adherend thickness and overlap lengths. Failure in the adherends was prevented by using high strength steel. The conclusion of their numerical analysis, supported by experiments, was that in the case of ductile adhesive and short overlap, failure occurs by global yielding. Conversely, if the overlap is long and the adhesive is not ductile, failure is due to an excess of strain localized at the ends of the overlap.

In conclusion of this Section it must be acknowledged that, nowadays, most of the numerical studies on the failure of the adhesive joints make use of CZM or XFEM, which are treated in subsequent chapters of this book, as previously stated. In this perspective, simulating plasticity can be mostly regarded as a part of modelling the post-yield behaviour of the joint, whilst the failure phase is simulated with CZM or XFEM.

5.3 MODEL SIZE REDUCTION (EFFICIENT MODELLING)

As stated previously, when a large structure containing bonded joints must be modelled, the need for limiting the number of nodes and elements becomes essential. Since the early applications of FEM to bonded joints, special formulations have been proposed to model the behaviour of the adhesive using only one element in the layer thickness, and including in the element formulation the main features in terms of stress state and deformability.

Carpenter (1980) developed two formulations of a special element based on the theories of Goland and Reissner (1944) or Ojalvo and Eidinoff (1978), respectively, in which the adhesive is described as a sort of two-node beam. Since the adherends are also described as beams or plates, having their mid-line or mid-surface at mid-height, the offset between the nodes of the adhesive and of the adherends is accounted for by means of a transformation matrix.

Several contributions to this line of research were already presented in the 1980s. Rao et al. (1982) proposed a six-node isoparametric element that modelled the adhesive as elastic springs. Similarly, Kuo (1984) developed a two dimensional spring element with 16 nodes, subjected only to shear stresses and strains, and applied it to evaluate SIF. Yadagiri et al. (1987) formulated a six-noded isoparametric element in two different versions (one stiff and one flexible) to model the adhesive layer including viscoelastic behaviour; standard eight-noded plane strain elements were used to model the adherends. A dedicated FEM program called VANIS was written on purpose. Amijima and Fuji (1987) wrote a computer program that implemented a simple one-dimensional model for the analysis of single lap joints, in which the adherends were modelled as conventional beams and the adhesive was an elastic layer subjected to normal and shear stresses and strains. Carpenter and Barsoum (1989) presented a two-noded and a four-noded element to represent the adhesive, suitable to be used with both plane stress or plane strain elements and beams or plates to model the adherends. The offset of the nodes was taken into account; moreover, the formulation prevented the presence of singularities at the overlap ends, thus avoiding the related convergence problems. Later, Lin and Lin (1993) presented an element formulation based on the Timoshenko beam to describe the behaviour of the single lap joint, also including the effects of adhesive thickness and dissimilar adherends. Tong and Sun (2003) created a special model to analyse bonded repairs of curved structures. The adherends were modelled with shells, and the adhesive with a pseudo-brick element. The formulation was implemented in an in-house FEM program called BPATCH.

In recent times, with the aim of obtaining the most efficient modelling technique, Stapleton et al. (2011) presented a model of a single lap joint in which an element alone was used to simulate the complete joint. This special element included the overlapping parts of the adherends, represented by beams or plates in cylindrical bending, and the adhesive, represented by a bed of normal and shear springs. Instead of the usual shape functions, the displacement field obtained from the minimum potential energy principle was implemented. The special element could be connected with standard beam elements at its end nodes to model the remainder of the structure. The adhesive behaviour was assumed to be linear elastic, but the progressive failure under the increasing load was reproduced by progressively shortening the bonded length. The response of the element was validated by comparison with standard detailed FEM modelling and, in the case of double cantilever beam specimens, with experimental results. In a further evolution of this approach, Stapleton et al. (2014) tested two methods to efficiently model adhesive cohesive failure. In the first method, one element is used to model the joint and the shape functions are recalculated to account for the progressive damage. In the second method, the joint element is, in turn, divided in a small number of sub-elements to describe the cracked condition. In both cases, the formulation is limited to a beam-like description of the joint and of the structure.

The main shortcoming of all the above mentioned approaches is that they remain confined to in-house programs and are difficult to include in general purpose, commercial, FEM codes. The widespread popularity of the latters in the industrial world, and the aim of exploiting their capability to solve large structures and complicated non-linear cases, together with the advantage of using popular graphical interfaces for pre-/post-processing, create the need for avoiding special elements and in-house FEM programs. A research line has been presented in recent years in this perspective. Castagnetti and Dragoni (2009) showed a comparison of two different approaches based on standard FEM to analyse the joints (Fig. 5.3). In the first approach, the joint is modelled using beams for the adherends and a single layer of square elements for the adhesive; the offset caused by the distance between the adhesive-adherend interface and the mid-height of the adherend is compensated by extending the adhesive thickness and adopting a corrected elastic modulus. In the second approach, the elements modelling the adhesive keep its actual height; the offset is taken into account by means of kinematic constraints among the nodes of the adhesive and of the adherend, similarly to the approach adopted by Carpenter (1980). The comparison with a detailed model showed that the first approach leads to large errors, while the second gives accurate results. It was also assessed that an optimal element length in the overlap direction is equal to the distance between the mid-planes of the adherends. The mesh size reduction obtained with this technique decreased the computational time of two orders of magnitude, compared to a detailed model. In subsequent works, Castagnetti et al. (2009, 2010) applied this technique to the post-elastic analysis of peel joints. Again, the adhesive layer was modelled with a single row of continuum elements, the adherends with beams, and the offset among the nodes was considered via the kinematic constraints. The materials behaviour was elasto-plastic, or simply elastic until failure in the case of a brittle adhesive, and the von Mises criterion was assumed. The comparison with both detailed modelling and experimental results showed that the reduced model gave good results in terms of stiffness and yield load. Also, the post-elastic behaviour was satisfactorily reproduced in case of a brittle adhesive. For these post-elastic cases, the optimal element length in the overlap direction was one quarter of the distance between the mid-planes of the adherends. The decrease in computational time using the reduced model was about 50 times in average.

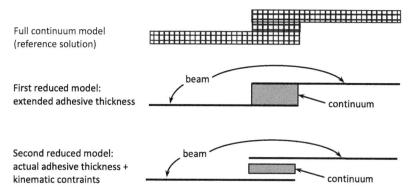

FIG. 5.3 Different modelling techniques for a lap joint used by Castagnetti and Dragoni (2009).

In a later work, Castagnetti et al. (2011) re-applied their reduction technique, based on kinematic constraining, to model the failure of the assembly of two square-section, thin-walled beams, joined by means of bonded overlapping plates. In this case, to improve the simulation of the post-elastic phase, the adhesive layer was modelled with a row of cohesive elements. The walls of the beams were modelled with shell elements. The comparison with experimental data, in terms of load-displacement curves, showed that the results of the simulation were accurate in the post-elastic phase as well.

5.4 TREATMENT OF THE SINGULARITIES

As anticipated in the previous sections, the end of the adherend/adhesive interface theoretically represents a singularity point; the problem is reconsidered here in more detail. This fact is due to two features that coexist in a bonded joint: one geometrical, the re-entrant corner, and one material-related, the adherend/adhesive interface.

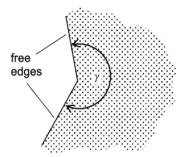

FIG. 5.4 Corner in a plate (re-entrant case, $\gamma > 180°$).

The problem related to the effect of a corner on the stresses was treated by Williams (1952) for the case of an elastic, homogeneous plate, loaded in its plane and subjected to various boundary conditions applied to the edges (Fig. 5.4). For the case in which both edges are free, relevant to describe the corner, the equation that rules the behaviour is

$$\sin(\lambda\gamma) = \pm\lambda\sin(\gamma) \tag{5.7}$$

where γ is the angle of the vertex (measured on the material side) and λ an eigenvalue. If a solution exists for $0 < \lambda < 1$, the stresses are singular at the vertex. This occurs for angles between 180° and 360°, that is, when the corner is re-entrant (it is worth to note that for $\gamma = 360°$, i.e. when the corner degenerates in a crack, $\lambda = 0.5$, which is the singularity order in fracture mechanics).

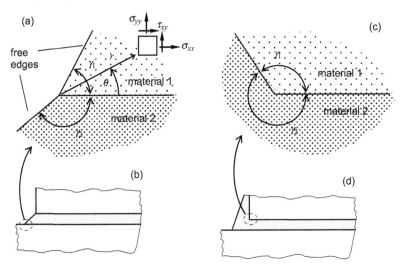

FIG. 5.5 Cases of singularity and related positions in bonded joints: (a), (b) interface end; (c), (d) interface corner.

After a preliminary study (Bogy 1968), the solution to the problem of the corner at the end of the interface (Fig. 5.5a) between different materials was obtained by Bogy (1971) in a more effective way by using the material constants defined by Dundurs (1969). Indeed, it was found that, considering materials 1, 2 of the bond interface (in practice, adherend and adhesive), having respectively the shear moduli G_1, G_2 and Poisson's ratios ν_1, ν_2, the singularity is ruled by the angles γ_1, γ_2 and by

$$\alpha = \frac{G_1(1-\nu_2) - G_2(1-\nu_1)}{G_1(1-\nu_2) + G_2(1-\nu_1)}; \quad \beta = \frac{1}{2}\frac{G_1(1-2\nu_2) - G_2(1-2\nu_1)}{G_1(1-\nu_2) + G_2(1-\nu_1)} \tag{5.8a,b}$$

which are the Dundurs constants (here reported in case of plane strain conditions). An intuitive interpretation of this fact is that the behaviour is not influenced by the four individual values of the elastic constants G_i, ν_i ($i = 1$, 2) of the two materials, but by their relative values defined by the two ratios α, β. The solution of the elastic problem is governed by the Bogy determinant

$$D(\gamma_1, \gamma_2, \alpha, \beta; p) = a\beta^2 + 2b\alpha\beta + c\alpha^2 + 2d\beta + 2e\alpha + f \tag{5.9}$$

where the coefficients a, b, c, d, e, f are functions of γ_1, γ_2 and of the auxiliary parameter p (see Bogy 1971 for the detailed expressions).

The values p_n of p satisfying the condition $D = 0$, i.e. the roots of equation (5.9), determine the singularity. In particular, the stress is singular if at least one root (its real part if the root is complex) exists in the range $]0,1[$. Thus, the generic stress component $\sigma_{ij}(i, j = x, y)$ is given by

$$\sigma_{ij} = \sum_n \frac{K_n}{r^{\lambda_n}} f_{ij,n}(\theta) \qquad (5.10)$$

where r, θ are the polar coordinates, $\lambda_n = 1 - p_n$ is the n-th singularity exponent, to which are associated the SIF K_n and the angular function $f_{ij,n}(\theta)$. In the commonest situation, only one singular term exists and dominates the stress distribution in the vicinity of the corner, and the stresses tend to infinity for $r \to 0$. Therefore, as in linear elastic fracture mechanics, the parameter that characterizes the stress distribution is not the maximum local stress (by definition unbounded) but the SIF, which defines how "intense" is the distribution near the corner.

An example of interface end singularity occurs in a lap joint when the adhesive spew does not exceed the thickness of the adhesive layer, as shown in Fig. 5.5b (note that in this case material 1 is the adhesive and material 2 is the lower adherend; γ_1 is the spew angle and $\gamma_2 = 180°$). Another case of singularity is the interface corner (Fig. 5.5c) studied by Bogy and Wang (1971), whose solution originates a different determinant but presents the same conceptual aspects as for the interface end; in particular, the stresses again tend to infinity for $r \to 0$. An example of this case occurs when the spew embeds the unloaded end of an adherend as shown in Fig. 5.5d (note that in this case material 1 is the upper adherend and material 2 is the adhesive; $\gamma_1 = 90°$ and $\gamma_2 = 270°$). By studying the determinants, the number of singular terms in equation (5.10) can be assessed; the usual case is having only one singular term, but for some geometry and pair of materials a second singular term can also arise.

The analytical studies on the singularity cited until here enable the assessment of whether the stress distribution is singular or not and the order (exponent λ) of the singularity, but cannot evaluate the SIF; this latter task must be carried out numerically. Some authors developed numerical solutions of the problem. For instance, Sawa et al. (2000) solved the elasticity equations by means of an Airy stress function expanded in the Fourier series, while Lazzarin et al. (2002) assumed a two-term expansion for the stress distribution and solved the set of differential equations by means of an ad hoc iterative procedure to treat the boundary-value problem as an initial-value problem. An attempt to evaluate the SIF in a simpler way was proposed by Goglio and Rossetto (2009), who developed a procedure based on the stresses calculated with the one-dimensional solution of Bigwood and Crocombe (1989). Very recently, Zappalorto et al. (2015) have developed an analytical solution, of remarkable mathematical complexity, for the three-dimensional stresses in anisotropic bi-material corners.

However, in most cases, the SIF is obtained by means of FEM modelling. In a pioneering work, Groth (1988) approached the study of interface corners through subsequent steps. For the sake of comparison, the singularity was assessed analytically with the determinant of Bogy and Wang (1971), which evidenced the existence of two singularity exponents λ_1, λ_2. A "global" FEM analysis of the complete test specimen was carried out, considering both linear and geometrically non-linear behaviour. Then, a "local", refined model was used to study the stresses in the interface corner, using as boundary conditions the previously obtained displacements from the global analysis. In the same paper (Groth 1988), it was pointed out that the results given by FEM in the vicinity of the singular vertex must be fitted with a two-term equation as

$$\sigma_y^{FEM} = \frac{Q_y^1}{\xi^{\lambda_1}} + \frac{Q_y^2}{\xi^{\lambda_2}} \tag{5.11}$$

where σ_y^{FEM} is the peel stress obtained by FEM, Q_y^1, Q_y^2 are the associated generalized SIF, and ξ is the non-dimensional radial distance from the vertex. The values of the SIF and the singularity exponents were obtained by the least squares method.

This is the procedure followed in practice – more frequently considering only one singular term, which is enough in most cases – by all authors, as Adams and Harris (1987) already mentioned in Section 5.1; Hattori (1991), to study single lap joints; and Lazzarin et al. (2002), to support and compare their numerical solution. Figure 5.6 shows an example of least squares fitting, in log-log scale, of the stresses obtained from a FE model to calculate the singularity exponent and the SIF. In the typical case, the points obtained from the first-second elements closest to the singularity vertex must be discarded, as well as the points "too far" from the vertex that are also influenced by the terms in eq. (5.10) following the first one. Thus, least square fitting must be carried out in an intermediate zone, where the points lie on a straight line. More recently, fitting of FEM results was also used by Goglio and Rossetto (2010) to assess the influence of the geometry of the joint on the singular stress distribution. Both general (overlap length, thickness) and local parameters (edge shape) were considered, and it was noticed that the most influencing factor is the edge angle: for instance, a change from 90° (square edge) to 60° (fillet edge) reduces the singularity exponent by a factor of more than two and the stress intensity by five times.

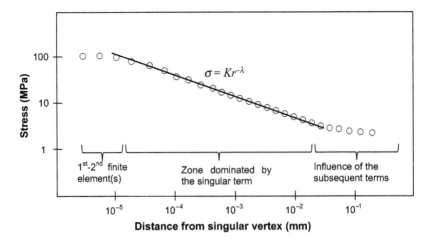

FIG. 5.6 Fitting of the FE stresses close to a singularity to assess exponent (λ) and SIF (K).

It must be pointed out that this approach requires an extreme local refinement of the mesh. Indeed, whilst the non-singular stresses in the adhesive layer can be correctly obtained with few elements in the thickness (normally less than ten), the element size at the corner must be reduced down to 10^{-4}-10^{-3} times the adhesive thickness to describe the singular stress distribution. An example is shown in Fig. 5.7. This leads to an order of magnitude of 10^5 elements and nodes in the model, even in a two-dimensional case. However, since the analysis is a linear elastic one, the related computational effort is relatively easy to sustain by present hardware and FEM software, and runs of this type require few minutes of CPU time.

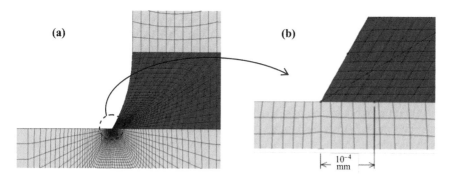

FIG. 5.7 Mesh refinement to evaluate the singular stress distribution: (a) overlap end; (b) detail view of the interface corner (from Goglio and Rossetto 2010, copyright Elsevier).

As an alternative, special FEM formulations have been developed to deal with the singularity. In particular, Barut et al. (2001) developed a "global" two-dimensional element by including, instead of the usual interpolation functions, the exact solution based on singular eigenfunctions similar to equation (5.10), in addition to non-singular terms. This element can represent an interface end (Fig. 5.5a) or corner (Fig. 5.5c) and is connected to the rest of the model by imposing continuity at the boundary. The formulation was implemented in the commercial FEM package Ansys. Thus, standard finite elements could be used to model the joint around the singular region, and the general pre/post-processing and solution capabilities of the package could be used.

A completely different use of FEM for the analysis of the singularity was presented by Wang and Rose (2000). The aim of this work was to obtain practical formulae for the stress intensity factor K, written in the form

$$K = [\sigma^* A(v) + \tau^* B(v)] h^{\lambda} \tag{5.12}$$

where σ^* and τ^* are the shear and peel stresses obtained from conventional one-dimensional solutions for the bonded joints, $A(v)$ and $B(v)$ are coefficients, v is the Poisson's ratio, h is the thickness of the adhesive, and λ is the singularity exponent. Assuming the adherends as rigid with respect to the adhesive, the authors determined polynomial expressions for $A(v)$ and $B(v)$ by interpolating FEM results in the two cases of square edge and spew fillet.

A more sophisticated approach to assess the SIF, in the case of a bonded butt joint, was presented by Akisanya (1997), which used the reciprocal work integral contour method, based on the Betti's reciprocity theorem. An advantage of this approach is that the SIF is obtained from FEM results far from the singularity. Thus, neither an extreme mesh refinement, nor the use of special elements, are required.

It should be noted that all the works reported until here in the present section were dedicated to the analytical and numerical assessment of the singularity parameters, namely the singularity exponent and the SIF. In an engineering perspective this is only half of the task, the remaining part being the assessment of a relevant strength parameter to be used for comparison. From the physical view point the stresses cannot tend to infinity, as they cause either failure or yield of the material (more likely in a modern, ductile adhesive), thus their peak is limited to a finite value. Moreover, the actual geometry of a joint never exhibits sharp corners, since some degree of roundness is always present. Therefore, the singular behaviour can be regarded as a mathematical idealization of the actual stress distribution. Nonetheless, it is clear that the interface ends and corners represent severe stress concentrators and must be considered with attention. To this aim, some attempts to relate the singular stress field to the failure of the joint have been carried out. Groth (1988) complemented his study cited above by applying the SIF to predict the failure of single lap joints. The theoretical assumption is that, as in linear elastic fracture mechanics, failure occurs whenever the SIF – evaluated, as previously stated, by means of a least squares fit of the FEM results – exceeds a critical value. When the case is influenced by two SIF, as in equation (5.11), an equivalent single factor can be conveniently defined, again by fitting FEM results. The critical value must be evaluated experimentally, by testing a set of specimens for failure under controlled conditions. A further difficulty is added by the fact that the SIF and its critical value depend on the singularity exponent λ, which in turn is influenced by the materials and also by the angles. Therefore, it is not possible to assess the critical value for an adhesive alone, nor for an adhesive/adherend pair. On the basis of his results, Groth (1998) concluded that a SIF-based criterion is suitable for brittle adhesives.

An interesting discussion on this aspect was presented by Akisanya and Meng (2003), which investigated the validity of the SIF as a suitable parameter to predict joint strength. The main conclusion of an elasto-plastic analysis, supported by experiments on butt joints, was that the criterion holds if the plastic zone near the singular corner is contained within the zone dominated by the elastic singular solution, and this is possible only if the adhesive layer is thick enough. This latter condition is often not verified for actual structural bonds, in which the adhesive layer is kept intentionally thin to achieve the best strength. Thus in these cases, the criterion is not applicable.

In conclusion, the relatively high ductility of modern adhesives enables them to sustain a non-negligible amount of plastic deformation before failure. Thus, a merely elastic study of the singularity is likely not capable of predicting the failure of the adhesive. Moreover, the absence of real sharp corners reduces the intensity of the actual stress field. However, the singularity is always present in FEM modelling of the adhesive joints. An analyst unaware of the problem could be misled by the related spurious stress values obtained and by the impossibility to achieve convergence by refining the mesh.

5.5 GEOMETRIC NON-LINEARITY

When an adhesive joins flexible parts, the displacement and rotation occurring under load can be not negligible; thus, the approximation of studying the equilibrium in an undeformed condition (as usually done in linear elasticity) is not acceptable. This effect is shown in Fig. 5.8 for a single lap joint. In the original, undeformed condition, the adherend centre lines are offset of d, equal to the sum of the half-thicknesses of the two adherends and the thickness of the adhesive (Fig. 5.8a). Under load, the rotation of the joint due to the curvature assumed by the adherends reduces the offset to d' (Fig. 5.8b). Therefore, under increasing load, the bending moment in the joint increases in a less than linear way.

This phenomenon was recognized and treated by Goland and Reissner (1944), who simplified the study by neglecting the thickness of the adhesive and assuming that the overlap behaves as a plate of thickness equal to the sum of the thicknesses of the adherends. They wrote the bending moment M_0 at the edges of the overlap in the form

$$M_0 = kT \frac{t}{2} \tag{5.13}$$

where T is the axial load, t is the thickness of each one of the two adherends (supposed to be equal) and k is the moment factor, given by the formula

$$k = \frac{\sinh(u_1 l)\cosh(u_2 c)}{\sinh(u_1 l)\cosh(u_2 c) + \dfrac{u_1}{u_2}\cosh(u_1 l)\sinh(u_2 c)} \tag{5.14}$$

In equation (5.14), l and c are respectively the unbonded length of each adherend and the half-overlap length, $u_i = \sqrt{T/D_i}$ ($i = 1, 2$), where D_1 is the bending stiffness of each adherend and D_2 the bending stiffness of the overlap.

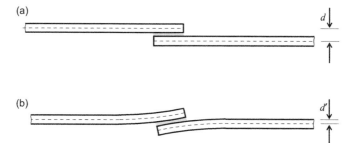

FIG. 5.8 Moment arm reduction in a single lap joint due to rotation under load: (a) unloaded condition, undeformed joint; (b) loaded condition, deformed joint.

Several attempts to improve this result were presented in subsequent times. Hart-Smith (1973), assuming the two adherends in the overlap uncoupled, obtained lower k values, tending to zero in the case of long overlaps. Oplinger (1994) adopted a different correction, obtaining k values similar to Goland and Reissner (1944) and different from Hart-Smith (1973) in the case of long overlaps. Recently, Zhao et al. (2010) proposed a new calculation of the k factor, based on the assumption of neglecting the deformation of the overlap region, which adds the advantage of being applicable also in case of non-identical adherends. Another contribution to the problem was added by Guo et al. (2006) that, by means of an analytical solution obtained as an extension of Goland and Reissner's (1944), studied the effect of the boundary conditions on the single lap joint. They found that the way the load is applied, in particular the presence of spacers to align the halves of the joint in the testing machine, is influential only if the grips are close to the overlap ends. Furthermore, the spacers are more influential in the case of compressive loading.

To study a geometrically non-linear problem by means of the FEM, the formulation of the method must be capable to follow the configuration change of the body from the reference to the current state. If the reference state is the initial state (0) the approach is called Total Lagrangian. If the reference state is the one preceding the next state (t) the approach is called Updated Lagrangian. In this context, Lagrangian is in contrast to the Eulerian approach, in which a control volume is assumed, and the flow of the material through it is considered (which is typical of fluid mechanics). The study is carried out incrementally: as the solution has been found at time t, the new solution at time $t + \Delta t$ is searched. Usually, the equilibrium is written by means of the principle of the virtual displacements

$$\int_{t+\Delta t_V} ({}^{t+\Delta t}\sigma_{ij}\,\delta^{t+\Delta t}\varepsilon_{ij})\,d^{t+\Delta t}V = {}^{t+\Delta t}\mathcal{L} \qquad (5.15)$$

where, at time $t + \Delta t$ (superscript preceding the symbols), V is the volume, σ_{ij} is the Cauchy stress, ε_{ij} is the engineering infinitesimal strain, $\mathcal{L}$ is the external virtual work; δ is the virtual "variation in" (strain in this case). Again, the tensor notation implies summation of the repeated indices over their variation range ($i, j = 1, 2, 3$). The basic difficulty is that the configuration in which equation (5.15) is written is still unknown (and it is not possible to approximate it with the undeformed state, as in linear analysis); therefore, suitable measures of stress and strain must be used.

The usual stress measure used in these problems is the 2nd Piola-Kirchhoff tensor that, at current time (t) and adopting as reference state the initial state (0), is defined as

$$ {}^{t}_{0}S_{ij} = \frac{{}^{0}\rho}{{}^{t}\rho}\,\frac{\partial^{0}x_i}{\partial^{t}x_m}\,{}^{t}\sigma_{mn}\,\frac{\partial^{0}x_j}{\partial^{t}x_n} \qquad (5.16)$$

where ρ is the density and x_i is a coordinate (it can be easily noticed that, if the current and reference state coincide, the 2nd Piola-Kirchhoff stress reduces to the Cauchy stress). The related strain measure is the Green-Lagrange tensor, which, considering again times (t) and (0), is defined as

$$ {}^{t}_{0}E_{ij} = \frac{1}{2}\left(\frac{\partial^{t}u_i}{\partial^{0}x_j} + \frac{\partial^{t}u_j}{\partial^{0}x_i} + \frac{\partial^{t}u_k}{\partial^{0}x_i}\frac{\partial^{t}u_k}{\partial^{0}x_j} \right) \qquad (5.17)$$

where u_i is the displacement component in the direction of x_i. The important fact is that the 2nd Piola-Kirchhoff stress and the Green-Lagrange strain are energetically conjugated; thus, adopting for instance the total Lagrangian formulation, it follows that equation (5.15) can be written as

$$\int_{t+\Delta t_V} ({}^{t+\Delta t}\sigma_{ij}\,\delta{}^{t+\Delta t}\varepsilon_{ij})\,d{}^{t+\Delta t}V = \int_{0_V} ({}^{t+\Delta t}_{0}S_{ij}\,\delta{}^{t+\Delta t}_{0}E_{ij})\,d^0V = {}^{t+\Delta t}\mathcal{L} \qquad (5.18)$$

Thanks to this transform, the integration can be carried out over a known state (^0V). If the updated Lagrangian formulation is adopted, in equations (5.18) index 0 is replaced by t. In any case, equation (5.18) is then linearized and implemented in the FEM discretization. A notable aspect in this class of problems is that, in addition to the usual stiffness matrix of the elements, related to material behaviour, a geometric stiffness matrix, originated from the non-linear part of the strain, also arises (for example, in a truss or beam element the geometric stiffness accounts for the stiffening effect due to the axial tension). Like in the case of the non-linear material behaviour cited in Section 5.2, a suitable solution method must be adopted: a common choice is the Newton-Raphson iterative scheme.

As FEM formulations for geometrically non-linear cases became available, they were readily applied to the study of bonded, flexible joints as well. Crocombe and Adams (1981) used a geometrically non-linear FEM model to study the peel test, in which the flexible strip that is peeled off undergoes large displacements and rotations. It was found that the maximum stress occurs at the bond end close to the flexible strip. The study also made use of fracture mechanics, including the presence of a crack at the adherend/adhesive interface, and the SIFs related to modes I and II were evaluated.

Reddy and Roy (1988) developed a bi-dimensional finite element for the geometrically non-linear analysis of bonded joints based on the updated Lagrangian formulation. The stress distribution under three different boundary conditions for the simple lap joints was assessed, evidencing the importance of the applied constraints. The response of a cantilevered joint subjected to a deflection equal to 50% of its span was calculated, which highlighted the need for the non-linear analysis to reproduce the phenomenon accurately. Tsai and Morton (1994) carried out a two-dimensional geometrically non-linear analysis to create a set of results on single lap joints to support a comparison among the theories of Goland and Reissner (1944), Hart-Smith (1973), and Oplinger (1994). They noticed that in the case of a short overlap the Hart-Smith solution is more suitable, whilst in the case of a long overlap, the Oplinger solution gives a better approximation. However, the Goland and Reissner solution is accurate enough in both cases of short or long overlaps.

Adopting the updated Lagrangian formulation, Andruet et al. (2001) developed special finite elements for the geometrically non-linear analysis of bonded joints, in two- and three-dimensional cases. In the two-dimensional case, the adherends were described by non-linear Bernoulli-Euler beams elements, the adhesive layer by a single (through-the-thickness) row of continuum quadrilateral elements, with offset nodes (as, for example, in Carpenter (1980) or Castagnetti and Dragoni (2009)). In the three-dimensional case, the adherends were described by nine node shell elements assuming the Bathe (2006) formulation, and the adhesive layer by brick elements with offset nodes, again using one single element through the adhesive thickness. The use of this formulation compared favourably with Tsai and Morton's (1994) and reduced considerably the computational effort with respect to a general purpose, non-linear, FEM code. Of course, as already stated in Section 5.3, the shortcoming of these special elements is that their usage is limited to in-house written FEM programs.

5.6 CONCLUSIONS

This chapter has reported in brief the principles and several cases of application of the FEM to the analysis of bonded joints. The topics are presented under the light of selected papers of the related scientific literature. Since several decades, the FEM has proven to be the natural tool to describe complex shapes and details of the joints that an analytical study cannot reproduce, either as geometry or as stress distribution. Moreover, the behaviour of the material, which in some cases (more frequently for the adhesive) may be non-linear, can be modelled with the FEM. Modelling plasticity of the adhesive is often a key instrument to predict the failure of the joints.

Practical problems in meshing the joints are due to the characteristic length – in other terms, the scale – of the zones which constitute a bonded structure. Typical values, as order of magnitude, are usually 10^{-1} mm for the adhesive thickness, 10^0 for the adherend thickness, 10^1-10^2 mm (or even 10^3 mm) for the overall dimension of the components. Therefore, to keep the model size and the consequent computational effort to a sustainable level, it is necessary to manage in an effective way the transition of mesh density from one zone to another, respecting the general requirements of FEM modelling in terms of acceptable element aspect ratio and internal angles.

Related to the latter point is the need for efficient modelling, to reduce mesh size, i.e. the number of nodes and elements in the model. Several formulations have been proposed to describe the response of the adhesive with a single row of elements. This approach leads to considerable savings, but has the shortcoming of being applicable only to in-house written programs. Alternatively, standard elements existing in commercial FEM codes have been adapted for the purpose. This offers the advantage (especially in an industrial context) that well known solvers, as well as pre-/post-processing graphical tools, can be used. In both cases, a crucial point is in managing the connections among the interface nodes of the adhesive and the mid-line nodes of the adherends, if modelled with beam or shell elements.

A peculiar aspect of the bonded joints is the stress singularity that occurs at the end or interface corners due to material mismatch and geometry. If a detailed study of the singularity is needed, it must be undertaken with a suitable approach, either by adopting an extreme local mesh refinement, or by using special elements that include the singularity in their shape functions. As the singular stress is a mathematical abstraction, due to the facts that in practice sharp corners do not exist and the stress is self-limited by plasticity, it is arguable and still subject to study whether a criterion based on the SIF can predict failure. Apart from this, the main practical consequence of the singularity from the numerical viewpoint is that local stresses do not converge by refining the mesh and this could be misleading for an analyst unaware of the problem. If the singularity is not of interest, the stresses at a proper distance from the singular point must be considered, for instance in the mid-thickness plane of the adhesive. Another typical feature of the adhesive joints is the geometrical non-linearity of the single lap joint related to the rotation of the overlap under loading. From the engineering viewpoint, this is a beneficial effect, since it reduces the stresses in the joint. Regarding the analysis, elements formulated to account for large displacements and rotations must be used to reproduce this behaviour. However, modern non-linear codes regularly include this feature.

Within the book, this chapter represents an introduction to the use of the FEM in the analysis of adhesive joints; it has mainly dealt with consolidated topics, although still of practical interest. More sophisticated or recent topics – namely fracture mechanics, CZM, damage mechanics and XFEM – are dealt with in the following chapters.

5.7 REFERENCES

Abdel Wahab, M. M. A. 2000. On the use of fracture mechanics in designing a single lap adhesive joint. Journal of Adhesion Science and Technology 14: 851-865.

Abdel Wahab, M. M., Ashcroft, I. A., Crocombe, A. D. and Smith, P. A. 2004. Finite element prediction of fatigue crack propagation lifetime in composite bonded joints. Composites: Part A 35: 213-222.

Adams, R. D. and Peppiatt, N. A. 1974. Stress analysis of adhesive-bonded lap joints. The Journal of Strain Analysis for Engineering Design 9: 185-196.

Adams, R. D. and Peppiatt, N. A. 1977. Stress analysis of adhesive bonded tubular lap joints. The Journal of Adhesion 9: 1-18.

Adams, R. D., Coppendale, J. and Peppiatt, N. A. 1978. Stress analysis of axisymmetric butt joints loaded in torsion and tension. The Journal of Strain Analysis for Engineering Design 13: 1-10.

Adams, R. D. and Harris, J. A. 1987. The influence of local geometry on the strength of adhesive joints. International Journal of Adhesion and Adhesives 7: 69-80.

Akisanya, A. R. 1997. On the singular stress field near the edge of bonded joints. Journal of Strain Analysis 32: 301-311.

Akisanya, A. R. and Meng, C. S. 2003. Initiation of fracture at the interface corner of bi-material joints. Journal of the Mechanics and Physics of Solids 51: 27-46.

Amijima, S. and Fuji, T. 1987. A microcomputer program for stress analysis of adhesive-bonded joints. International Journal of Adhesion and Adhesives 7: 199-204.

Andruet, R. H., Dillard, D. A. and Holzer, S. M. 2001. Two- and three-dimensional geometrical nonlinear finite elements for analysis of adhesive joints. International Journal of Adhesion and Adhesives 21: 17-34.

Barut, A., Guven, I. and Madenci, E. 2001. Analysis of singular stress fields at junctions of multiple dissimilar materials under mechanical and thermal loading. International Journal of Solids and Structures 38: 9077-9109.

Bathe, K. J. 1996. Finite Element Procedures. Prentice Hall, Upper Saddle River, NJ, USA.

Bigwood, D. A. and Crocombe, A. D. 1989. Elastic analysis and engineering design formulae for bonded joints. International Journal of Adhesion and Adhesives 9: 229-242.

Bogy, D. B. 1968. Edge-bonded dissimilar orthogonal elastic wedges under normal and shear loading. Journal of Applied Mechanics 35: 460-466.

Bogy, D. B. 1971. Two edge-bonded elastic wedges of different materials and wedge angles under surface tractions. Journal of Applied Mechanics 38: 377-386.

Bogy, D. B. and Wang, K. C. 1971. Stress singularities at interface corners in bonded dissimilar isotropic elastic materials. International Journal of Solids and Structures 7: 993-1005.

Campilho, R. D. S. G., Banea, M. D., Neto, J. A. B. P. and da Silva, L. F. M. 2013. Modelling adhesive joints with cohesive zone models: Effect of the cohesive law shape of the adhesive layer. International Journal of Adhesion and Adhesives 44: 48-56.

Carpenter, W. C. 1980. Stresses in bonded connections using finite elements. International Journal for Numerical Methods in Engineering 15: 1659-1680.

Carpenter, W. C. and Barsoum, R. 1989. Two finite elements for modeling the adhesive in bonded configurations. The Journal of Adhesion 30: 25-46.

Castagnetti, D. and Dragoni, E. 2009. Standard finite element techniques for efficient stress analysis of adhesive joints. International Journal of Adhesion and Adhesives 29: 125-135.

Castagnetti, D., Dragoni, E. and Spaggiari, A. 2009. Efficient post-elastic analysis of bonded joints by standard finite element techniques. Journal of Adhesion Science and Technology 23: 1459-1476.

Castagnetti, D., Dragoni, E. and Spaggiari, A. 2010. Failure analysis of bonded T-peel joints: Efficient modelling by standard finite elements with experimental validation. International Journal of Adhesion and Adhesives 30: 306-312.

Castagnetti, D., Dragoni, E. and Spaggiari, A. 2011. Failure analysis of complex bonded structures: Experimental tests and efficient finite element modeling by tied mesh method. International Journal of Adhesion and Adhesives 31: 338-346.

Crocombe, A. D. and Adams, R. D. 1981. Peel analysis using the finite element method. The Journal of Adhesion 12: 127-139.

Crocombe, A. D. and Adams, R. D. 1982. An elasto-plastic investigation of the peel test. The Journal of Adhesion 13: 241-267.

Crocombe, A. D. 1989. Global yielding as a failure criterion for bonded joints. International Journal of Adhesion and Adhesives 9: 145-153.

da Silva, L. F. M., das Neves, P. J. C., Adams, R. D. and Spelt, J. K. 2009. Analytical models of adhesively bonded joints—part I. Literature survey. International Journal of Adhesion and Adhesives 29: 319-330.

de Moura, M. F. S. F., Campilho, R. D. S. G. and Gonçalves, J. P. M. 2008. Crack equivalent concept applied to the fracture characterization of bonded joints under pure mode I loading. Composites Science and Technology 68: 2224-2230.

de Souza Neto, E. A., Peric, D. and Owen, D. R. J. 2008. Computational Methods for Plasticity – Theory and Applications. John Wiley & Sons, Chichester, West Sussex, UK.

Dundurs, J. 1969. Discussion of the paper 'Edge-bonded dissimilar orthogonal elastic wedges under normal and shear loading' by D. Bogy. Journal of Applied Mechanics 36: 650-652.

Gleich D. M., van Tooren, M. J. L. and de Haan, P. A. J. 2000. Shear and peel stress analysis of an adhesively bonded scarf joint. Journal of Adhesion Science and Technology 14: 879-893.

Gleich D. M., van Tooren, M. J. L. and Beukers, A. 2001. Analysis and evaluation of bondline thickness effects on failure load in adhesively bonded structures. Journal of Adhesion Science and Technology 15: 1091-1101.

Goglio, L., Rossetto, M. and Dragoni, E. 2008. Design of adhesive joints based on peak elastic stresses. International Journal of Adhesion and Adhesives 28: 427-435.

Goglio, L. and Rossetto, M. 2009. Evaluation of the singular stresses in adhesive joints. Journal of Adhesion Science and Technology 23: 1441-1457.

Goglio, L. and Rossetto, M. 2010. Stress intensity factors in bonded joints: Influence of the geometry. International Journal of Adhesion and Adhesives 30: 313-321.

Goland, M. and Reissner, E. 1944. The stresses in cemented joints. Journal of Applied Mechanics 11: A17-27.

Groth, H. L. 1988. Stress singularities and fracture at interface corners in bonded joints. International Journal of Adhesion and Adhesives 8: 107-112.

Guo, S., Dillard, D. A. and Plaut, R. H. 2006. Effect of boundary conditions and spacers on single-lap joints loaded in tension and compression. International Journal of Adhesion and Adhesives 26: 629-638.

Hamoush, S. A. and Ahmad, S. H. 1989. Fracture energy release rate of adhesive joints. International Journal of Adhesion and Adhesives 9: 171-178.

Han, X., Crocombe, A. D., Anwar, S. N. R., Hu, P. and Li, W. D. 2014. The Effect of a hot–wet environment on adhesively bonded joints under a sustained load. The Journal of Adhesion 90: 420-436.

Hart-Smith, L. J. 1973. Adhesive-bonded single-lap joints. Technical Report NASA CR-11236.

Hassanipour, M. and Oechsner, A. 2011. Implementation of a pressure sensitive yield criterion for adhesives into a commercial finite element code. The Journal of Adhesion 87: 1125-1147.

Hattori, T. 1991. A stress-singularity-parameter approach for evaluating the adhesive strength of single lap joints. JSME International Journal 34: 326-331.

He, X. 2011. A review of finite element analysis of adhesively bonded joints. International Journal of Adhesion & Adhesives 31: 248-264.

Karachalios, E. F., Adams, R. D. and da Silva, L. F. M. 2013. Single lap joints loaded in tension with high strength steel adherends. International Journal of Adhesion and Adhesives 43: 81-95.

Katnam, K. B., Crocombe, A. D., Khoramishad, H. and Ashcroft, I. A. 2011. The static failure of adhesively bonded metal laminate structures: A cohesive zone approach. Journal of Adhesion Science and Technology 25: 1131-1157.

Kuo, A. S. 1984. A two-dimensional shear spring element. AIAA Journal 22: 1460-1464.

Lang, T. P. and Mallick, P. K. 1998. Effect of spew geometry on stresses in single lap adhesive joints. International Journal of Adhesion & Adhesives 18: 167-177.

Lang, T. P. and Mallick, P. K. 1999. The effect of recessing on the stresses in single lap adhesively bonded joints. International Journal of Adhesion & Adhesives 19: 257-271.

Lazzarin, P., Quaresimin, M. and Ferro, P. 2002. A two-term stress function approach to evaluate stress distributions in bonded joints of different geometries. Journal of Strain Analysis 37: 385-398.

Lin, C.-C. and Lin, Y.-S. 1993. A finite element model of single-lap adhesive joints. International Journal of Solids and Structures 30: 1679-1692.

Mubashar, A., Ashcroft, I. A., Critchlow, G. W. and Crocombe, A. D. 2009. Moisture absorption–desorption effects in adhesive joints. International Journal of Adhesion & Adhesives 29: 751-760.

Mubashar, A., Ashcroft, I. A., Critchlow, G. W. and Crocombe, A. D. 2011. Strength prediction of adhesive joints after cyclic moisture conditioning using a cohesive zone model. Engineering Fracture Mechanics 78: 2746-2760.

Noda, N.-A. and Lan, X. 2012. Stress intensity factors for an edge interface crack in a bonded semi-infinite plate for arbitrary material combination. International Journal of Solids and Structures 49: 1241-1251.

Ojalvo, I. U. and Eidinoff, H. L. 1978. Bond thickness effects upon stresses in single lap adhesive joints. AIAA Journal 16: 204-211.

Oplinger, D. W. 1994. Effects of adherend deflections in single lap joints. International Journal of Solids and Structures 31: 2565-2587.

Pirondi, A. and Nicoletto, G. 2004. Fatigue crack growth in bonded DCB specimens. Engineering Fracture Mechanics 71: 859-871.

Pradhan, S. C., Iyengar, N. G. R. and Kishore, N. N. 1995. Finite element analysis of crack growth in adhesively bonded joints. International Journal of Adhesion & Adhesives 15: 33-41.

Quaresimin, M. and Ricotta, M. 2006. Life prediction of bonded joints in composite materials. International Journal of Fatigue 28: 1166-1176.

Raghava, R., Caddell, R. M. and Yeh, G. S. Y. 1973. The macroscopic yield behaviour of polymers. Journal of Materials Science 8: 225-232.

Rao, B. N., Rao, Y. V. K. S. and Yadagiri, S. 1982. Analysis of composite bonded joints. Fibre Science and Technology 17: 77-90.

Reddy, J. N. and Roy, S. 1988. Non-linear analysis of adhesively bonded joints. International Journal of Non-Linear Mechanics 23: 97-112.

Rispler, A. R., Tong, L., Steven, G. P. and Wisnom, M. R. 2000. Shape optimisation of adhesive fillets. International Journal of Adhesion & Adhesives 20: 221-231.

Sancaktar, E. and Narayan, K. 1999. Substrate volume and stress gradient concepts in mechanical adhesion: Analysis of single straight sections. Journal of Adhesion Science and Technology 13: 237-271.

Sawa, T., Liu, J., Nakano, K. and Tanaka, J. 2000. A two-dimensional stress analysis of single-lap adhesive joints of dissimilar adherends subjected to tensile loads. Journal of Adhesion Science and Technology 14: 43-66.

Sheppard, A., Kelly, D. and Tong, L. 1998. A damage zone model for the failure analysis of adhesively bonded joints. International Journal of Adhesion & Adhesives 18: 385-400.

Stapleton, S. E, Waas, A. M. and Bednarcy, B. A. 2011. Modeling progressive failure of bonded joints using a single joint finite element. AIAA Journal 49: 1740-1749.

Stapleton, S. E., Pineda, E. J., Gries, T. and Waas, A. M. 2014. Adaptive shape functions and internal mesh adaptation for modelling progressive failure in adhesively bonded joints. International Journal of Solids and Structures 51: 3252-3264.

Tong, L. and Sun, X. 2003. Adhesive elements for stress analysis of bonded patch to curved thin walled structures. Computational Mechanics 30: 143-154.

Tsai, M. Y. and Morton, J. 1994. An evaluation of analytical and numerical solutions to the single-lap joint. International Journal of Solids and Structures 31: 2537-2563.

Wang, C. H. and Rose, L. R. F. 2000. Compact solutions for the corner singularity in bonded lap joints. International Journal of Adhesion & Adhesives 20: 145-154.

Williams, M. L. 1952. Stress singularities resulting from various boundary conditions in angular corners of plates in extension. Journal of Applied Mechanics 19: 526-528.

Wooley, G. R. and Carver, D. R. 1971. Stress concentration factors for bonded lap joints. Journal of Aircraft 8: 817-820.

Xu, W. and Li, G. 2010. Finite difference three-dimensional solution of stresses in adhesively bonded composite tubular joint subjected to torsion. International Journal of Adhesion & Adhesives 30: 191-199.

Yadagiri, S., Reddy, C. P. and Reddy, T. S. 1987. Viscoelastic analysis of adhesively bonded joints. Computers and Structures 27: 445-454.

Zappalorto, M., Carraro, P. A. and Quaresimin, M. 2015. Analytical solution for the three-dimensional stress field in anisotropic composite bimaterial corners. Composite Structures 122: 127-138.

Zhao, X., Adams, R. D. and da Silva, L. F. M. 2010. A new method for the determination of bending moments in single lap joints. International Journal of Adhesion & Adhesives 30: 63-71.

Zienkiewicz, O. C. and Cheung, Y. K. 1967. The Finite Element Method in Structural and Continuum Mechanics. McGraw-Hill, London.

6

C H A P T E R

Cohesive Zone Modelling for Static Applications

Raul D.S.G. Campilho

6.1 INTRODUCTION

Adhesively-bonded joints are extensively used in several fields of engineering, such as automotive, aeronautical and space structures, as an easy method to join components, assuring at the same time the design requirements for the structure (Lee et al. 2010). General capabilities of this jointing method involve more uniform stress fields than fastening or riveting, fluid sealing, high fatigue resistance and the possibility to join different materials on account of corrosion prevention and accommodation of different thermal expansion of the adherends (Banea and da Silva 2009, Campilho et al. 2012). The techniques for strength prediction of bonded joints have also improved. Initially, theoretical methods (mainly closed-form) were proposed for stress distributions in the adhesive for simple geometries such as the single or double-lap joint and failure estimation was carried out by comparison of the maximum stresses with the material strengths (Lai et al. 1996). Some decades later, the Finite Element Method (FEM) initiated its incursion in the analysis of adhesively-bonded joints (e.g. the work of Wooley and Carver (1971)), by consideration of stress/strain or fracture mechanics criteria for failure prediction (Tsai and Morton 1994). Even though these analyses were promising, they had few limitations: stress/strain predictions depend on the mesh size at the critical regions, while fracture criteria such as the Virtual Crack Closure Technique (VCCT) are restricted to Linear Elastic Fracture Mechanics (LEFM) and need an initial crack.

Cohesive Zone Models (CZM) have been used in the last decades for the strength prediction of adhesive joints, as an add-on to FEM analyses that allows simulation of damage growth within bulk regions of continuous materials or interfaces between different materials (Feraren and Jensen 2004, da Silva and Campilho

Departamento de Engenharia Mecânica, Instituto Superior de Engenhariado Porto, Instituto Politécnicodo Porto, Rua Dr. António Bernardino de Almeida, 431, 4200-072 Porto, Portugal.
E-mail: raulcampilho@gmail.com

2012). Compared to the conventional FEM, a much more accurate prediction is achieved, since different shapes can be developed for the cohesive laws, depending on the nature of the material or interface to be simulated. The triangular and trapezoidal CZM shapes are most commonly used for strength prediction of typical structural materials. For the application of this technique, traction-separation laws with a pre-defined shape are established at the failure paths and the values of strain energy release rate in tension and shear (G_I and G_{II}, respectively) along the fracture paths and respective critical values or toughness (G_{Ic} and G_{IIc}) are required. The cohesive strengths in tension and shear (t_n^0 and t_s^0, respectively) are equally needed and they relate to damage initiation, i.e. end of the elastic behaviour and beginning of damage. Different techniques are nowadays available for the definition of the cohesive parameters (G_{Ic}, G_{IIc}, t_n^0 and t_s^0) such as the property identification technique, the direct method and the inverse method. These methods usually rely on the Double-Cantilever Beam (DCB), End-Notched Flexure (ENF), or single-lap specimens, generally with good results (Li et al. 2005, Blackman et al. 2006, Biel and Stigh 2008, Banea et al. 2010, Campilho et al. 2011a). The property identification technique consists of the separated calculation of each one of the cohesive law parameters by suitable tests, while in the inverse method, the CZM parameters are estimated by iterative fitting of the FEM prediction with experimentally measured data (typically the load-displacement, P-δ, curve) up to an accurate representation. The direct method gives the precise shape of the CZM laws of a specific material or interface (Pandya and Williams 2000). This is done by differentiation of G_I (tension) or G_{II} (shear) with respect to the relative opening of the crack (δ_n for tension or δ_s for shear).

Carlberger and Stigh (2010) estimated the CZM law shapes of a thin adhesive layer in tension and shear with DCB and ENF tests, respectively, for an adhesive thickness, t_A, between 0.1 and 1.6 mm. The cohesive laws were found by a direct method based on the differentiation of the G_I/G_{II} vs. δ_n/δ_s data. The CZM shapes and respective parameters significantly varied with t_A, ranging from a rough triangular shape for the smaller values of t_A to a trapezoidal shape for bigger values of t_A. Ji et al. (2010) studied the influence of t_A on t_n^0 and G_{Ic} for a brittle epoxy adhesive, by using the DCB specimen and the direct method for parameter estimation. The analysis methodology relied on the measurement of G_{Ic} by an analytical technique proposed by Andersson and Stigh (2004) that required the relative rotation between the adherends of the DCB specimen. Derivation of the tensile cohesive stress (t_n) – δ_n laws was quickly obtained by differentiation of the G_I – δ_n data (Sørensen 2002, Zhu et al. 2009) and clearly showed a reduction of t_n^0 and increase of G_{Ic} with bigger values of t_A. On the other hand, the influence of the deviation between the parameterized approximations and the real cohesive laws on the output of the simulations, is also largely dependent on parameters such as the adherend thickness (t_P) or the stiffness of the adherends. Pinto et al. (2009) evaluated the tensile strength of single-lap joints between adherends with different values of t_A and materials, with parameterized trapezoidal shapes being used in the numerical (CZM) simulations. The P-δ data revealed that the use of accurate CZM shapes for the adhesive is much more important with stiff adherends, since the differential deformation between the adherends is minimized and the adhesive is evenly loaded along the entire bond length. Conversely, compliant adherends give rise to substantial shear and peel peak stress gradients, providing relatively insensitive results to the shape of the cohesive law. The results of Ridha et al. (2011) modelled a thin adhesive layer of a high elongation epoxy adhesive in scarf repairs on composite panels. Linear, exponential, and trapezoidal softening was considered to model plasticity of the adhesive, with the use of linear softening resulting in under prediction of the actual strengths of the repairs by nearly 20%, because of premature softening at the bond edges after the peak strength that is not consistent with the actual behaviour of the adhesive.

This chapter gives a general description of CZM applied to the strength prediction of bonded joints under static conditions. An overview of the technique is initially given, followed by the most common techniques for estimation of the cohesive laws, essential as input in the numerical simulations. Convergence problems are relevant in the CZM analysis of bonded joints due to the sometimes abrupt failure process, and are specifically addressed in this chapter. Finally, two case studies are presented in this chapter regarding the application of CZM techniques to adhesively-bonded joints, which enable a better perception regarding the accuracy of this technique. The first case study investigates the effect of the cohesive parameters of the adhesive layer, while the second one addresses the influence of the cohesive law shape of the adhesive layer; both on the accuracy of the strength predictions.

6.2 COHESIVE ZONE MODELLING

The conventional techniques for strength prediction of bonded joints such as continuum mechanics or LEFM-based techniques endured success some decades ago, but they are limited to small-scale yielding ahead of the crack tip. However, for modern toughened adhesives the plastic zones that develop in the adhesive layer can be larger than t_p (Ji et al. 2010). CZM were developed in the late 1950's/early 1960's by Barenblatt (1959) and Dugdale (1960) for static applications. CZM are based on spring or more typically cohesive elements (Feraren and Jensen 2004), connecting two-dimensional (2D) or three-dimensional (3D) solid elements of structures. The CZM laws can be easily incorporated in conventional FEM software to simulate crack propagation in various materials, including adhesively bonded joints (Ji et al. 2010). This method is based on the establishment of cohesive laws to model interfaces or finite regions.

The CZM laws are established between paired nodes of cohesive elements, and they can be used to connect superimposed nodes of elements representing different materials or different plies in composites, to simulate a zero thickness interface (local approach; Fig. 6.1a; (Pardoen et al. 2005)), or they can be applied directly between two non-contacting materials to simulate a thin strip of finite thickness between them, e.g. to simulate an adhesive bond (continuum approach; Fig. 6.1b; (Xie and Waas 2006)). Some numerical works of strength prediction of bonded joints by CZM take advantage of the local approach (Campilho et al. 2005, Liljedahl et al. 2006). With this methodology, the plastic dissipations of the adhesive bond are simulated by the solid finite elements, whilst the cohesive elements simulate damage growth (Fig. 6.1a). The "intrinsic fracture energy" should be considered for the CZM laws instead of the fracture toughness (G_c), while the plastic dissipations of ductile materials take place at the solid elements representative of the adhesive bond (Liljedahl et al. 2006). Thus, G_c is the sum of these two components. The effects of external and internal constraints on the plastic dissipations of an adhesive bond are thus accounted for in the local approach. Application of the continuum approach (Fig. 6.1b) involves the replacement of the entire adhesive bond by a single row of cohesive elements with the representative behaviour of the adhesive bond (Kafkalidis and Thouless 2002, Campilho et al. 2008a). Opposite to the local approach, the CZM elements' stiffness represents the adhesive layer stiffness in each mode of loading. This approach has been widely used in the simulation of bonded joints, with accurate results after proper calibrations are undertaken for the CZM laws (Campilho et al. 2009b). The main disadvantage of this approach is that CZM become dependent on the joint geometry, more specifically of t_P and t_A, because these largely affect the size of the fracture process zone (FPZ) and plasticity ahead of the crack tip (Ji et al. 2010).

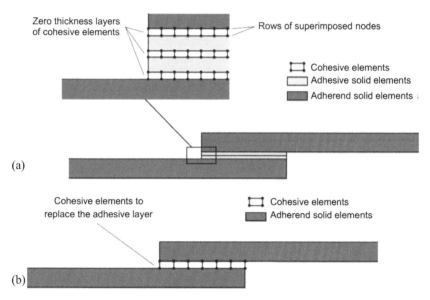

FIG. 6.1 Cohesive elements to simulate zero thickness failure paths – local approach (a) and to model a thin adhesive bond between the adherends – continuum approach (b) in an adhesive joint.

CZM simulate stress evolution and subsequent softening up to complete failure, which corresponds to material degradation. The CZM laws are usually represented by linear relations at each one of the loading stages (Yang and Thouless 2001). Figure 6.2 represents the 2D triangular CZM model actually implemented in Abaqus® (Providence, RI, USA) for static damage growth. The 3D additionally includes the tearing component. More details on the 3D CZM is available in (Campilho et al. 2008b) or in Abaqus® (2013).

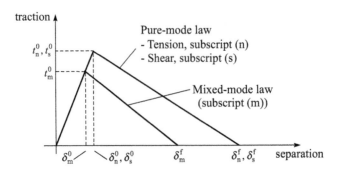

FIG. 6.2 Triangular CZM law (adapted from Abaqus® (2013)).

The subscripts n and s relate to pure normal (tension) and shear behaviours, respectively. t_n and t_s are defined as current stresses in tension and shear, respectively, δ_n^0 and δ_s^0 are the peak strength displacements, and δ_n^f and δ_s^f the failure displacements (defined by G_{Ic} or G_{IIc}, respectively, as these represent the area under the CZM laws). In the mixed-mode CZM law, t_m^0 is the mixed-mode cohesive strength, δ_m^0 the corresponding displacement, and δ_m^f the mixed-mode failure displacement. Under mixed loading, stress and/or energetic criteria are used to combine the pure mode laws. With this procedure, the complete failure response of structures can be simulated (Zhu et al. 2009). Typically, continuum mechanics criteria are used for the onset of damage and energy criteria for propagation (Chen et al. 2011). This allows the simulation of onset and non-self-similar growth of damage without user intervention and not requiring an initial flaw, unlike what occurs with conventional fracture mechanics criteria. On the other hand, compared to continuum mechanics techniques, CZM are mesh independent if enough integration points are simultaneously under softening during the failure process (Campilho et al. 2009d, Campilho et al. 2011b). This technique also allows the combining of multiple failure possibilities, and the knowledge of the damage onset site is not required as input, since damages initiates at any CZM element when the damage onset criterion is attained. The main limitation of CZM is that cohesive elements must exist at the planes where damage is prone to occur, which, in several applications, can be difficult to know in advance. However, in bonded joints that damage propagation is restricted to the adhesive layer or the adhesive/adherend interfaces, which turns the analysis procedure easier. Developed CZM include triangular (Alfano and Crisfield 2001), linear-parabolic (Allix and Corigliano 1996), polynomial (Chen 2002), exponential (Chandra et al. 2002), and trapezoidal laws (Campilho et al. 2008a). The triangular, linear-exponential, and trapezoidal shapes are the most commonly used CZM shapes (Fig. 6.3). In the trapezoidal law, δ_n^s and δ_s^s are the stress softening onset displacements.

CZM can also be adapted to simulate ductile adhesive layers by using trapezoidal laws (Campilho et al. 2009b). Although it is always advised to use the most suitable CZM shape and to perform accurate parameter estimations, few works showed acceptable predictions for small variations to the optimal CZM parameters and shapes (Liljedahl et al. 2006, Biel and Stigh 2008). Nonetheless, the effect of the

CZM law shape on the strength predictions may significantly vary depending on the structure geometry and post-elastic behaviour of the materials.

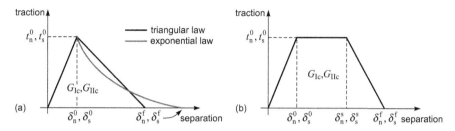

FIG. 6.3 Different shapes of pure mode CZM laws: triangular or linear-exponential (a) and trapezoidal (b).

The CZM law effects became evident in the experimental and FEM study of Pinto et al. (2009), whose objective was the strength comparison of single-lap joints with similar and dissimilar adherends and values of t_p bonded with the adhesive 3M DP-8005®. The accurate shape of the CZM law was considered fundamental for the strength prediction and $P\text{-}\delta$ response of the structure when using stiff adherends. Under these conditions, peel stresses are minimal and, due to the large longitudinal stiffness, shear stresses distribute more evenly along the bond length. Thus, the $P\text{-}\delta$ curve is very similar in shape to the chosen shear CZM law. On the other hand, compliant adherends led to large shear and peel stress gradients. Since this implies different damage states along the adhesive layer, using an inaccurate CZM law gives adhesive stresses that are over predicted at some elements and under predicted at others. Thus, by using compliant adherends the overall behaviour gave smaller errors. Ridha et al. (2011) considered scarf repairs on composite panels bonded with the high elongation epoxy adhesive FM® 300M (Cytec). CZM laws with linear, exponential, and trapezoidal softening were compared, and linear degradation resulted in under predictions of the repairs strength of nearly 20%, on account of excessive plastic degradation at the bond edges that was not observed in the real joints. Regarding the application of CZM for strength prediction of adhesive bonds, trapezoidal laws are recommended for ductile adhesives (Feraren and Jensen 2004, Campilho et al. 2010b), and this is particularly critical when considering stiff adherends, due to the practical absence of differential deformation effects in these components along the overlap (Alfano 2006, Pinto et al. 2009). In contrast, triangular CZM are efficient for brittle materials that do not plasticize by a significant amount after yielding (Campilho et al. 2011b), and also for the intralaminar fracture of composite adherends in bonded structures, due to their intrinsic brittleness (Xie et al. 2006). For adhesives that exhibit a relatively brittle behaviour in tension while showing large plastic flow in shear, the proper selection of the CZM parameters and also the minimization of the constant stress (plastic flow) region in the tensile law result on a good representation of the adhesive behaviour. The material/interfacial behaviour that the CZM law is simulating should always be the leading decision factor to select the most appropriate shape. Despite this fact, other issues should be taken into account (da Silva and Campilho 2012). In fact, the CZM law shape also influences the iterative solving procedure and the time required to attain the solution of a given engineering problem: larger convergence difficulties in the iterative solving procedure usually take place for trapezoidal rather than triangular CZM laws due to the more abrupt change of stiffness in the cohesive elements during stress softening. Actually, for a fixed value of the material properties G_{Ic} and G_{IIc}, the larger the constant stress length of the trapezoidal law, the bigger is the descending slope. Additionally, exponential and trapezoidal CZM are more difficult to formulate and implement in FEM software.

The CZM parameter effects are also detailed in several works. The main joint geometry parameters that affect the CZM parameters of adhesive layers are t_A and t_P, which emphasizes the importance of the t_A and t_P consistency between the fracture tests and the structures to be simulated (Bascom and Cottington 1976, Chai 1986, Leffler et al. 2007). In the work of Carlberger and Stigh (2010), the CZM parameters in tension and shear were determined for a thin layer of adhesive using the DCB and ENF test configurations, respectively, considering $0.1 \leq t_A \leq 1.6$ mm. It was concluded that the CZM parameters significantly vary with t_A, namely an increase of G_{Ic} and G_{IIc} with this parameter. Corroboration of the adhesive restraining effects was equally accomplished by Ji et al. (2010), which studied the influence of t_A on t_n^0 and G_{Ic} for a brittle epoxy adhesive, by using the DCB specimen and the direct method for parameter estimation. Results clearly showed a reduction of t_n^0 and increase of G_{Ic} with bigger values of t_A. On the other hand, a few studied showed variations of G_{Ic} and G_{IIc} by modification of t_P. In the work of Mangalgiri et al. (1987), symmetric and unsymmetric DCB specimens were experimentally tested with different values of t_P (by considering 8, 16, or 24 plies of carbon-fibre adherends). The static tests showed a large improvement of G_{Ic} between composites with 8 and 16 plies. Devitt et al. (1980) equally used the DCB test to investigate this effect, and found a 9% increase in the value of G_{Ic} of bonded joints made of glass-epoxy composites by duplicating the number of plies of the adherends. From these studies, it is clear that the differences take place at relatively low t_P values. Since most bonded joints are made between thin adherends/sheets, the understanding of how t_P affects the fracture toughness is highly relevant.

6.2.1 Estimation of the Cohesive Parameters

CZM is a powerful technique for the strength prediction of bonded joints and to account for the largely nonlinear fracture behaviour of modern adhesively bonded joints. However, the cohesive parameters for a given application require accurate estimation procedures under identical restraining conditions and respective validation in order to accurately simulate the failure process (Carlberger and Stigh 2010). Despite this fact, standardized methods for the definition of t_n^0 and t_s^0 are not yet available (Lee et al. 2010). Recently, several researchers focused on the cohesive parameters estimation (t_n^0, t_s^0, G_{Ic} and G_{IIc}) and mainly three techniques were proposed with this purpose: the property identification technique, the inverse method, and the direct method, which significantly differ in terms of complexity and accuracy. The few of these works that validated with mixed-mode experiments the estimated pure mode CZM typically made use of DCB, ENF, or single lap specimens, generally with good results (Kafkalidis and Thouless 2002, Zhu et al. 2009).

The property identification method involves the individual estimation of all parameters relevant to the CZM laws by suitable tests. According to previous tests (Pandya and Williams 2000), this technique is most critical when using bulk tests, due to the known deviations between the properties of adhesives as bulk and thin layers. This is mainly caused by two factors: the restraining effect of the surrounding adherends to the adhesive layer, and the typical mixed-mode crack propagation in adhesive bonds. Actually, in bulk materials, cracks tend to grow perpendicularly to the direction of maximum principal stress (Chai 1992). In thin bonds, cracks are forced to follow the bond length path since, as the adhesive is typically weaker and more compliant than the components to be joined, failure is often cohesive within the adhesive. The inverse technique is based on an iterative fitting analysis to match the numerical results with experimental data on bonded joints, such as the P-δ curve, allowing the inverse determination of CZM laws with parameterized shape for the specific geometrical conditions of the experimental specimens. It should be mentioned that for these two methods (property identification and inverse), a simplified shape for the CZM laws should be initially selected (e.g. bilinear or trilinear), based on the knowledge of the material behaviour to be simulated (Campilho et al. 2009b). In opposition to these two techniques, the direct method outputs the precise shape of the CZM laws. This is done by initially evaluating G_{Ic} or G_{IIc} by proper fracture characterization tests up to crack initiation, typically by J-integral-based methods, and then by differentiating the G_I-δ_n or G_{II}-δ_s curves, which provides the directly measured CZM laws (Pandya and Williams 2000). However, for subsequent use by FEM strength prediction of adhesively-bonded structures, it is common practice to build a simplified approximation for easier implementation.

Disregarding the considered method to define the CZM laws, most likely deviations will exist between the predicted CZM laws and the actual behaviour of the adhesive layer (Leffler et al. 2007). However, between the three techniques, the inverse and direct methods are considered to be more accurate because the adhesive layer is characterized as a thin layer, i.e., under identical circumstances to the real applications (Pardoen et al. 2005). The direct method is particularly accurate in the sense that it provides the exact CZM laws of the adhesive, whereas the inverse method starts from a simplified assumed CZM shape. Between all CZM parameters, G_c is the most important on account of its large influence on the strength predictions (Campilho et al. 2012). The estimation of this parameter is regularly performed by LEFM techniques, which generally work well for brittle and moderately ductile adhesives. For highly ductile adhesives, LEFM techniques are not appropriate, and the J-integral is a viable option. However, this method is more complicated to apply, since it requires additional data such as the adherends rotation during the test (Zhu et al. 2009). Independent of the parameter determination method, the CZM parameters invariably depend on t_A and t_P. Thus, the specimens used to characterize the CZM laws of the adhesive should have identical t_A and t_P values to the structures to be simulated (Leffler et al. 2007).

6.2.2 Convergence Issues

The use of CZM is intrinsically linked to convergence issues that may occur in the incremental iterative process at the time of reaching the solution of the FEM system of equations. Actually, in a given increment, the solution is attained when the iterative procedure gives an error between the left and right sides of the fundamental equation under a software pre-defined or used-defined value (Abaqus® 2013). Usually, the main factors contributing to convergence difficulties are the propagation of damage by CZM formulations (which depend on the CZM law shape) and respective instability, the consideration of plastic constitutive behaviours for the solid elements, the adhesive plasticity, the adherends stiffness, the orthotropic nature of composite materials, coarse meshes at or near stress concentration/damage propagation sites and structural geometries with large stress concentrations, and singular regions or intersecting cracks (Hamitouche et al. 2008), more specifically if two or more of these features exist concurrently within an analysis. In FEM models, the increments usually run smoothly up to the damage uptake by the cohesive elements along the failure paths after which convergence problems appear due to the abrupt variation of the stiffness matrix between consecutive increments.

Trapezoidal laws give more convergence problems because of the more abrupt change in the stiffness of the cohesive elements during softening. Gradual softening and reduction of the mesh size at the fracture regions cannot be enough to surpass difficulties and attain total separation. Convergence problems may also occur during unstable crack propagation, when the available value of strain energy release rate (G) is higher than the respective G_c (Abaqus® 2013), which can be prevented by smaller step incrementations. Isotropic plasticity in the adherends or orthotropic elasticity for composite adherends creates larger difficulties during CZM damage growth and, under these conditions, elastic-perfectly plastic approximations or approximation to isotropic equivalent properties, respectively, can assist a smoother fracture without compromising the results. On the same line of adherends plasticity, compliant adherends, such as polymers, usually endure significant deformations and element distortions near the damage propagation sites, making it difficult to reduce iterative errors during crack growth with implicit solving techniques such as in Abaqus®/Standard. There is no ultimate solution to improve convergence for this scenario, although the local reduction of the mesh size, or the use of geometric non-linearities in the FEM software may help. As regards adhesive behaviour, brittle adhesives make the CZM softening process more abrupt, notwithstanding the CZM law shape that, added to the smaller FPZ at the crack tip, results in larger convergence difficulties. Depending on the bonded system, the increase of G_{Ic} and/or G_{IIc} is sometimes enough to achieve surface separation, although a small over-prediction of the structural strength is naturally associated (up to 10% in single lap joints for 100% increase of G_{Ic} and G_{IIc} (Campilho et al. 2012)). All the other issues, such as coarse meshing, singularities, or intersecting cracks, are partially solved by a bigger mesh refinement.

Transversely to all these scenarios, the reduction of the cohesive elements' initial stiffness up to t_n^0 or t_s^0 is equally a possibility for the reduction of convergence problems by the reduction of the adjacent elements' distortion and stress concentrations. However, this must be performed in such a manner that the structure's overall stiffness is not compromised (for both local and continuum approaches) and, additionally for adhesive bonds modelled by a continuum CZM, bearing in mind that the constitutive behaviour of the bond is being altered. Interpenetration of the crack faces can also occur for small stiffness values and should be prevented as well (Khoramishad et al. 2010). Following extensive studies, the value of 10^6 N/mm^3 was recommended by Gonçalves et al. (2000) for a triangular static CZM (local approach modelling) for accurate convergence and minimization of convergence difficulties during the nonlinear procedure. Nonetheless, values down to 5×10^4 N/mm^3 were used without compromising the global behaviour of bonded repairs on composite structures (Campilho et al. 2008b). Smaller values are not recommended on account of stiffness accuracy issues and numerical ill-conditioning problems (Khoramishad et al. 2010).

Apart from the proposed solutions, FEM software packages usually hold a few techniques to aid the iterative process. Abaqus®, as an example, gives the user three main possibilities (Abaqus® 2013): (1) viscous regularization, (2) automatic stabilization, and (3) non-default solution controls.

Using viscous regularization in Abaqus®/Standard improves the convergence of fracture problems, by the implementation of viscous regularization within the constitutive equations, which allows stresses to be outside the limits of the CZM laws. A small value for the viscosity parameter, compared to the characteristic time increment, usually improves the convergence rate of simulations during the softening regime, without compromising the accuracy of the results.

Consideration of automatic stabilization is another possibility to improve convergence for unstable problems (e.g., due to buckling or material softening). If the instability is localized, a local transfer of strain energy takes place from one part of the model to the neighbouring regions, and global solution methods may not work (Abaqus® 2013). These difficulties have to be solved either dynamically or with the introduction of (artificial) damping. Abaqus®/Standard uses an automatic algorithm for the stabilization of locally unstable quasi-static problems by the automatic addition of volume-proportional damping to the model.

The use of non-default solution controls consists of the user modification of the nonlinear equation solution accuracy and time increment adjustment to actively affect the incremental solving procedure and solver parameters to force convergence. However, caution is required, since, if less strict convergence criteria are imposed, the results may be accepted as converged, although they are not sufficiently close to the exact system solution. The line search algorithm is also within the scope of solution controls, allowing the detection of divergent equilibrium iterations in strongly nonlinear problems solved by the Newton algorithms of Abaqus®/Standard. A scale factor is then applied to the computed solution correction, which helps to prevent divergence.

The work of Khoramishad et al. (2010), relating to the development of a fatigue CZM for single-lap joints, included a preliminary study on the viscous regularization technique of Abaqus® as a requirement for the attainment of static failure, for further progress to fatigue damage implementation. The authors emphasized on the necessity of not changing the overall structural response by the inclusion of viscous damping in the models. As a result, the optimal value of this parameter was found by a parametric study with decreasing levels of implemented viscous damping coefficient, μ (Fig. 6.4).

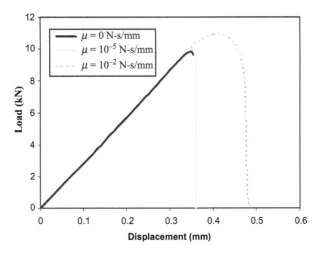

FIG. 6.4 The effect of μ on the predicted P-δ curve of a single-lap joint (Khoramishad et al. 2010).

Figure 6.4 shows the result of a non-optimized value of μ (10^{-2} N·s/mm), as the response of the bonded joints are largely affected, and the final optimized results by consideration of $\mu = 10^{-5}$ N·s/mm since this quantity allowed surpassing the divergent behaviour of the solution although it did not change the overall results.

6.3 CASE STUDIES

Two case studies are presented in this chapter regarding the application of CZM techniques to adhesively-bonded joints: (1) the effect of the cohesive parameters of the adhesive layer and (2) the effect of the cohesive law shape of the adhesive layer; both based on the accuracy of the strength predictions.

6.3.1 Effect of the Cohesive Parameters of the Adhesive Layer

In this study, the influence of the cohesive law parameters of a triangular CZM used to model a thin adhesive layer in bonded joints is studied, to estimate their effect on the predictions. Some conclusions are established to provide important data for the proper selection of the cohesive parameters' estimation technique and expected accuracy of the simulation results.

6.3.1.1 Characterization of the Materials

The adherends were fabricated from unidirectional carbon-epoxy pre-preg (SEAL®Texipreg HS 160 RM; Legnano, Italy) with 0.15 mm thickness plies and $[0]_{16}$

lay-up. Table 6.1 specifies the elastic ply properties, modelled as elastic orthotropic (Campilho et al. 2005). The adhesive Araldite® 2015 (Basel, Switzerland) was characterized in previous works (Campilho et al. 2011c), whose properties, shown in Table 6.2, were used to construct the triangular cohesive laws (the initial yield stress was calculated for a plastic strain of 0.2%).

Table 6.1 Elastic orthotropic properties of a unidirectional carbon-epoxy ply aligned in the fibres direction (x-direction; y and z are the transverse and through-the-thickness directions, respectively) (Campilho et al. 2005).

$E_x = 1.09E + 05$ MPa	$v_{xy} = 0.342$	$G_{xy} = 4315$ MPa
$E_y = 8819$ MPa	$v_{xz} = 0.342$	$G_{xz} = 4315$ MPa
$E_z = 8819$ MPa	$v_{yz} = 0.380$	$G_{yz} = 3200$ MPa

Table 6.2 Properties of the adhesive Araldite® 2015 (Campilho et al. 2011c).

Property	
Young's modulus, E [GPa]	1.85 ± 0.21
Poisson's ratio, $v*$	0.33
Tensile yield strength, σ_y [MPa]	12.63 ± 0.61
Tensile failure strength, σ_f [MPa]	21.63 ± 1.61
Tensile failure strain, ε_f [%]	4.77 ± 0.15
Shear modulus, G [GPa]	0.56 ± 0.21
Shear yield strength, τ_y [MPa]	14.6 ± 1.3
Shear failure strength, τ_f [MPa]	17.9 ± 1.8
Shear failure strain, γ_f [%]	43.9 ± 3.4

Bulk tests under tension were selected to characterize the adhesive in tension and Thick Adherend Shear Tests (TAST) were chosen for shear characterization. The bulk specimens were manufactured by the indications of the NF T 76-142 French standard, to prevent the creation of voids. Thus, the 2 mm thick plates were fabricated in a sealed mould, followed by precision machining to produce the dogbone shape described in the standard. The TAST characterization of the adhesive was carried out according to the 11003-2:1999 ISO standard, considering DIN Ck 45 steel adherends. More details about the fabrication and testing procedures can be found in the work of Campilho et al. (2011c).

6.3.1.2 Joint Fabrication and Testing

Figure 6.5 represents the joint geometry. The characteristic dimensions were defined as (in mm): overlap length $L_O = 10\text{-}80$, width $b = 15$, total length between gripping points $L_T = 240$, $t_P = 2.4$, and $t_A = 0.2$. The bonding surfaces were prepared by manual abrasion with 220 grit sandpaper, followed by wiping with acetone. The desired value of t_A was achieved with a dummy adherend and a 0.2 mm spacer under the upper adherend, jointly with the application of pressure with grips. Tabs were glued at the specimen edges for correct alignment in the testing machine. For proper adhesive curing at room temperature, testing was carried out one week after fabrication. Eight different values of L_O were evaluated (10, 20, 30, 40, 50, 60, 70, and 80 mm). The joints were tested in an Instron® 4208 (Norwood, MA, USA) electro-mechanical testing machine with a 100 kN load cell, at room temperature and under displacement control (0.5 mm/min). The testing machine grips displacement was considered to build the $P\text{-}\delta$ curves. For each value of L_O, six specimens were tested, with at least four valid results.

6.3.1.3 Numerical Analysis

6.3.1.3.1 Modelling Conditions

The numerical analysis in the FEM package Abaqus® aimed to check the accuracy of its triangular CZM embedded formulation to predict the strength of adhesively-bonded single-lap joints, and to evaluate the impact of cohesive parameter misjudgements on the strength predictions, either caused by intrinsic limitations of the data reduction techniques, or by different restraining scenarios between the characterization tests and the structures to be simulated (e.g. t_A or t_P inconsistencies). The triangular CZM formulation was chosen for this analysis because of its simplicity, large use for investigation purposes, and availability in Abaqus® including a mixed-mode formulation, which is absolutely necessary to model the single-lap joints used in this case study. However, other CZM shapes are available, such as the trapezoidal, which, for this particular case, would more faithfully simulate ductile adhesive behaviour (Campilho et al. 2009b). The numerical analysis considered geometrical non-linear effects (Campilho et al. 2008b, Campilho et al. 2009b) with the elastic orthotropic properties of Table 6.1 to simulate the adherends and the triangular cohesive model presented in the following section for the adhesive. The single-lap joint meshes were built without symmetry conditions (Fig. 6.5).

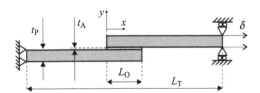

FIG. 6.5 Geometry and characteristic dimensions of the single-lap specimens.

Figure 6.6 shows the mesh for the $L_O = 10$ mm joint. The meshes of the numerical models were built automatically considering bias effects with smaller elements near the overlap edges, as these regions are known to be theoretically singular regions with major peel and shear peak stresses (Panigrahi and Pradhan 2007). The mesh size was adjusted in all models for a similar element size at the overlap edges (edge of ≈ 0.1 mm), thus allowing for the accurate capture of these stress variations (Campilho et al. 2005). Two-dimensional numerical approximations were considered to simulate the joints, with 4-node plane-strain elements (CPE4 from Abaqus®) and with COH2D4 4-node cohesive elements, compatible with the CPE4 elements (Campilho et al. 2009c). The joints were fully restrained (i.e. clamped) at one of the edges to simulate real clamping conditions in the machine grips, and the other edge was subjected to a tensile displacement concurrently with transverse restraining (Goyal et al. 2008) (Fig. 6.5). For the analysis, a single row of cohesive elements was considered for the adhesive (Campilho et al. 2011b) incorporating a damage model between each set of paired nodes, as defined further in Section 6.3.1.3.2. The proposed modelling technique is currently implemented within the Abaqus® CAE suite and will be briefly described below.

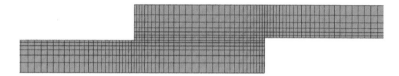

FIG. 6.6 Detail of the mesh for the L_0 = 10 mm model.

6.3.1.3.2 CZM Formulation

CZM simulate the elastic loading, initiation of damage and further propagation due to local failure within a material. CZM are based on a relationship between stresses and relative displacements (in tension or shear) connecting paired nodes of the cohesive elements (Fig. 6.2), to simulate the elastic behaviour up to t_n^0 (tension) or t_s^0 (shear) and subsequent softening, to account for the gradual degradation of material properties up to complete failure (Campilho et al. 2009a). Generically speaking, the shape of the cohesive laws can be adjusted to conform to the behaviour of the material or interface they are simulating (Campilho et al. 2009b). The values of G_I and G_{II}, representing the areas under the traction-separation laws in tension or shear, respectively, are equalled to G_{Ic} for tension or G_{IIc} for shear, allowing for the definition of δ_n^f and δ_s^f, respectively. Under pure tension or shear, damage propagation occurs at a specific pair of nodes when the stresses are released in the respective traction-separation law. Under the mixed mode, stress and energetic criteria are often used to combine tension and shear (Campilho et al. 2008a). In this case study, a continuum-based approach, i.e. using the cohesive elements to model solids rather than interfaces, was considered to model the finite value of t_A. The cohesive layer is assumed to be under one direct component of strain (through-the-thickness) and one transverse shear strain, computed directly from the element kinematics. The membrane strains are assumed as zero, which is appropriate for thin and compliant layers between stiff adherends. The traction-separation laws assume an initial linear elastic behaviour followed by linear evolution of damage (Fig. 6.2). Elasticity is defined by an elastic constitutive matrix (**K**) relating the current stresses (**t**) and strains (**ε**) in tension and shear across the interface (subscripts n and s, respectively) (Abaqus® 2013)

$$\mathbf{t} = \begin{Bmatrix} t_n \\ t_s \end{Bmatrix} = \begin{bmatrix} K_{nn} & K_{ns} \\ K_{ns} & K_{ss} \end{bmatrix} \cdot \begin{Bmatrix} \varepsilon_n \\ \varepsilon_s \end{Bmatrix} = \mathbf{K}\boldsymbol{\varepsilon} \qquad (6.1)$$

The matrix **K** contains the stiffness parameters of the adhesive layer, given by the relevant elastic moduli. A suitable approximation for thin adhesive layers is provided with $K_{nn} = E$, $K_{ss} = G$, $K_{ns} = 0$ (Campilho et al. 2008a, Campilho et al. 2010a). E and G are the tensile and shear elastic moduli. Damage initiation can be specified by different criteria. In this case study, the quadratic nominal stress criterion was selected for the initiation of damage, previously tested for accuracy (Campilho et al. 2009b), expressed as (Abaqus® 2013).

$$\left\{ \frac{\langle t_n \rangle}{t_n^0} \right\}^2 + \left\{ \frac{t_s}{t_s^0} \right\}^2 = 1 \qquad (6.2)$$

$\langle\ \rangle$ are the Macaulay brackets, emphasizing that a purely compressive stress state does not initiate damage (Jing et al. 2009). After t_m^0 is attained (Fig. 6.2) by the fulfilment of equation (6.2), the material stiffness initiates a degradation process. Complete separation and δ_m^f (Fig. 6.2) are predicted by a linear power law form of the required energies for failure in the pure modes (Abaqus® 2013)

$$\frac{G_I}{G_{Ic}} + \frac{G_{II}}{G_{IIc}} = 1 \qquad (6.3)$$

Table 6.3 shows the values introduced in Abaqus® for the adhesive layer damage laws (Campilho et al. 2011b). These properties were established from the data of Table 6.2, considering the average values of the experiments.

Table 6.3 Properties of the adhesive Araldite® 2015 for CZM modelling (Campilho et al. 2009, Campilho et al. 2011).

Property	
E [GPa]	1.85
G [GPa]	0.56
t_n^0 [MPa]	21.63
t_s^0 [MPa]	17.9
G_{Ic} [N/mm]	0.43
G_{IIc} [N/mm]	4.70

6.3.1.4 Results and Discussion

6.3.1.4.1 Joint Strength

All the joints experienced a cohesive failure of the adhesive layer. Figure 6.7 reports the maximum load (P_m) as a function of L_O, showing an increase of P_m at a slightly decreasing rate with L_O (Jain and Mai 1999, Reis et al. 2005), although the P_m-L_O plot is nearly linear. As previously discussed, this occurs from the high stiffness of the adherends and ductility of the adhesive (John et al. 1991, Hu and Soutis 2000). The absence of a strength plateau in the P_m-L_O curve is justified by the high strength of the carbon-epoxy (i.e., the tensile strength of the laminates was not attained for the tested range of L_O values), and by the ductility of the adhesive that allowed a progressively larger redistribution of stresses in the adhesive layer up to the largest value of L_O, initiating at the loci of peak stresses, i.e., the overlap edges (Davis and

Bond 1999, McGeorge 2010). In fact, since the fracture was always abrupt, only with only a negligible crack growth before P_m for the bigger values of L_O, it can be concluded that the adhesive plasticity always held up crack initiation at the overlap edges up to P_m, keeping these regions at the peak strength, while stresses increased at the inner regions (Hu and Soutis 2000).

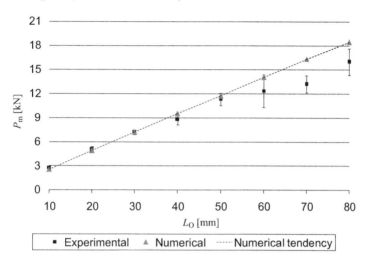

FIG. 6.7 Experimental and numerical comparison between the P_m values as a function of L_O.

6.3.1.4.2 Cohesive Parameters Influence

In this section, the influence of per cent variations of G_{Ic}, G_{IIc}, t_n^0, t_s^0, and their combined effect, on the value of P_m/P_m^0 of the joints is numerically assessed (P_m^0 represents P_m for the initial parameters of Table 6.3). Percentile variations of the initial properties between -80 to $+100\%$ were considered, whilst the non-mentioned cohesive parameters in all analyses were kept unchanged (values of Table 6.3). The influence of each parameter on the damage laws is depicted in Fig. 6.8 ((a) for G_{Ic} and the tensile law; (b) for G_{IIc} and the shear law) and Fig. 6.9 ((a) for t_n^0 and the tensile law; (b) for t_s^0 and the shear law).

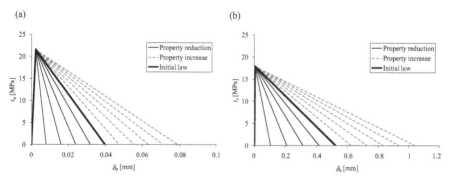

FIG. 6.8 Cohesive laws for values of G_{Ic} (a) and G_{IIc} (b) ranging from -80 to $+100\%$ of the initial ones, in increments of 20%.

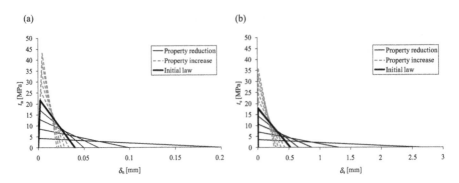

FIG. 6.9 Cohesive laws for values of t_n^0 (a) and t_s^0 (b) ranging from –80 to +100% of the initial ones, in increments of 20%.

Varying G_{Ic} or G_{IIc} changes the slope of the decaying portion of the respective cohesive law, while t_n^0 or t_s^0 remain identical. The modification of t_n^0 or t_s^0, by keeping the respective value of G_{Ic} or G_{IIc} unchanged, greatly changes the softening behaviour and value of δ_n^f or δ_s^f, respectively. As discussed in detail in the following section, the fluctuations of P_m/P_m^0 with G_{Ic} and G_{IIc} are justified by the variations of δ_n^f or δ_s^f of the damage laws by the modifications of these parameters (Fig. 6.8) and the actual values of δ_n or δ_s along the entire bondline when P_m is attained, which determine the loads transmitted by the adhesive layer according to the established cohesive laws. On the other hand, the influence of t_n^0 and t_s^0 on P_m/P_m^0 will mainly depend on the value of t_m^0 (Fig. 6.2), which is attained when equation (6.2) is fulfilled, and whose value significantly changes by the modification of t_n^0 and t_s^0 in the same equation. Actually, the variation of t_m^0, P_m/P_m^0 will be affected to an extent that depends on the values of δ_n and δ_s along the entire adhesive layer at the time of failure. Ridha et al. (2011) addressed the strength of scarf repairs in composite panels by CZM modelling. An analysis on the cohesive parameters influence was conducted by testing percentile variations of –50 and 50% of the initial values and simultaneous variations of t_n^0/t_s^0 or G_{Ic}/G_{IIc}. The results showed that reducing these parameters has a large influence on the strength predictions, whilst positive variations gave small variations on the strength of the repairs.

Fracture Toughness

Figures 6.10, 6.11, and 6.12 describe the influence of percentile variations of G_{Ic}, G_{IIc}, and G_{Ic} plus G_{IIc}, respectively, on P_m/P_m^0. Figure 6.10, relating to G_{Ic}, shows a significant under-prediction of P_m/P_m^0 for reductions of G_{Ic} (maximum of $\approx 35.5\%$ for an 80% reduction of the initial G_{Ic} and $L_O = 40$ mm), occurring by the smaller values of δ_n^f (Fig. 6.8a) in the tensile cohesive law, which results in premature failure at the overlap edges. A slight reduction of the reported under-prediction is found near $L_O = 80$ mm, since, for bigger values of L_O, the peak values of δ_n focus at a smaller normalized region at the overlap edges. The gradual increase of P_m/P_m^0 from $L_O = 30$ mm to $L_O = 10$ mm is accredited to the smaller values of δ_n at the overlap edges with the reduction of L_O, which leads to smaller actual values of δ_n in the tensile cohesive law when P_m is attained (Fig. 6.8a), rendering any under-prediction of G_{Ic} less preponderant. Over-predicting G_{Ic} gives minor improvements of P_m/P_m^0 (maximum of ≈4.9% for L_O = 50 mm) (Ridha et al. 2011), equally smaller for shorter overlaps, due to the corresponding reduction of δ_n values. The negligible influence of the G_{Ic} over predictions, when compared to the under-predictions, is also closely related to the sole attainment of large δ_n values (bigger than δ_n^f for the initial parameters; Fig. 6.8a) at the overlap edges, which renders any increase of G_{Ic} above its initial value not significant (Ridha et al. 2011).

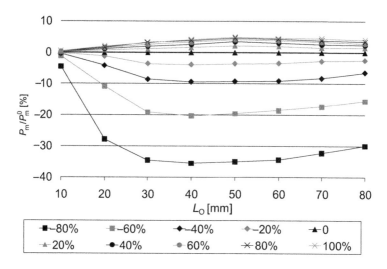

FIG. 6.10 Percentile variation of P_m/P_m^0 with G_{Ic} values ranging from −80 to +100% of the initial ones.

Figure 6.11 corresponds to G_{IIc} and depicts a significant difference to the data of Fig. 6.10 (G_{Ic}), as P_m/P_m^0 varies nearly proportionally with L_O for under-predictions of G_{IIc}. This is related to the more uniform values of δ_s along the bondline, compared to δ_n. As a result, for small values of L_O, the value of P_m/P_m^0 corresponds to a state of stress in which all the cohesive elements of the adhesive are very close to t_s^0 (Fig. 6.2 and Fig. 6.8b). Thus, any modification to the shear cohesive law at $\delta_s > \delta_s^0$ does not reflect by a large amount on P_m/P_m^0. The increase of L_O steadily increases the gradients of δ_s along the bondline, associating P_m/P_m^0 to an increasing portion of the overlap with $\delta_s > \delta_s^0$ (at the overlap edges). As a result, the softening shape of the shear damage law becomes progressively more preponderant with L_O. The maximum reduction of P_m/P_m^0 of ≈34.8%, was found for L_O = 80 mm. On the other hand, identically to the G_{Ic} data, over-predicting G_{IIc} only causes a maximum P_m/P_m^0 improvement of ≈5.7% (L_O = 80 mm) (Ridha et al. 2011), because of the occurrence of large values of δ_s (bigger than δ_s^f for the initial parameters; Fig. 6.8b) at a restricted region at the overlap edges.

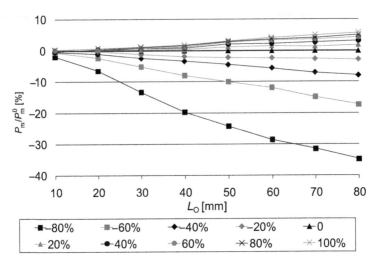

FIG. 6.11 Percentile variation of P_m/P_m^0 with G_{IIc} values ranging from −80 to +100% of the initial ones.

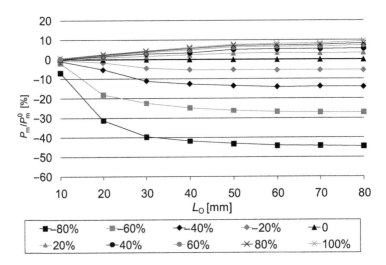

FIG. 6.12 Percentile variation of P_m/P_m^0 with G_{Ic} and G_{IIc} values ranging from −80 to +100% of the initial ones.

The combined modification of G_{Ic} and G_{IIc} (Fig. 6.12) gives a reduced influence on P_m/P_m^0 between −20 and +100% of the initial values (maximum of ≈9.3% for $L_O = 80$ mm), and large reductions from −40 to −80% that attain its maximum, of ≈44.5%, for $L_O = 80$ mm. The value of P_m/P_m^0 increases from $L_O = 40$ to $L_O = 10$ mm, owing to the combined effect of G_{Ic} (Fig. 6.10) and G_{IIc} (Fig. 6.11). The bigger deviations of P_m/P_m^0, compared to Fig. 6.10 and Fig. 6.11, also relate to the joint influence of G_{Ic} and G_{IIc} on the failure process.

Cohesive Strength

The influence of t_n^0 and t_s^0 on P_m/P_m^0 is shown in Fig. 6.13, Fig. 6.14, and Fig. 6.15 for t_n^0, t_s^0, and t_n^0 plus t_s^0, respectively. The variation of these parameters also affects δ_n^f and δ_s^f (Fig. 6.9) to keep G_{Ic} or G_{IIc} constant.

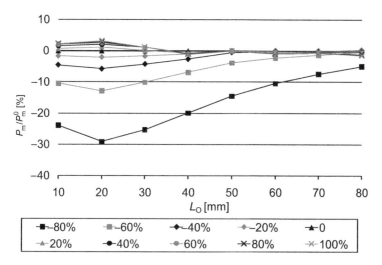

FIG. 6.13 Percentile variation of P_m/P_m^0 with t_n^0 values ranging from −80 to +100% of the initial ones.

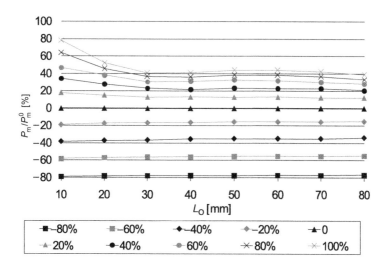

FIG. 6.14 Percentile variation of P_m/P_m^0 with t_s^0 values ranging from −80 to +100% of the initial ones.

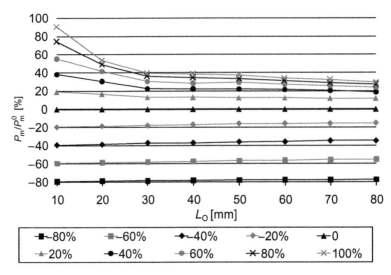

FIG. 6.15 Percentile variation of P_m/P_m^0 with t_n^0 and t_s^0 values ranging from −80 to +100% of the initial ones.

Figure 6.13 displays a larger influence of t_n^0 on P_m/P_m^0 for the smaller values of L_O, for under and over predictions of t_n^0, due to the concentration of peel δ_n values at a larger normalized region at the overlap edges. With the increase of L_O, the concentration of peel δ_n values occurs over a smaller normalized region, giving a less significant influence of t_n^0 on the global behaviour of the joints. Figure 6.13 also reports a much lesser influence on P_m/P_m^0 by over predicting t_n^0 than under-predicting it (Ridha et al. 2011). In both these scenarios, these variations are closely related to the attainment of t_m^0 (Fig. 6.2 and equation (6.2)). Actually, the improvement of P_m/P_m^0 by over-predicting t_n^0 (maximum of ≈3.2% for L_O = 20 mm) is related to the smaller influence of t_n stresses in equation (6.2). Conversely, the under prediction of t_n^0 is largely more preponderant on P_m/P_m^0 (maximum of ≈29.1% for L_O = 20 mm), owing to a premature occurrence of t_m^0 (Fig. 6.2) by the larger influence of t_n on the failure process (equation (6.2)).

Figure 6.14 depicts a nearly proportional percentile reduction of P_m/P_m^0 with t_s^0 notwithstanding the value of L_O (maximum of ≈79.1% for L_O = 10 mm). Actually, the transmission of loads through the adhesive is accomplished mainly by shear (Shin and Lee 2003). By reducing t_s^0, t_m^0 diminishes nearly proportionally at almost the entire overlap (equation (6.2)), and P_m/P_m^0 follows the same tendency. This is valid either for small values of L_O, for which P_m/P_m^0 relates to small gradients of δ_s, and big values of L_O, corresponding to large δ_s variations. On the other hand, the improvement of P_m/P_m^0 with over-predictions of t_s^0 is only close to proportional for L_O = 10 mm (maximum of ≈78.7% for L_O = 10 mm and t_s^0 increase of +100%), due to the evenness of δ_s values along the overlap that result of a value of P_m/P_m^0 almost exclusively depending on t_s^0. For bigger values of L_O, owing to the enlarging δ_s gradients along the overlap, increasing t_s^0 results in higher load transfer for $\delta_s < \delta_s^0$

(δ_s^0 is the shear relative displacement at softening onset; Fig. 6.2) and smaller or eventually nil load transfer for $\delta_s > \delta_s^0$ (Fig. 6.9b). As a result of these conflicting variations along the overlap when P_m/P_m^0 is attained, the improvement of P_m/P_m^0 is limited.

The combined influence of t_n^0 and t_s^0 (Fig. 6.15) is identical to the sole effect of t_s^0 (Fig. 6.14), but with a slightly bigger impact on P_m/P_m^0, especially for the smaller values of L_O (maximum improvement of $\approx 90.0\%$ and reduction of 79.4%, for the respective variations of t_n^0 and t_s^0 equal to $+100\%$ and -80%), because of the greater importance of shear stresses on the joint strength than peel ones (Shin and Lee 2003), and to a larger influence of t_s^0 on the results (Fig. 6.14).

Combination of Cohesive Strength and Fracture Toughness

Figure 6.16 reports on the combined influence of similar percentile variations of G_{Ic}, G_{IIc}, t_n^0 and t_s^0 on P_m/P_m^0. The results show that the relationship is typically linear for under-predictions of the cohesive parameters, following the overall tendency of Fig. 6.15 (t_n^0 and t_s^0 have a higher influence on P_m/P_m^0 than G_{Ic} and G_{IIc}; Fig. 6.12).

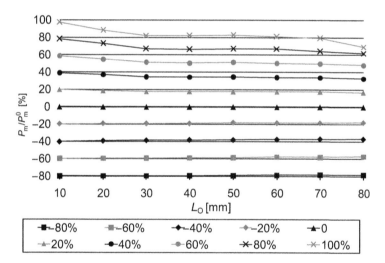

FIG. 6.16 Percentile variation of P_m/P_m^0 with G_{Ic}, G_{IIc}, t_n^0 and t_s^0 values ranging from -80 to $+100\%$ of the initial ones.

With regard to the increase of the cohesive parameters, the relationship is nearly proportional for $L_O = 10$ mm, but quickly diminishes for bigger values of L_O (maximum deviation for the $L_O = 80$ mm joint: 100% improvement of the cohesive properties gives only a $\approx 69.6\%$ increase of P_m/P_m^0). This trend also resembles the results of Fig. 6.15, relating to t_n^0 and t_s^0, but the increase was larger, since G_{Ic} and G_{IIc} were increased as well (as previously discussed, the fracture parameters play an important role for large values of L_O).

6.3.1.5 Conclusions of the Study

This case study aimed at the evaluation of the cohesive parameters influence of a triangular CZM, used to approximate the behaviour of a thin adhesive layer in bonded joints, on the value of P_m/P_m^0, after validation of the methodology, with experiments. The results allowed a critical perception of the effect of these parameters on the numerical predictions. Owing to the range of available techniques for the CZM definition, with dissimilar degrees of complexity and expected accuracy of the quantitative estimations, such a study is highly important for the selection of the most suitable method. A few other practical scenarios were emphasized for which the knowledge of the influence of such parameters deviations is particularly useful. The single-lap geometry was chosen covering a wide range of values of L_O, allowing testing different loading conditions, from a practically even distribution of shear stresses (small values of L_O) to a condition of large shear stress gradients and more localized influence of peel stresses (large values of L_O). The quantitative results presented in this case study are solely applicable to the particular set of geometric and material properties selected, but they can qualitatively be extrapolated for different bonded geometries and materials. For the conditions tested, under-predicting G_{Ic} and/or G_{IIc} is highly detrimental to the accuracy (maximum under-prediction of $\approx 44.5\%$ for $L_O = 80$ mm, by reducing G_{Ic} and G_{IIc} by -80% of the initial values), except for extremely small values of L_O. On the other hand, the over-prediction of G_{Ic} and/or G_{IIc} only slightly affects the results (maximum over-prediction of $\approx 9.3\%$ for $L_O = 80$ mm; G_{Ic} and G_{IIc} improvement of $+100\%$). Over-predictions of t_n^0 are almost inconsequent (maximum of $\approx 3.2\%$ for the $L_O = 20$ mm joint and $+100\%$ improvement), but moderate variations are expected if this parameter is under-predicted, especially for small values of L_O (maximum of $\approx 29.1\%$ for $L_O = 20$ mm and -80% reduction). Opposing to t_n^0, t_s^0 largely influences the results with a nearly proportional relation between the under-prediction of P_m/P_m^0 and the percentile variation of t_s^0 (maximum of $\approx 79.1\%$ for $L_O = 10$ mm and -80% reduction). For over-predictions of t_s^0, the improvement of P_m/P_m^0 is not so notorious, especially for large values of L_O (maximum of $\approx 78.7\%$ for $L_O = 10$ mm and t_s^0 increase of $+100\%$). The combined effect of t_s^0 and t_n^0 is close to that of t_n^0 (maximum improvement of $\approx 90.0\%$ and reduction of 79.4%, for the respective variations of t_n^0 and t_s^0 equal to $+100\%$ and -80%). The simultaneous variation of G_{Ic}, G_{IIc}, t_n^0 and t_s^0 gives values of P_m/P_m^0 in close proportion with the parameter percentile variations, except for over-predictions and large values of L_O. In these circumstances, the improvement is not so significant, with a maximum deviation for $L_O = 80$ mm (over-prediction of $\approx 69.6\%$ and $+100\%$ properties improvement).

6.3.2 Effect of the Cohesive Law Shape of the Adhesive Layer

This case study addresses the influence of the CZM shape (triangular, linear-exponential or trapezoidal) used to model a thin adhesive layer in single-lap adhesive joints, for an estimation of its influence on strength prediction under different material conditions. The study provides guidelines on the possibility to use a CZM shape that may not be the most suited for a particular adhesive, but that may be more straightforward to use/implement and have less convergence problems (e.g. triangular shaped CZM), thus attaining the solution faster.

6.3.2.1 Characterization of the Materials

Unidirectional carbon-epoxy pre-preg (SEAL® Texipreg HS 160 RM) with 0.15 mm thickness was considered for the composite adherends of the single-lap joints, with the $[0]_{16}$ lay-up. Table 6.1 presents the elastic properties of a unidirectional lamina, modelled as elastic orthotropic in the FEM analysis (Campilho et al. 2005). Two epoxy adhesives were considered. The adhesive Araldite® AV138 is a two-part (resin + hardener) brittle and high strength adhesive suited to bond a large variety of materials such as metals or polymers/polymer composites. The adhesive Araldite® 2015 is equally a two-part structural adhesive, showing a smaller ultimate strength than the previous one, but allowing large plastic flow prior to failure, which is an important feature for bonded joints, as it allows redistribution of stresses at stress concentration regions, which usually takes place because of the sharp edges at the overlap ends and also joint asymmetry/distinct deformation of the adherends along the overlap. The adhesives were previously characterized regarding E, G, the failure strengths in tension and shear (corresponding to t_n^0 and t_s^0) and the values of G_{Ic} and G_{IIc}. For the characterization of the mechanical properties, bulk tests were performed to characterize the adhesives in tension, and TAST were chosen for shear characterization. It should be pointed out that the cohesive strengths of thin adhesive layers and the bulk strengths of adhesives are different quantities (Ji et al. 2010). This is because bulk adhesives are homogeneous materials cracking perpendicularly to the maximum principal stress direction, while adhesives as thin layers are highly constrained between stiff adherends, and damage growth under these conditions occurs under mixed-mode (tension plus shear) and along the predefined path of the bonding direction. In this case study, the cohesive strengths of the adhesives were assumed as equal to their bulk quantities as an approximation. The good correspondence that was observed by the comparisons to the experimental data allowed the assumption that a fair approximation was attained, and corroboration of the use of these properties.

The bulk specimens were manufactured following the NF T 76-142 French standard, to prevent the creation of voids. Thus, 2 mm thick plates were fabricated in a sealed mould, followed by precision machining to produce the dogbone shape described in the standard. The TAST characterization of the adhesive was carried out according to the 11003-2:1999 ISO standard, considering DIN Ck 45 steel adherends. More details about the fabrication and testing procedures can be found in the reference (Campilho et al. 2011c). Characterization of the adhesives regarding the elastic constants, strengths and strains in tension and shear, was previously conducted in the work of da Silva et al. (2008) (Araldite® AV138) and by the chapter author in a previous work (Campilho et al. 2009c) (Araldite® 2015). The values of G_{Ic} and G_{IIc} for the AV138 were determined by the authors in Campilho et al. (2011b) by numerical fitting procedures. The authors also estimated in a previous work the values of G_{Ic} and G_{IIc} for the 2015 study (Campilho et al. 2009c), by DCB (G_{Ic}) and ENF tests (G_{IIc}) using different fracture mechanics data reduction methods. The relevant mechanical properties of these adhesives, which were used to construct the cohesive laws, are summarized in Table 6.4 (the initial yield stress was calculated for a plastic strain of 0.2%). The large difference between G_{Ic} and G_{IIc} observed in Table 6.4 for the 2015 study is typical of ductile structural adhesives, which show a significantly larger plastic flow in shear than in tension (Neto et al. 2012).

Table 6.4 Properties of the adhesives Araldite® AV138 and 2015 (da Silva et al. 2008, Campilho et al. 2009c, Campilho et al. 2011b).

Property	AV138	2015
Young's modulus, E [GPa]	4.89 ± 0.81	1.85 ± 0.21
Poisson's ratio, v^*	0.35	0.33
Tensile yield strength, σ_y [MPa]	36.49 ± 2.47	12.63 ± 0.61
Tensile failure strength, σ_f [MPa]	39.45 ± 3.18	21.63 ± 1.61
Tensile failure strain, ε_f [%]	1.21 ± 0.10	4.77 ± 0.15
Shear modulus, G [GPa]	1.56 ± 0.01	0.56 ± 0.21
Shear yield strength, τ_y [MPa]	25.1 ± 0.33	14.6 ± 1.3
Shear failure strength, τ_f [MPa]	30.2 ± 0.40	17.9 ± 1.8
Shear failure strain, γ_f [%]	7.8 ± 0.7	43.9 ± 3.4
Toughness in tension, G_{Ic} [N/mm]	0.20 [a]	0.43 ± 0.02
Toughness in shear, G_{IIc} [N/mm]	0.38 [a]	4.70 ± 0.34

* manufacturer's data
[a] estimated in Campilho et al. (2011b)

6.3.2.2 Joint Fabrication and Testing

The single-lap joint geometry and characteristic dimensions are represented in Fig. 6.5. The following dimensions were considered (in mm): L_O = 10-80, b = 15, L_T = 240, t_p = 2.4 and t_A = 0.2. Eight different values of L_O were evaluated (10, 20, 30, 40, 50, 60, 70 and 80 mm). The joints were fabricated by the following steps: (1) the surfaces to be bonded were roughened by manual abrasion with 220 grit sandpaper and cleaned with acetone, (2) the joints were bonded in an apparatus for the correct alignment, and the desired value of t_A was achieved during assembly with a dummy adherend and a 0.2 mm calibrated spacer under the upper adherend, jointly with the application of pressure with grips, and (3) tabs were glued at the specimen edges for a correct alignment in the testing machine. The reported method for the joints assembly assured the precision of the obtained t_A values, to reduce test data scatter to a minimum. The joints were left to cure at room temperature for one week to assure complete curing, and the excess adhesive at the bonding region was then removed by precision milling to provide square edges at the overlap edges. Tensile testing of the joints was carried out in an Instron® 4208 electro-mechanical testing machine with a 100 kN load cell, at room temperature and under displacement control (0.5 mm/min). The testing machine grips displacement was considered to build the P-δ curves. For each value of L_O, six specimens were tested, with at least four valid results.

6.3.2.3 Numerical Analysis

6.3.2.3.1 Modelling Conditions

The FEM software Abaqus® was considered for this study, to evaluate the modelling accuracy of its CZM embedded formulation when stipulating different CZM shapes to model the adhesive layer in single-lap joints. A geometrically non-linear static analysis was performed (Campilho et al. 2008b, Campilho et al. 2009b), modelling the adherends with the elastic orthotropic properties of Table 6.1. Figure 6.6 depicts an example of FEM mesh for the $L_O = 10$ mm joint. The meshes for all FEM models were automatically created by the software considering bias effects, with smaller sized elements near the overlap edges and in the thickness direction near the adhesive. Actually, it is known that the overlap edges are theoretically singularity spots with large stress variations (Panigrahi and Pradhan 2007). To provide identical modelling conditions, the FEM elements size in all models was made equal at the overlap edges (approximately 0.2×0.2 mm elements), thus allowing the accurate capture of stress variations (Campilho et al. 2005). The joints were simulated with two-dimensional FEM models, using 4-node plane-strain elements (CPE4 from Abaqus®) and COH2D4 4-node cohesive elements, compatible with the CPE4 elements (Campilho et al. 2009c). Boundary conditions included clamping the joints at one of the edges, to reproduce the testing machine gripping, while the opposite edge was pulled in tension together with lateral restraining (Fig. 6.5). The adhesive layer was modelled with a single row of cohesive elements (Campilho et al. 2011b) and a damage model between each set of paired nodes with varying CZM shape, as defined in the following section. This technique is implemented in Abaqus® CAE and will be briefly described for the different types of cohesive laws evaluated.

6.3.2.3.2 CZM Formulation

CZM reproduce the elastic loading up to a peak load, damage onset and crack growth due to local failure. CZM are typically founded on a relationship between stresses/cohesive tractions and relative displacements (in tension or shear) that connect homologous nodes of the cohesive elements, to simulate the elastic behaviour up to t_n^0 (tension) or t_s^0 (shear) and subsequent stiffness reduction, related to the progressive material degradation up to final failure (Turon et al. 2007, Campilho et al. 2009a). In this case study, triangular, linear-exponential and trapezoidal shapes were evaluated (Fig. 6.3 schematically represents these three CZM shapes with the associated nomenclature). The linear-exponential law is linear up to t_n^0 or t_s^0, and afterwards undergoes an exponential softening up to failure. This shape is an approximation of the full-exponential law (Chandra et al. 2002), providing, in this case, a more abrupt stress drop than the triangular law, after the peak loads are achieved. G_I and G_{II} are the areas under the CZM laws in tension or shear, respectively. The definition of the normal or shear maximum relative displacements (δ_n^f and δ_s^f, respectively) is carried out by making $G_I = G_{Ic}$ for tension or $G_{II} = G_{IIc}$ for shear. The initial linear elastic behaviour in the CZM laws (notwithstanding their shape) is defined by $\mathbf{K}$ relating $\mathbf{t}$ and $\boldsymbol{\varepsilon}$ (Abaqus® 2013)

$$\mathbf{t} = \begin{Bmatrix} t_n \\ t_s \end{Bmatrix} = \begin{bmatrix} K_{nn} & K_{ns} \\ K_{ns} & K_{ss} \end{bmatrix} \cdot \begin{Bmatrix} \varepsilon_n \\ \varepsilon_s \end{Bmatrix} = \mathbf{K}\boldsymbol{\varepsilon} \tag{6.4}$$

$\mathbf{K}$ contains the adhesive stiffness parameters. As for the previous case study, these parameters were equalled to $K_{nn} = E$, $K_{ss} = G$ and $K_{ns} = 0$ (Campilho et al. 2010a). For all of the three CZM shapes, initiation of damage was evaluated by the following quadratic nominal stress criterion, previously tested for accuracy (Campilho et al. 2009b), and expressed as (Abaqus® 2013)

$$\left\{ \frac{\langle t_n \rangle}{t_n^0} \right\}^2 + \left\{ \frac{t_s}{t_s^0} \right\}^2 = 1 \tag{6.5}$$

Thus, initiation of damage is coupled between tension and shear (Li et al. 2006). After the criterion of equation (6.5) is met, the material stiffness initiates a degradation process. However, from this point on, an uncoupled tensile/shear behaviour was used, in which the tensile and shear behaviours of the CZM elements are independent up to failure. This choice was made because of the Abaqus® unavailability of mixed-mode coupling criteria for the trapezoidal CZM formulation.

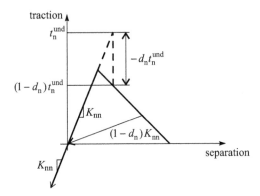

FIG. 6.17 Definition of the damage variable in tension, d_n, in Abaqus® (extrapolation is possible for d_s).

The softening regions of the CZM laws are defined in Abaqus® by specification of the damage variable (d_n for tension or d_s for shear), as a function of $\delta_n - \delta_n^0$ (tension) or $\delta_s - \delta_s^0$ (shear), i.e., as a function of the effective displacement beyond damage initiation. This is described by the following formulae. Figure 6.17 pictures the definition of d_n for the triangular law, although it can be extrapolated to d_s (Abaqus® 2013)

$$t_n = (1 - d_n) t_n^{und}$$
$$t_s = (1 - d_s) t_s^{und} \tag{6.6}$$

where t_n^{und} and t_s^{und} are the current cohesive tractions in tension and shear, respectively, without stiffness degradation. In this expression, $d_{n,s} = 0$ for n undamaged material (in the elastic region) and $d_{n,s} = 1$ for fully damaged material. By this principle, the generic expression (in tension or shear) of $d_{n,s}$ for the triangular law takes the form (Abaqus® 2013)

$$d_{n,s} = \frac{\delta_{n,s}^f (\delta_{n,s} - \delta_{n,s}^0)}{\delta_{n,s} (\delta_{n,s}^f - \delta_{n,s}^0)} \tag{6.7}$$

For the linear-exponential law, the expression of $d_{n,s}$ gives (Abaqus® 2013)

$$d_{n,s} = 1 - \frac{\delta_{n,s}^0}{\delta_{n,s}} \left(1 - \frac{1 - e^{\left(-\alpha \left(\frac{\delta_{n,s} - \delta_{n,s}^0}{\delta_{n,s}^f - \delta_{n,s}^0} \right) \right)}}{1 - e^{(-\alpha)}} \right) \tag{6.8}$$

where α is a non-dimensional parameter, related to a specific material, that establishes the rate of damage evolution with $\delta_{n,s}$ (for $\alpha = 0$ a triangular law is attained). In this case study, $\alpha = 7$ was chosen to provide a significant difference to the triangular shape, by a significantly faster degradation after $t_{n,s}^0$ is reached. For the trapezoidal law, the value of $d_{n,s}$ is divided into the constant stress region ($\delta_{n,s}^0 < \delta \leq \delta_{n,s}^s$; Fig. 6.3) and softening region ($\delta_{n,s}^s < \delta \leq \delta_{n,s}^f$; Fig. 6.3) as follows (Abaqus® 2013)

$$\begin{vmatrix} d_{n,s} = 1 - \dfrac{\delta_{n,s}^0}{\delta_{n,s}} & \text{if} & \delta_{n,s}^0 < \delta \leq \delta_{n,s}^s \\[3mm] d_{n,s} = 1 - \dfrac{m\delta_{n,s} + b}{K_{nn,ss}\delta_{n,s}} & \text{if} & \delta_{n,s}^s < \delta \leq \delta_{n,s}^f \end{vmatrix} \tag{6.9}$$

The values of m and b relate to the straight line equation of the decaying portion of the CZM law with respect to the $t_{n,s} - \delta_{n,s}$ plot origin, given by (Abaqus® 2013)

$$m = \frac{-t_{n,s}^0}{\delta_{n,s}^f - \delta_{n,s}^s}; \qquad b = t_{n,s}^0 - m\delta_{n,s}^s \tag{6.10}$$

The values of $\delta_{n,s}^f$ are found by consideration of the area under the $t_{n,s} - \delta_{n,s}$ plot to be equal to $G_{Ic,IIc}$.

Table 6.5 Properties of the adhesives Araldite® AV138 and 2015 for CZM modelling.

Property	AV138	2015
E [GPa]	4.89	1.85
G [GPa]	1.56	0.56
t_n^0 [MPa]	39.45	21.63
t_s^0 [MPa]	30.2	17.9
G_{Ic} [N/mm]	0.20	0.43
G_{IIc} [N/mm]	0.38	4.70

On the other hand, several techniques are available for the definition of $\delta_{n,s}^s$ (trapezoidal law), such as pre-established ratios between $\delta_{n,s}^s$ and $\delta_{n,s}^f$ (Yang et al. 1999), use of experimental failure strain data (Campilho et al. 2008a), or pre-established decaying slope up to $\delta_{n,s}^f$ (e.g. identical slope between the tensile and shear CZM laws, if only tensile data is available) (Carlberger and Stigh 2007). It this case study, the first approach was adopted, considering $\delta_{n,s}^s/\delta_{n,s}^f = 0.8$. Table 6.5 shows the considered values for the adhesive layer CZM laws, estimated from the experimental data of Table 6.4 and considering the average values of the experiments. Figure 6.18 details the CZM laws with different shapes for the adhesives AV138 and 2015 in tension (a) and shear (b).

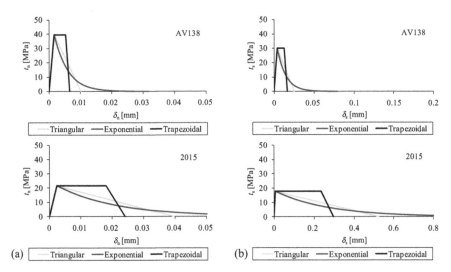

FIG. 6.18 CZM laws in tension (a) and shear (b) for both the adhesives tested.

6.3.2.4 Results and Discussion

6.3.2.4.1 Joint Strength

All the joints experienced a cohesive failure of the adhesive layer. Previously to the analysis, a mesh dependency study was carried out to ascertain if the selected mesh refinement was enough to ensure convergence to the right solution. This analysis considered the joints with both adhesives and considering $L_O = 10$ and 80 mm, which give the most significant difference in the adhesive layer stress state. Increasing refinements were considered, with element lengths at the overlap edges of 0.05, 0.1, 0.2 and 0.4 mm. Maximum deviations of 0.3% were found relatively to the average value of P_m between the four mesh sizes (independent analysis for each adhesive). This behaviour was expected, since, in CZM modeling, an energetic criterion is used for damage propagation, based on the input values of G_{Ic} and G_{IIc}. Since the energy required for crack growth is averaged over the damaged area, results are mesh independent provided that a minimum refinement is used, more specifically if a minimum of three to four elements are undergoing the softening process at the damage front (Kafkalidis and Thouless 2002, Campilho et al. 2008a).

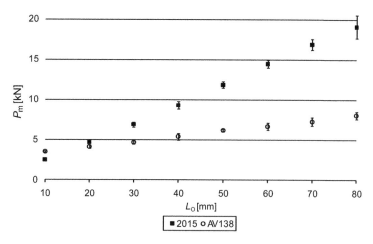

FIG. 6.19 Experimental plot of the P_m-L_O values for the adhesives AV138 and 2015.

Figure 6.19 reports P_m as a function of L_O for both adhesives tested, showing a nearly linear increase of P_m with L_O. The non-existence of a limiting P_m value in the P_m-L_O curves is justified by the high strength of the carbon-epoxy adherends (i.e., the tensile strength of the laminates was not attained for the tested L_O values up to failure in the adhesive layer). Actually, the maximum value of longitudinal axial stresses (σ_{xx}) was found for the joints with L_O = 80 mm (610.7 MPa for the joints with Araldite® 2015 and 393.9 MPa for the joints with Araldite® AV138), being much smaller in magnitude to the tensile strength of the employed composite, of ≈ 2000 MPa (Campilho et al. 2009b). Analyzing Fig. 6.19 in more detail, for L_O = 10 mm, the AV138 shows a larger P_m value than the 2015, which is accredited to the bigger adhesive strength (Table 6.5), and to the fact that shear stresses, which rule the failure process, are nearly constant over the overlap for very short overlaps (Campilho et al. 2012). As a result, failure depends almost exclusively on the adhesive strengths, while the fracture toughness (much bigger for the 2015) becomes irrelevant. For increasing values of L_O, the 2015 shows a steeper increase of P_m than the AV138, because it is extremely ductile (Table 6.4) and the joints fail with a significant degree of plastic flow in the adhesive layer (Adams et al. 1997). In fact, adhesive plasticization takes place at the overlap edges where stresses peak, together with redistribution of stresses in the adhesive layer towards the inner overlap regions (Davis and Bond 1999, McGeorge 2010). Because of this issue, a nearly proportional relationship exists between P_m and L_O in Fig. 6.19. To further corroborate this fact, fracture was always abrupt in the test specimens, with only a negligible crack growth before P_m for the bigger values of L_O, which shows that the adhesive plasticity always held up crack initiation at the overlap edges up to P_m, keeping these regions at the peak strength, while stresses increased at the inner regions (Hu and Soutis 2000). On the other hand, the AV138 shows a steady but significantly smaller improvement of P_m with L_O, due to its brittleness, testified by the corresponding values of $G_{Ic,IIc}$ (Table 6.4). Adding to this, peel peak stresses at the overlap ends progressively increase in magnitude with L_O (Campilho et al.

2012), as they gradually concentrate in a smaller region because of more localized bending of the adherends at the overlap edges. This, added to the reduced allowance of adhesive plastic flow (small value of G_{Ic}), gives a lesser advantage in the single-lap joints strength with the increase of L_O.

6.3.2.4.2 Cohesive Law Shape Effects

The CZM law shape influence on strength predictions was carried out considering triangular, linear-exponential and trapezoidal CZM, for a perception of the influence of this choice on the accuracy of the FEM simulations under different material/geometrical conditions. Initial emphasis was given to the size of the predicted length of the process zone at P_m (immediately before fracture), for a better understanding of the failure processes and differences between the FEM models with distinct CZM shapes for the adhesive layer. To this end, the shear process zone was considered, as it is the most significant to the failure process, and it was measured for the joints with L_O = 10 and 80 mm, which represent the limiting scenarios of L_O. Only the results for the Araldite® 2015 are presented, because any difference between models is more easily detected on account of the adhesive ductility, although the conclusions are identical to the Araldite® AV138. For L_O = 10 mm, the process zone extents at P_m were as follows (averaged over L_O): trapezoidal law – 79.1%, triangular law – 88.8% and linear-exponential law – 100%. For L_O = 80 mm, the following data was obtained: trapezoidal law – 90.2%, triangular law – 94.5% and linear-exponential law – 100%. These results are in agreement with the plotted laws used to model the adhesive (Fig. 6.18), showing that the extent of damage is largest for the linear-exponential law, followed by the triangular and trapezoidal laws, by the respective order. However, as will be discussed further in this section, this does not necessarily imply bigger values of predicted P_m by the linear-exponential law, as the triangular and, more specifically, the trapezoidal laws, allow bigger transmission of loads at the initial stages of damage (Fig. 6.18).

Single-Lap Joints with the Ductile Adhesive

Figure 6.20 reports the percentile deviation (Δ) between the experimental and FEM P_m values for the adhesive 2015 (averaged by the respective experimental P_m values). The slight inconsistent trend of the P_m-L_O plots is related to the calculation process to average experimental data, giving natural oscillations.

The results show that the trapezoidal law best approximates the best the experimental data. The percentile errors between the experimental and FEM data are negligible, with a maximum of 1.9% for L_O = 80. These results are consistent with previous observations for these types of adhesives (Campilho et al. 2008a, Ridha et al. 2011). Thus, the large plastic flow of the adhesive at a constant level of stresses after the peak strength is attained is captured by the FEM simulations, by using damage definitions that correspond to a constant level of stresses at the end of the elastic region. The use of a triangular law showed to consistently underestimate P_m, with a clear tendency for bigger discrepancies with larger values of L_O (Δ = –2.2% for L_O = 10 mm, growing steadily for bigger L_O values; Δ = –5.5% for L_O = 80 mm). The described tendency is justified in light of the typical stress distributions (namely

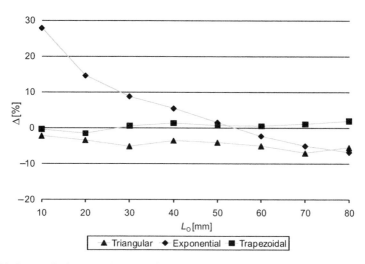

FIG. 6.20 Percentile deviation between the experimental and FEM P_m values for the adhesive 2015.

shear stresses) for single-lap joints. As a fact, for small values of L_O, the nearly constant level of shear stresses between overlap ends (Campilho et al. 2012) makes the CZM law shape practically irrelevant because at the time P_m is attained, the adhesive is evenly loaded in all its length. In the FEM analyses, this corresponds to a scenario in which the stress levels are close to $t_{n,s}^0$ along the entire bond, which renders the softening shape of the CZM law not so important. With bigger values of L_O the stress gradients increase (Campilho et al. 2012) and the deviation to the experimental data enlarges as well. Despite the variations in the experimental results, the triangular law still manages to predict P_m with an acceptable accuracy, which is an important feature to mention, as it is the easiest CZM law to use in terms of implementation, time of calculation, CZM parameter definition, and availability in commercial FEM codes. The use of more compliant adherends would reduce this deviation even further because of bigger stress gradients along the overlap (Campilho et al. 2011c). On the other hand, adherends such as steel would increase this deviation. The linear-exponential CZM gave opposite results for the range of L_O values evaluated. For small values of L_O, P_m was numerically overestimated (maximum Δ of 27.9% for L_O = 10 mm). The Δ values consistently reduced and approached the experimental results for L_O = 50 mm. From this point, under predictions of P_m were obtained with exponential softening (reaching $\Delta = -6.8\%$ for L_O = 80 mm). Analysis of the FEM results showed that the overestimation of P_m for the smaller L_O values is due to the following motives:

1. With the reduction of L_O, peel peak stresses develop at a larger normalized region of L_O (Campilho et al. 2012). With the increase of L_O, peel peak stresses concentrate at smaller normalized regions of L_O. This difference makes the preponderance of peel stresses not negligible for small L_O values. The over estimation of P_m for small L_O values is thus linked to the bigger value of δ_n^f for the linear-exponential law (Fig. 6.18a), which leads to failure at the overlap

edges at higher values of P_m. The peel stresses extension (normalized over the overlap) rapidly diminishes with the increase of L_O, reducing the error of the CZM predictions with the linear-exponential law.

2. With the reduction of L_O, owing to the bigger value of δ_s^f for the linear-exponential law induced by the steeper reduction of t_s after t_s^0 is attained, and also to a state of approximately constant shear stresses (Campilho et al. 2012), the CZM elements of the inner overlap region at the time of failure show smaller degradation (i.e., higher transmitted loads), and thus the predicted P_m values artificially increase.

Single-Lap Joints with the Brittle Adhesive

Figure 6.21 provides an identical comparison for the adhesive AV138, in which the oscillations are due to the aforementioned experimental variations. A large discrepancy can be readily observed in which regards the order of magnitude of Δ, since for the AV138 the maximum deviation is near 3%, compared to the approximate 30% for the 2015. On the other hand, the results of all the three CZM configurations follow the same tendency for the entire range of L_O values. This is related to the brittleness of the AV138, especially when compared to the large ductility of the 2015, which can be testified in Fig. 6.18 by the disparity in the $\delta_{n,s}^f$ values. Actually, for the shear behaviour (Fig. 6.18b), δ_s^f for the 2015 is more than one order of magnitude higher than for the AV138. As a result of this difference, the CZM shape of the AV138 is much less influential because the region under softening is negligible to that of the 2015.

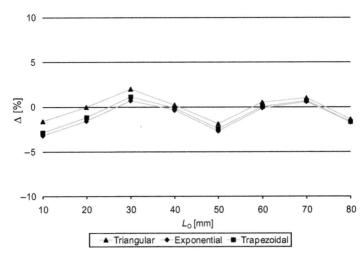

FIG. 6.21 Percentile deviation between the experimental and FEM P_m values for the adhesive AV138.

This can be observed in Fig. 6.22, which compares joints with $L_O = 80$ mm bonded with the 2015 (a) and AV138 (b) when P_m is attained (trapezoidal CZM). The stiffness degradation (SDEG) corresponds to d_s, with SDEG = 0 relating to the undamaged material and SDEG = 1 to complete failure. Since the region of influence of the CZM laws for the AV138 is restricted to a small portion of the overlap, any differences in shape have a reduced effect. The same tendency between all three CZM shapes is also a result of this, although a slight reduction of Δ between the three shapes occurs with the increase of L_O, with negligible variations for the bigger values of L_O. This variation can be accredited to the increasing degree of stress gradients in the adhesive bond, both peel and shear (Campilho et al. 2012), which further reduces the bond length under softening, where the differences between the three CZM shapes appear. Under brittle conditions, all the CZM shapes revealed to be accurate in predicting the measured response of the joints, although the best results (especially for small values of L_O) were found with the triangular law (maximum value of Δ of -1.9% for $L_O = 10$ mm). Compared to these and the experiments, the trapezoidal results showed a slight under prediction (maximum $\Delta = -2.9\%$ for $L_O = 10$ mm). The linear-exponential CZM further under predicts P_m (maximum $\Delta = -3.2\%$ for $L_O = 10$ mm), although following the very same trend of the previously reported data.

(a)   (b)

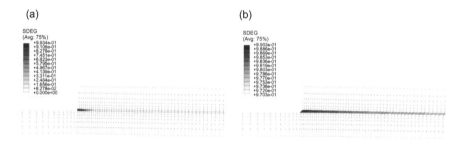

FIG. 6.22 Stiffness degradation for the joints with $L_O = 80$ mm bonded with the 2015 (a) and AV138 (b) when P_m is attained (trapezoidal CZM).

6.3.2.5 Conclusions of the Study

The main purpose of this case study was to evaluate the influence of the CZM shape used to model a thin adhesive layer in single-lap joints on the strength predictions, for different geometry/adhesive combinations. With this purpose, single-lap joints were bonded with a brittle and ductile adhesive and tested under tension, considering a large range of L_O values, which allowed to test different bond solutions in which regards to stress distributions (short overlaps are usually related to small shear stress gradients, while large overlaps give rise to large stress concentrations). The experimental results initially showed a markedly different trend for both adhesives as a function of L_O, since the brittle adhesive resulted in a smaller improvement of P_m with L_O, as the joints failed soon after the attainment of the adhesive strengths at the overlap ends. Oppositely, the joints bonded with the ductile adhesive showed a major strength improvement with L_O on account of failure ruled by allowance of large plastic flow in the adhesive layer. Regarding the different CZM shapes, these showed a significant influence on the results for the joints bonded with the 2015. These were more precisely modeled by the trapezoidal CZM that captured the adhesive plastic flow at the end of the elastic region, whilst the triangular CZM under predicted P_m up to $\Delta = -5.5\%$ for $L_O = 80$ mm. The linear-exponential CZM showed over predictions of P_m for short overlaps (up to 27.9%) and under predictions for long overlaps (up to -6.8%). For the AV138, the triangular CZM showed to be the most suited, although the results were very close between all CZM shapes tested (maximum deviations of -1.9%, -2.9% and -3.2% for the triangular, trapezoidal, and linear-exponential CZM, respectively). As a result of this study, some conclusions were established to properly select the CZM shape for a given adhesive, depending on its characteristics, but the importance of using the most suited CZM shape will invariably depend on the required precision and on CZM availability/easiness to use. Actually, triangular CZM are more widespread in commercial software, they are more straightforward to formulate, and give results faster on account of easier convergence. Overall, it was found that the influence of the CZM shape can be neglected when using brittle adhesives without compromising too much the accuracy, whilst for ductile adhesives this does not occur. Additionally, the smaller the value of L_O and the adhesive ductility, the greater is the influence of the CZM shape. In the end, any use of a CZM shape not suited to the material/interface to be simulated has to be balanced in these issues and expected variations in accuracy.

6.4 CONCLUSIONS

This chapter focused on CZM for static applications. CZM simulate the macroscopic damage along the fracture paths by the specification of a traction-separation response between paired nodes on either side of pre-defined crack paths. In most of the CZM, the traction–separation relations for the interfaces are such that, with increasing interfacial separation, the traction across the interface reaches a maximum (crack initiation), then it decreases (softening), and finally the crack propagates, permitting a total de-bond. The whole failure response and crack propagation can thus be modelled. A CZM simulates the fracture process, extending the concept of continuum mechanics by including a zone of discontinuity modelled by cohesive zones, thus using both strength and energy parameters to characterize the debonding process. This allows the approach to be of much more general utility than conventional fracture mechanics. The method is also mesh insensitive, provided that enough integration points undergo softening simultaneously. Studies demonstrated that it is possible to experimentally determine the appropriate CZM parameters of an adhesive bond, and to incorporate them into FEM analyses for excellent predictive capabilities. However, CZM present a limitation, as it is necessary to know beforehand the critical zones where damage is prone to occur, and to place the cohesive elements accordingly. Also, for ductile materials, the shape of the traction-separation law must be modified, which may give additional convergence problems. Different CZM shapes were addressed in this chapter. Discussions were included regarding the commonly used techniques for the estimation of the CZM parameters and convergence issues that usually take place when modelling damage by CZM. Two case studies were finally presented, regarding the effect of the CZM parameters and law shapes on the joints' strength, which showed that, although this technique is highly accurate, it requires a careful definition of the CZM laws for the best results.

6.5 REFERENCES

Abaqus® 2013. Documentation. D. Systèmes. Vélizy-Villacoublay.

Adams, R. D., Comyn, J. and Wake, W. C. 1997. Structural Adhesive Joints in Engineering. Chapman & Hall, London.

Alfano, G. and Crisfield, M. A. 2001. Finite element interface models for the delamination analysis of laminated composites: Mechanical and computational issues. International Journal for Numerical Methods in Engineering 50: 1701-1736.

Alfano, G. 2006. On the influence of the shape of the interface law on the application of cohesive-zone models. Composites Science and Technology 66: 723-730.

Allix, O. and Corigliano, A. 1996. Modeling and simulation of crack propagation in mixed-modes interlaminar fracture specimens. International Journal of Fracture 77: 111-140.

Andersson, T. and Stigh, U. 2004. The stress–elongation relation for an adhesive layer loaded in peel using equilibrium of energetic forces. International Journal of Solids and Structures 41: 413-434.

Banea, M. D. and da Silva, L. F. M. 2009. Adhesively bonded joints in composite materials: An overview. Proceedings of the Institution of Mechanical Engineers, Part L: Journal of Materials Design and Applications 223: 1-18.

Banea, M. D., da Silva, L. F. M. and Campilho, R. D. S. G. 2010. Temperature dependence of the fracture toughness of adhesively bonded joints. Journal of Adhesion Science and Technology 24: 2011-2026.

Barenblatt, G. I. 1959. The formation of equilibrium cracks during brittle fracture. General ideas and hypothesis. Axisymmetrical cracks. Journal of Applied Mathematics and Mechanics 23: 622-636.

Bascom, W. D. and Cottington, R. L. 1976. Effect of temperature on the adhesive fracture behavior of an elastomer-epoxy resin. The Journal of Adhesion 7: 333-346.

Biel, A. and Stigh, U. 2008. Effects of constitutive parameters on the accuracy of measured fracture energy using the DCB-specimen. Engineering Fracture Mechanics 75: 2968-2983.

Blackman, B. R. K., Brunner, A. J. and Williams, J. G. 2006. Mode II fracture testing of composites: A new look at an old problem. Engineering Fracture Mechanics 73: 2443-2455.

Campilho, R. D. S. G., de Moura, M. F. S. F. and Domingues, J. J. M. S. 2005. Modelling single and double-lap repairs on composite materials. Composites Science and Technology 65: 1948-1958.

Campilho, R. D. S. G., de Moura, M. F. S. F. and Domingues, J. J. M. S. 2008a. Using a cohesive damage model to predict the tensile behaviour of CFRP single-strap repairs. International Journal of Solids and Structures 45: 1497-1512.

Campilho, R. D. S. G., de Moura, M. F. S. F., Domingues, J. J. M. S. and Morais, J. J. L. 2008b. Computational modelling of the residual strength of repaired composite laminates using a cohesive damage model. Journal of Adhesion Science and Technology 22: 1565-1591.

Campilho, R. D. S. G., de Moura, M. F. S. F., Barreto, A. M. J. P., Morais, J. J. L. and Domingues, J. J. M. S. 2009a. Fracture behaviour of damaged wood beams repaired with an adhesively-bonded composite patch. Composites Part A: Applied Science and Manufacturing 40: 852-859.

Campilho, R. D. S. G., de Moura, M. F. S. F., Pinto, A. M. G., Morais, J. J. L. and Domingues, J. J. M. S. 2009b. Modelling the tensile fracture behaviour of CFRP scarf repairs. Composites Part B: Engineering 40: 149-157.

Campilho, R. D. S. G., de Moura, M. F. S. F., Ramantani, D. A., Morais, J. J. L. and Domingues, J. J. M. S. 2009c. Buckling behaviour of carbon-epoxy adhesively-bonded scarf repairs. Journal of Adhesion Science and Technology 23: 1493-1513.

Campilho, R. D. S. G., de Moura, M. F. S. F., Ramantani, D. A., Morais, J. J. L. and Domingues, J. J. M. S. 2009d. Tensile behaviour of three-dimensional carbon-epoxy adhesively bonded single- and double-strap repairs. International Journal of Adhesion and Adhesives 29: 678-686.

Campilho, R. D. S. G., de Moura, M. F. S. F., Ramantani, D. A., Morais, J. J. L. and Domingues, J. J. M. S. 2010a. Buckling strength of adhesively-bonded single and double-strap repairs on carbon-epoxy structures. Composites Science and Technology 70: 371-379.

Campilho, R. D. S. G., de Moura, M. F. S. F., Ramantani, D. A., Morais, J. J. L., Barreto, A. M. J. P. and Domingues, J. J. M. S. 2010b. Adhesively bonded repair proposal for wood members damaged by horizontal shear using carbon-epoxy patches. The Journal of Adhesion 86: 649-670.

Campilho, R. D. S. G., Banea, M. D., Chaves, F. J. P. and da Silva, L. F. M. 2011a. eXtended Finite Element Method for fracture characterization of adhesive joints in pure mode I. Computational Materials Science 50: 1543-1549.

Campilho, R. D. S. G., Banea, M. D., Pinto, A. M. G., da Silva, L. F. M. and de Jesus, A. M. P. 2011b. Strength prediction of single- and double-lap joints by standard and extended finite element modelling. International Journal of Adhesion and Adhesives 31: 363-372.

Campilho, R. D. S. G., Pinto, A. M. G., Banea, M. D., Silva, R. F. and da Silva, L. F. M. 2011c. Strength improvement of adhesively-bonded joints using a reverse-bent geometry. Journal of Adhesion Science and Technology 25: 2351-2368.

Campilho, R. D. S. G., Banea, M. D., Neto, J. A. B. P. and da Silva, L. F. M. 2012. Modelling of single-lap joints using cohesive zone models: Effect of the cohesive parameters on the output of the simulations. The Journal of Adhesion 88: 513-533.

Carlberger, T. and Stigh, U. 2007. An explicit FE-model of impact fracture in an adhesive joint. Engineering Fracture Mechanics 74: 2247-2262.

Carlberger, T. and Stigh, U. 2010. Influence of layer thickness on cohesive properties of an epoxy-based adhesive—An experimental study. The Journal of Adhesion 86: 816-835.

Chai, H. 1986. On the correlation between the mode I failure of adhesive joints and laminated composites. Engineering Fracture Mechanics 24: 413-431.

Chai, H. 1992. Experimental evaluation of mixed-mode fracture in adhesive bonds. Experimental Mechanics 32: 296-303.

Chandra, N., Li, H., Shet, C. and Ghonem, H. 2002. Some issues in the application of cohesive zone models for metal–ceramic interfaces. International Journal of Solids and Structures 39: 2827-2855.

Chen, J. 2002. Predicting progressive delamination of stiffened fibre-composite panel and repaired sandwich panel by decohesion models. Journal of Thermoplastic Composite Materials 15: 429-442.

Chen, Z., Adams, R. D. and da Silva, L. F. M. 2011. Prediction of crack initiation and propagation of adhesive lap joints using an energy failure criterion. Engineering Fracture Mechanics 78: 990-1007.

da Silva, L. R. M., da Silva, R. A. M., Chousal, J. A. G. and Pinto, A. M. G. 2008. Alternative methods to measure the adhesive shear displacement in the thick adherend shear test. Journal of Adhesion Science and Technology 22: 15-29.

da Silva, L. F. M. and Campilho, R. D. S. G. 2012. Advances in Numerical Modelling of Adhesive Joints. Springer, Heidelberg.

Davis, M. and Bond, D. 1999. Principles and practices of adhesive bonded structural joints and repairs. International Journal of Adhesion and Adhesives 19: 91-105.

Devitt, D. F., Schaperv, R. A. and Bradley, W. L. 1980. A method for determining the mode I delamination fracture toughness of elastic and viscoelastic composite materials. Journal of Composite Materials 14: 270-285.

Dugdale, D. S. 1960. Yielding of steel sheets containing slits. Journal of the Mechanics and Physics of Solids 8: 100-104.

Feraren, P. and Jensen, H. M. 2004. Cohesive zone modelling of interface fracture near flaws in adhesive joints. Engineering Fracture Mechanics 71: 2125-2142.

Gonçalves, J. P. M., de Moura, M. F. S. F., de Castro, P. M. S. T. and Marques, A. T. 2000. Interface element including point-to-surface constraints for three-dimensional problems with damage propagation. Engineering Computations 17: 28-47.

Goyal, V. K., Johnson, E. R. and Goyal, V. K. 2008. Predictive strength-fracture model for composite bonded joints. Composite Structures 82: 434-446.

Hamitouche, L., Tarfaoui, M. and Vautrin, A. 2008. An interface debonding law subject to viscous regularization for avoiding instability: Application to the delamination problems. Engineering Fracture Mechanics 75: 3084-3100.

Hu, F. Z. and Soutis, C. 2000. Strength prediction of patch-repaired CFRP laminates loaded in compression. Composites Science and Technology 60: 1103-1114.

Jain, L. K. and Mai, Y.-W. 1999. Analysis of resin-transfer-moulded single-lap joints. Composites Science and Technology 59: 1513-1518.

Ji, G., Ouyang, Z., Li, G., Ibekwe, S. and Pang, S.-S. 2010. Effects of adhesive thickness on global and local Mode-I interfacial fracture of bonded joints. International Journal of Solids and Structures 47: 2445-2458.

Jing, J., Gao, F., Johnson, J., Liang, F. Z., Williams, R. L. and Qu, J. 2009. Simulation of dynamic fracture along solder–pad interfaces using a cohesive zone model. Engineering Failure Analysis 16: 1579-1586.

John, S. J., Kinloch, A. J. and Matthews, F. L. 1991. Measuring and predicting the durability of bonded carbon fibre/epoxy composite joints. Composites 22: 121-127.

Kafkalidis, M. S. and Thouless, M. D. 2002. The effects of geometry and material properties on the fracture of single lap-shear joints. International Journal of Solids and Structures 39: 4367-4383.

Khoramishad, H., Crocombe, A. D., Katnam, K. B. and Ashcroft, I. A. 2010. Predicting fatigue damage in adhesively bonded joints using a cohesive zone model. International Journal of Fatigue 32: 1146-1158.

Lai, Y.-H., Dwayne Rakestraw, M. and Dillard, D. A. 1996. The cracked lap shear specimen revisited—a closed form solution. International Journal of Solids and Structures 33: 1725-1743.

Lee, M. J., Cho, T. M., Kim, W. S., Lee, B. C. and Lee, J. J. 2010. Determination of cohesive parameters for a mixed-mode cohesive zone model. International Journal of Adhesion and Adhesives 30: 322-328.

Leffler, K., Alfredsson, K. S. and Stigh, U. 2007. Shear behaviour of adhesive layers. International Journal of Solids and Structures 44: 530-545.

Li, S., Thouless, M. D., Waas, A. M., Schroeder, J. A. and Zavattieri, P. D. 2005. Use of a cohesive-zone model to analyze the fracture of a fiber-reinforced polymer–matrix composite. Composites Science and Technology 65: 537-549.

Li, S., Thouless, M. D., Waas, A. M., Schroeder, J. A. and Zavattieri, P. D. 2006. Mixed-mode cohesive-zone models for fracture of an adhesively bonded polymer–matrix composite. Engineering Fracture Mechanics 73: 64-78.

Liljedahl, C. D. M., Crocombe, A. D., Wahab, M. A. and Ashcroft, I. A. 2006. Damage modelling of adhesively bonded joints. International Journal of Fracture 141: 147-161.

Mangalgiri, P. D., Johnson, W. S. and Everett, R. A. 1987. Effect of adherend thickness and mixed mode loading on debond growth in adhesively bonded composite joints. The Journal of Adhesion 23: 263-288.

McGeorge, D. 2010. Inelastic fracture of adhesively bonded overlap joints. Engineering Fracture Mechanics 77: 1-21.

Neto, J. A. B. P., Campilho, R. D. S. G. and da Silva, L. F. M. 2012. Parametric study of adhesive joints with composites. International Journal of Adhesion and Adhesives 37: 96-101.

Pandya, K. C. and Williams, J. G. 2000. Measurement of cohesive zone parameters in tough polyethylene. Polymer Engineering & Science 40: 1765-1776.

Panigrahi, S. K. and Pradhan, B. 2007. Three dimensional failure analysis and damage propagation behavior of adhesively bonded single lap joints in laminated FRP composites. Journal of Reinforced Plastics and Composites 26: 183-201.

Pardoen, T., Ferracin, T., Landis, C. M. and Delannay, F. 2005. Constraint effects in adhesive joint fracture. Journal of the Mechanics and Physics of Solids 53: 1951-1983.

Pinto, A. M. G., Magalhães, A. G., Campilho, R. D. S. G., de Moura, M. F. S. F. and Baptista, A. P. M. 2009. Single-lap joints of similar and dissimilar adherends bonded with an acrylic adhesive. The Journal of Adhesion 85: 351-376.

Reis, P. N. B., Antunes, F. J. V. and Ferreira, J. A. M. 2005. Influence of superposition length on mechanical resistance of single-lap adhesive joints. Composite Structures 67: 125-133.

Ridha, M., Tan, V. B. C. and Tay, T. E. 2011. Traction–separation laws for progressive failure of bonded scarf repair of composite panel. Composite Structures 93: 1239-1245.

Shin, K. C. and Lee, J. J. 2003. Bond parameters to improve tensile load bearing capacities of co-cured single and double lap joints with steel and carbon fiber-epoxy composite adherends. Journal of Composite Materials 37: 401-420.

Sørensen, B. F. 2002. Cohesive law and notch sensitivity of adhesive joints. Acta Materialia 50: 1053-1061.

Tsai, M. Y. and Morton, J. 1994. An evaluation of analytical and numerical solutions to the single-lap joint. International Journal of Solids and Structures 31: 2537-2563.

Turon, A., Dávila, C. G., Camanho, P. P. and Costa, J. 2007. An engineering solution for mesh size effects in the simulation of delamination using cohesive zone models. Engineering Fracture Mechanics 74: 1665-1682.

Wooley, G. R. and Carver, D. R. 1971. Stress concentration factors for bonded lap joints. Journal of Aircraft 8: 817-820.

Xie, D. and Waas, A. M. 2006. Discrete cohesive zone model for mixed-mode fracture using finite element analysis. Engineering Fracture Mechanics 73: 1783-1796.

Xie, D., Salvi, A. G., Sun, C., Waas, A. M. and Caliskan, A. 2006. Discrete cohesive zone model to simulate static fracture in 2D triaxially braided carbon fiber composites. Journal of Composite Materials 40: 2025-2046.

Yang, Q. D., Thouless, M. D. and Ward, S. M. 1999. Numerical simulations of adhesively-bonded beams failing with extensive plastic deformation. Journal of the Mechanics and Physics of Solids 47: 1337-1353.

Yang, Q. D. and Thouless, M. D. 2001. Mixed-mode fracture analyses of plastically-deforming adhesive joints. International Journal of Fracture 110: 175-187.

Zhu, Y., Liechti, K. M. and Ravi-Chandar, K. 2009. Direct extraction of rate-dependent traction–separation laws for polyurea/steel interfaces. International Journal of Solids and Structures 46: 31-51.

7

CHAPTER

Parameter Identification in Cohesive Zone Modelling

Raul D.S.G. Campilho[1,*], Marcelo Costa[2],
Guilherme Viana[2] and Lucas F.M. da Silva[3]

7.1 INTRODUCTION

Adhesive bonding is a material joining process in which an adhesive, placed between the adherend surfaces, solidifies to produce an adhesive bond. Adhesively bonded joints are an increasing alternative to mechanical joints in engineering applications and provide many advantages over conventional mechanical fasteners. Among these advantages are lower structural weight, lower fabrication cost and improved damage tolerance. In fact, adhesive bonding has found applications in various areas from high technology industries such as aeronautics, aerospace and automotive to traditional industries such as construction, sports and packaging. There are several reference books dealing with adhesive joints such as those of Kinloch (1987) and da Silva et al. (2011). Bonded joints are frequently expected to sustain static or cyclic loads for considerable periods of time without any adverse effect on the load-bearing capacity of the structure. A lack of suitable predictive techniques (either analytical or numerical) has resulted in a tendency to 'overdesign' adhesive joints. Safety considerations often require that adhesively bonded structures, particularly those employed in primary load-bearing applications, include mechanical fasteners (e.g. bolts) as an additional safety precaution. These practices result in heavier and more costly components. The development of reliable design and predictive methodologies can be expected to result in a more efficient use of adhesives. There are two basic mathematical approaches for the analyses of adhesively bonded joints: closed-form analyses (analytical methods) and numerical methods (i.e. Finite

[1] Departamento de Engenharia Mecânica, Instituto Superior de Engenharia do Porto, Instituto Politécnico do Porto, Rua Dr. António Bernardino de Almeida, 431, 4200-072 Porto, Portugal.
[2] INEGI – Pólo FEUP, Rua Dr. Roberto Frias, s/n, 4200-465 Porto, Portugal.
[3] Departamento de Engenharia Mecânica, Faculdade de Engenharia, Universidade do Porto, Rua Dr. Roberto Frias, 4200-465 Porto, Portugal.
* Corresponding author: raulcampilho@gmail.com

Element or FE analyses). However, the analytical analysis of adhesive joints can be highly complex if composite adherends are used, if the adhesive deforms plastically or if there is an adhesive fillet. In those cases, several differential equations of high complexity might be obtained (non-linear and non-homogeneous). For those cases, numerical methods are more adequate. The FE method, the boundary element (BE) method and the finite difference (FD) method are the three major numerical methods for solving partial differential equations in science and engineering. The FE is, by far, the most common technique used in the context of adhesively bonded joints. Adams and co-workers are among the first to have used the FE method for analysing adhesive joint stresses (Adams and Peppiatt 1974). One of the first reasons for the use of the FE method was to assess the influence of the spew fillet. The joint rotation, the adherends and adhesive plasticity are other aspects that are easier to treat with a FE analysis. The use of the BE method is still very limited in the analysis of adhesive joints. The FD method is especially used for solving complex governing differential equations in closed-form models. The FE is, obviously, the most important method for strength prediction of bonded joints and several approaches to failure analysis exist: continuum mechanics, fracture mechanics and the more recent damage mechanics and eXtended Finite Element Method (XFEM).

The FE found application in the analysis of adhesively-bonded joints some decades later than the analytical approaches (e.g. the work of Carver and Wooley (1971)), by consideration of stress/strain or fracture mechanics criteria for failure prediction (Tsai and Morton 1994). Even though these analyses were promising, they had few limitations: stress/strain predictions depend of the mesh size at the critical regions, while fracture criteria such as the Virtual Crack Closure Technique (VCCT) are restricted to Linear Elastic Fracture Mechanics (LEFM) and require an initial crack. Cohesive Zone Models (CZM) have been used for the strength prediction of adhesive joints, as an add-on to FE analyses that allows simulation of damage growth within bulk regions of continuous materials or interfaces between different materials (Feraren and Jensen 2004, da Silva and Campilho 2011). Compared to conventional FE, a much more accurate prediction is achieved, since different shapes can be developed for the cohesive laws, depending on the nature of the material or interface to be simulated. The triangular and trapezoidal CZM shapes are most commonly used for strength prediction of typical structural materials. For the application of this technique, traction-separation laws with a pre-defined shape are established at the failure paths and the values of energy release rate in tension and shear (G_I and G_{II}, respectively) along the fracture paths and respective critical values or toughness (G_{Ic} and G_{IIc}) are required. The cohesive strengths in tension and shear (t_n^0 and t_s^0, respectively) are equally needed and they relate to damage initiation, i.e. end of the elastic behaviour and beginning of damage. Different techniques are nowadays available for the definition of the cohesive parameters (G_{Ic}, G_{IIc}, t_n^0 and t_s^0) such as the property identification technique, the direct method and the inverse method. These methods usually rely on the Double-Cantilever Beam (DCB), End-Notched Flexure (ENF) or single-lap specimens, generally with good results (Li et al. 2005, Banea et al. 2010, Campilho et al. 2011a). The property identification technique consists of the separated calculation of each one of the cohesive law parameters by suitable tests, while in the inverse method, the CZM parameters are estimated by iterative fitting of the FE prediction with experimentally measured data (typically the load-displacement, $P\text{-}\delta$, curve) up to an accurate representation. Both these approaches begin with the assumption of a CZM shape to simulate a specific material, which approximately replicates it in terms of post-elastic behaviour (Campilho et al. 2009a). On the other hand, the direct method gives the precise shape of the CZM laws of a specific material or interface, since these are estimated from the experimental data of fracture tests such as the DCB or ENF (Pandya and Williams 2000). This is done by differentiation of G_I (tension) or G_{II} (shear) with respect to the relative opening of the crack (δ_n for tension or δ_s for shear). Nonetheless, it is usual to convert the obtained shape in an approximated parameterized shape for introduction in the FE software. This chapter addresses in detail the aforementioned CZM parameter identification methods with selected examples of application, after a state-of-the-art review regarding strength prediction techniques for bonded joints.

7.2 STRENGTH PREDICTION BY NUMERICAL TECHNIQUES

The FE method is a numerical analysis procedure that provides an approximate solution to problems in various fields of engineering. Ashcroft (2011) gives a description of the method applied to adhesive joints. The FE method is based on the idea of building a complicated object with simple blocks or dividing a complicated object into small and manageable pieces. The first efforts to use piecewise continuous functions defined over triangular domains appeared in applied mathematics literature with the work of Courant (1943). Advances in the aerospace industry and the development of computers in the 1950's and 1960's saw further development and computerization of these methods (Turner et al. 1956). This was the direct pre-cursor of nearly all current commercial FE analysis methods. The term FE was introduced by Clough (1960). Zienkiewicz and Cheung (1967) described the applicability of the method to general field applications. In the 1980's, generalised software packages designed to run on powerful computers were developed and improved techniques for the analysis of non-linear problems were established. The FE method is now used in practically all fields of engineering analysis such as structural, heat transfer, material transport (such as diffusion), fluid mechanics and electro-magnetics. Recent FE programs offer the possibility to multi-physics problems (coupling of different analysis classes, such as thermo-structural or hygro-thermo-structural problems).

To predict the joint strength, one must have the stress distribution and a suitable failure criterion. The stress distribution can be obtained by a FE analysis or a closed-form model. For complex geometries and elaborate material models, the FE method is preferable. One of the simplest failure models is that based on a stress or strain limit state, i.e. based on a continuum mechanics approach (Adams et al. 1997). Fracture mechanics principles can also be used within a FE analysis. This can be based on either the stress intensity factor or energy approaches (Shahin and Taheri 2008). An extension to this approach is damage modelling with cohesive elements (CZM technique), which also allows simulating damage of a material ahead of a crack. The technique is a combination of continuum mechanics and fracture mechanics. One of the most important advantages of CZM is related to their ability to simulate onset and non-self-similar growth of damage. No initial crack is needed and damage propagation takes place without user intervention. They do not depend on an initial flaw, unlike conventional fracture mechanics approaches. Usually, CZM are based on spring or cohesive elements (Feraren and Jensen 2004), connecting plane or two-dimensional (2D) or three-dimensional (3D) solid elements of structures. Those elements are placed at the planes where damage is prone to occur, which, in several applications, can be difficult to identify a priori. However, an important characteristic of bonded assemblies is that damage propagation is restricted to well defined planes, i.e., at or near the interfaces between the adhesive and the adherends, or inside the adhesive, which allows overcoming this limitation. CZM are based on a softening relationship between stresses and relative displacements between the crack faces, thus simulating a gradual degradation of material properties. The shape of the softening laws can be adjusted to conform to the behaviour of the material or interface they are simulating. Thus, they can be adapted to simulate ductile adhesive layers, which can be modelled with trapezoidal laws (Campilho et al. 2008, 2009a). The areas under the traction-separation laws in each mode are equalled to the respective value of critical strain energy release rate or fracture toughness (G_c) of the adhesive layer or interface. Under pure mode, damage propagation occurs in a given integration point when the stresses are released in the respective traction-separation law. Under mixed-mode, energetic criteria are often used to combine pure modes I and II (2D analyses) or I, II and III (3D analyses), thus simulating the typical mixed-mode behaviour inherent to bonded assemblies. The complete P-δ curve and respective failure mode can be obtained with this methodology. Another method for modelling cracks in materials is the XFEM, which uses enriched shape functions to represent a discontinuous displacement field. The main advantage of XFEM is that the crack may initiate at any point in the material and propagate based on loading conditions. No remeshing is required as the crack can grow within an element, and it does not need to follow element boundaries.

7.3 DAMAGE MECHANICS

Advanced modelling techniques are required that comprise accurate failure predictions, surpassing the aforementioned limitations associated to the continuum and fracture mechanics approaches, to effectively model damage evolution within a material or structure with bonded components (Liljedahl et al. 2006). Structural damage during loading can be found in the form of micro-cracks over a finite volume or interfacial region between bonded components, such that load transfer is locally reduced, globally resulting on a drop of applied load for a given value of δ applied to the structure. Figure 7.1 reports on a typical uni-axial stress-strain (σ-ε) diagram up to failure for a ductile material or structure. A FE model built solely with solid continuum elements comprising the elastic and plastic constitutive behaviours of each one of the materials wrongly gives as modelling output the abcd′ curve because of generalised plasticization without damage evolution, while a damage and failure based model can actually provide the real abcd curve, by allowing damage to grow through the simulation of material stiffness degradation between point c (damage onset) and point d (complete failure).

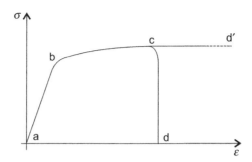

FIG. 7.1 Typical uni-axial σ-ε response of a ductile material or structure.

Damage mechanics permits the simulation of step-by-step damage and fracture at a pre-defined crack path or arbitrarily within a finite region up to complete structural failure (Duan et al. 2004). However, this is still an innovative field under intense development, regarding more accurate modelling techniques, reliable and simple parameter determination methods, increase of robustness and elimination of convergence issues (Liljedahl et al. 2006), and it is also under heavy implementation in commercial FE software packages such as Abaqus® (Campilho et al. 2011b). The use of CZM coupled to conventional FE analyses is the most widespread damage mechanics-based method of predicting static or fatigue damage uptake in structures (Yang and Thouless 2001). Damage modelling can be separated into local or continuum approaches. In the local approach, damage is confined to a zero volume line or a surface, allowing the simulation of an interfacial failure between materials, e.g. between the adhesive bond and the adherend (Yang et al. 2001), the interlaminar failure of stacked composites (Turon et al. 2007) or the interface between solid phases of materials (Chandra et al. 2002). By the continuum approach, damage is modelled over a finite region, within solid finite elements of structures to simulate a bulk failure (Song et al. 2006) or along an adhesive bond to model a cohesive fracture of the adhesive bond (Kafkalidis and Thouless 2002). Under the continuum assumption, thin adhesive bonds to join structural members are a large field of application of CZM (Campilho et al. 2008), and the single row of cohesive elements used to model the bulk strip of adhesive makes it impossible to differentiate thickness-wise effects or concentrations of stresses towards the interface, providing an equivalent behaviour of the bond. Differences exist regarding the definition of the CZM laws, namely in which concerns their initial stiffness, between the local approach modelling (the initial stiffness can use the penalty function method) and continuum CZM modelling (the initial stiffness must accurately reproduce the deformation behaviour of the thin material strip). CZM modelling is described in detail in the next Section, prior to focusing on the cohesive parameters' estimation.

7.4 COHESIVE ZONE MODELLING

Although the computer implementation of LEFM techniques endured great success some decades ago, they are restricted to small-scale yielding beyond the crack tip. Since for modern toughened adhesives the plastic zones developing along the adhesive bond can be larger than the adherends thickness, t_p (Ji et al. 2010), a large effort was undertaken beginning in the late 1950's/early 1960's by the independent studies of Barenblatt (1959) and Dugdale (1960). The concept of cohesive zone was proposed by these authors to describe damage under static loads at the cohesive process zone ahead of the apparent crack tip. CZM were largely refined and tested since then to simulate crack initiation and propagation in cohesive and interfacial failure problems or composite delaminations. CZM are based on spring or more typically cohesive elements (Feraren and Jensen 2004), connecting 2D or 3D solid elements of structures. An important feature of CZM is that they can be easily incorporated in conventional FE software to model the fracture behaviour in various materials, including adhesively bonded joints (Ji et al. 2010). CZM are based on the assumption that one or multiple fracture interfaces/regions can be artificially introduced in structures, in which damage growth is allowed by the introduction of a possible discontinuity in the displacement field. The technique consists on the establishment of traction-separation laws (addressed as CZM laws) to model interfaces or finite regions.

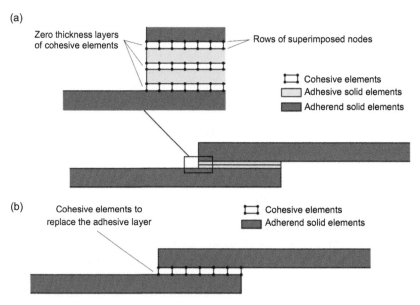

FIG. 7.2 Cohesive elements to simulate zero thickness failure paths – local approach (a) and to model a thin adhesive bond between the adherends – continuum approach (b) in an adhesive joint.

The CZM laws are established between paired nodes of cohesive elements, and they can be used to connect superimposed nodes of elements representing different materials or different plies in composites, to simulate a zero thickness interface (local approach; Fig. 7.2a; (Pardoen et al. 2005)), or they can be applied directly between two non-contacting materials to simulate a thin strip of finite thickness between them, e.g. to simulate an adhesive bond (continuum approach; Fig. 7.2b; (Xie and Waas 2006)).

A few works on CZM techniques use the local approach (Liljedahl et al. 2006, Turon et al. 2007, Campilho et al. 2005). With this methodology, the plastic dissipations in the adhesive bond are simulated by the solid finite elements, whilst the applicability of cohesive elements is restricted to damage growth simulation (Fig. 7.2a). The CZM laws usually present an extremely high initial stiffness (penalty function method), not to change the structures' global stiffness. Placement of the cohesive elements at different growth planes in the transverse direction of the adhesive bond is also feasible, capturing the respective stress gradients and concentration towards the singular regions (Campilho et al. 2005). In the local approach for bonded joints simulation, the adhesive is modelled as an elasto-plastic continuum by solid FE elements (Pardoen et al. 2005) and the "intrinsic fracture energy" is considered for the CZM laws instead of G_c, relating to the required dissipation of energy to create a new surface, while the plastic dissipations of ductile materials take place at the solid elements representative of the adhesive bond (Liljedahl et al. 2006). Thus, G_c is the sum of these two energy components, increasing by inclusion of the plastic dissipation of materials in the models. Under these assumptions, damage growth is ruled by the work of separation of the fracture surfaces instead of G_c, due to the dissipated energy by the continuum elements. The effects of external and internal constraints on the plastic dissipations of an adhesive bond are thus accountable for in the local approach. On the other hand, compared to the continuum approach, to be described in the following, more parameters and computations are needed (Ji et al. 2010).

CZM have also been used to simulate the behaviour of adhesive bonds by a continuum approach (Fig. 7.2b), by the replacement of the entire adhesive bond by a single row of cohesive elements with the representative behaviour of the adhesive bond (Campilho et al. 2008, Kafkalidis and Thouless 2002). The initial stiffness of the cohesive elements, unlike happens for the local approach, represents the adhesive layer stiffness in each mode of loading, and the global behaviours of the adherends are fully correlated by these elements. Due to the evident simplicity of this approach, it has been widely used in the damage growth simulation of bonded joints, giving accurate results providing that accurate calibrations are undertaken for the CZM laws (Campilho et al. 2009a). Despite the computationally efficiency of continuum CZM modelling of bonded joints, few limitations exist: (1) the physical meaning of the fracture process was somehow lost, because real cohesive separations are usually accompanied with localized plastic behaviours across the adhesive interlayer, even for brittle adhesives, represented with this method by averaged equivalent properties and (2) CZM become dependent of the structures' geometry, more specifically of t_P and the adhesive thickness (t_A), because these largely affect the size of the fracture process zone (FPZ) and plasticity around the crack tip, thus making the CZM laws dependent on these parameters (Ji et al. 2010).

From this discussion, it is evident that CZM provide a macroscopic reproduction of damage along a given path, disregarding the microscopic phenomena on the origin of failure, by the specification of a traction-relative displacement (t-δ) response between paired nodes on either side of the pre-defined crack path, by specification of large scale parameters ruling the crack growth process such as G_{Ic} or G_{IIc} (Zhao et al. 2011). CZM thus simulate stress evolution and subsequent softening up to complete failure, to account for the gradual degradation of material properties. The traction-separation laws are typically represented by linear relations at each one of the loading stages (Yang and Thouless 2001), although one or more stages can be defined differently for a more accurate representation of the materials behaviour. Figure 7.3 represents the 2D triangular CZM model actually implemented in Abaqus® for static damage growth, which is described here in detail. The 3D version is similar, but it also includes the tearing component. More details on the 3D CZM implemented in Abaqus® can be found in the work of Campilho et al. (2008) or in Abaqus® (2009).

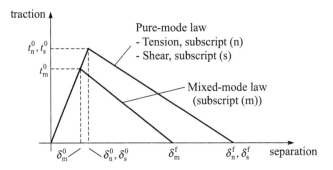

FIG. 7.3 Triangular CZM law (adapted from Abaqus® (2009)).

Regarding the nomenclature used throughout this work, the subscripts n and s relate to pure normal (tension) and shear behaviours, respectively. t_n and t_s are defined as current stresses in tension and shear, respectively, δ_n^0 and δ_s^0 are the peak strength displacements, and δ_n^f and δ_s^f the failure displacements. The values of δ_n^f and δ_s^f are defined by G_{Ic} or G_{IIc}, respectively, as these represent the area under the CZM laws. As for the mixed-mode CZM law (Fig. 7.3), t_m^0 is the mixed-mode cohesive strength, δ_m^0 the corresponding displacement, and δ_m^f the mixed-mode failure displacement. Under pure mode loading, the respective t-δ response attains its peak at the cohesive strength (t_n^0 or t_s^0), corresponding to damage initiation by the induced reduction of stiffness of the cohesive element. Softening follows and, when the values of t are completely cancelled, the crack propagates up to the adjacent set of paired nodes in the failure path, permitting the gradual debonding between crack faces. Under mixed loading (i.e., when two or three modes of loading are simultaneously present), stress and/or energetic criteria are often used to combine the pure mode laws, thus simulating the typical mixed-mode behaviour inherent to bonded assemblies. By the mentioned principles, the complete failure response of structures can be simulated (Zhu et al. 2009). For the estimation of the cohesive law parameters, a few data reduction techniques, described further in detail, are available (e.g. the property identification technique, the inverse method and the direct method) that enclose varying degrees of complexity and expected accuracy of the results (Liljedahl et al. 2006, Sørensen and Jacobsen 2003).

A CZM thus extends the concepts of continuum mechanics described in Section 2.2 by including a zone of discontinuity when the pre-defined CZM path is fulfilled, using characteristic parameters of the continuum mechanics approaches for the onset of damage and energy parameters of fracture approaches for propagation (Banea and da Silva 2009, Chen et al. 2011). This allows the CZM to be of much more general utility than conventional fracture mechanics, because of the simulation of onset and non-self-similar growth of damage without user intervention, and non-dependence on an initial flaw, unlike conventional fracture mechanics approaches. Compared to continuum mechanics approaches, CZM are mesh independent provided that sufficient integration points are simultaneously under softening at the crack front (Campilho et al. 2009b, 2011b). Actually, when the mixed-mode softening initiates at a given set of paired nodes, material degradation initiates, leading to a stress reduction and consequent stress redistribution to the neighbouring sets of nodes. If this is performed smoothly by sufficiently small load increments, then a stable propagation takes place, cancelling any mesh influence. CZM have been largely used in recent years to simulate the behaviour of structures up to failure, as they allow to include in the numerical models multiple failure possibilities, within different bulk regions of materials or between materials interfaces, e.g. at the adhesive bonding interfaces. The knowledge of the loci of damage occurring in a structure is not required as an input parameter, as the CZM globally searches damage initiation sites at specified failure paths satisfying the established criteria. However, cohesive elements must exist at the planes where damage is prone to occur, which, in several applications, can be difficult to know in advance. Although a large amount of CZM failure paths may be introduced in a structure to simulate different fractures (Campilho et al. 2009a), it is not feasible to introduce cohesive elements between every field element, even in a moderate size mesh. However, an important feature in bonded assemblies is that damage propagation is restricted to well defined planes, i.e., at or near the adhesive/adherend interfaces between the adhesive and the adherends, or cohesively in the adhesive bond, which allows overcoming this limitation (Campilho et al. 2008). Several CZM law shapes have been presented in the literature, depending of the nature of the material or interface to be simulated. The triangular, exponential and trapezoidal shapes, as shown in Fig. 7.4, are the most commonly used for strength prediction of typical materials. For the represented trapezoidal CZM, δ_n^s and δ_s^s are the stress softening onset displacements.

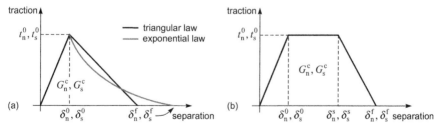

FIG. 7.4 Different shapes of pure mode CZM laws: triangular or exponential (a) and trapezoidal (b).

7.4.1 Triangular Cohesive Zone Model

The triangular CZM law, described in detail in this Section, is the most commonly used due to its simplicity, reduced number of parameters to be determined, and generally acceptable results for most real conditions (Liljedahl et al. 2006). However, generically speaking, the shape of the cohesive laws can be adjusted to conform to the behaviour of the material or interface they are simulating for more precise results (Campilho et al. 2009a). The current Section describes the static triangular 2D CZM implemented in Abaqus® v6.14 (Fig. 7.3). Different shape CZM are based on this formulation, typically differing on the softening simulation.

Under pure mode, damage propagation occurs at a specific set of paired nodes when the values of t_n or t_s are released in the respective CZM law. Under mixed-mode, stress and energetic criteria are often used to combine tension and shear (Campilho et al. 2008). Cohesive elements are assumed to be under one direct component of strain (tension) and one transverse shear strain, which are computed directly from the element kinematics. The membrane strains are assumed as zero, which is appropriate for thin and compliant bonds between stiff adherends. Strength evolution is initially defined by a constitutive matrix relating the current stresses, $\mathbf{t}$, and strains, $\boldsymbol{\varepsilon}$, in tension and shear across the cohesive elements (subscripts n and s, respectively) (Abaqus® 2009)

$$\mathbf{t} = \begin{Bmatrix} t_n \\ t_s \end{Bmatrix} = \begin{bmatrix} K_{nn} & K_{ns} \\ K_{ns} & K_{ss} \end{bmatrix} \cdot \begin{Bmatrix} \varepsilon_n \\ \varepsilon_s \end{Bmatrix} = \mathbf{K}_{COH}\boldsymbol{\varepsilon} \tag{7.1}$$

The matrix $\mathbf{K}_{COH}$ contains the stiffness parameters of the adhesive bond (K_{nn}, K_{ss} and K_{ns}), whose definition depends on whether the local or continuum approach is being used. In the local approach, used to simulate zero thickness fractures, the $\mathbf{K}_{COH}$ parameters are chosen as an extremely large value (penalty function method) for the cohesive elements not to interfere with the structure deformations, given that this task is to be performed solely by the continuum FE elements (Campilho et al. 2005). For continuum CZM modelling of bulk thin strips, and more specifically for adhesive bonds, a suitable approximation is provided with $K_{nn} = E$, $K_{ss} = G$, $K_{ns} = 0$ (Abaqus® 2009); E and G are the longitudinal and transverse elastic moduli, respective. Under these conditions, the model response for the adhesive bond accurately reproduces its deformation behaviour. A few user implemented models in the literature (Campilho et al. 2008) specify equation (7.1) directly in terms of $\mathbf{t}$-$\boldsymbol{\delta}$ relationship; $\boldsymbol{\delta}$ representing the vector of relative displacements including δ_n and δ_s. With this formulation, K_{nn} and K_{ss} are defined as the ratio between E or G and t_A. Damage initiation under mixed-mode conditions can be specified by different criteria. In this description, the quadratic nominal stress criterion for the initiation of damage is presented, previously tested for accuracy (Campilho et al. 2009a), and expressed as (Abaqus® 2009)

$$\left\{ \frac{\langle t_n \rangle}{t_n^0} \right\}^2 + \left\{ \frac{t_s}{t_s^0} \right\}^2 = 1 \tag{7.2}$$

$\langle \rangle$ are the Macaulay brackets, emphasizing that a purely compressive stress state does not initiate damage. After t_m^0 is attained (Fig. 7.3) by the fulfilment of equation (7.2), the material stiffness initiates a softening process. This is simulated by the energy being released in a cohesive zone ahead of the crack tip (FPZ). This region is where the material undergoes damage by different ways (Pereira and de Morais

2003), e.g. micro-cracking, extensive plasticity and fibre bridging (e.g. for composite adherends). Numerically, this is implemented by a damage parameter whose values vary from zero (undamaged) to unity (complete loss of stiffness) as the material deteriorates. Complete separation (δ_m^f in Fig. 7.3) is predicted by a linear power law form of the required energies for failure in the pure modes (Abaqus® 2009)

$$\frac{G_{\mathrm{I}}}{G_{\mathrm{Ic}}} + \frac{G_{\mathrm{II}}}{G_{\mathrm{IIc}}} = 1 \qquad (7.3)$$

7.4.2 Determination of the Cohesive Parameters

CZM analyses offer a powerful means to account for the largely nonlinear fracture behaviour of modern adhesively bonded joints, but the CZM parameters require careful calibrations by experimental data and respective validation in order to accurately simulate the failure process (Carlberger and Stigh 2010). Despite this fact, standardized methods for the definition of t_n^0 and t_s^0 are not yet available (Lee et al. 2010). In recent years, many works were published regarding the definition of the CZM parameters (t_n^0, t_s^0, G_{Ic} and G_{IIc}) and a few data reduction techniques are currently available (e.g. the property identification technique, the inverse method and the direct method) that enclose varying degrees of complexity and expected accuracy of the results. The few of these works that validated with mixed-mode experiments the estimated pure mode CZM typically made use of DCB, ENF or single lap specimens, generally with good results (Kafkalidis and Thouless 2002, Zhu et al. 2009).

The property identification method consists of the isolated definition of all the cohesive law parameters by suitable tests, and it is particularly critical if bulk tests are used, owing to reported deviations between the bulk and thin adhesive bond cohesive properties (Pandya and Williams 2000). This is caused by the strain constraining effect of the adherends in bonded assemblies, and also by the typical mixed-mode crack propagation in adhesive bonds. Actually, in bulk materials cracks tend to grow perpendicularly to the direction of maximum principal stress (Chai 1992). In thin bonds, cracks are forced to follow the bond length path since, as the adhesive is typically weaker and more compliant than the components to be joined, failure is often cohesive within the adhesive. The inverse method consists of the trial and error fitting analysis to experimental data on bonded joints, such as the $P\text{-}\delta$ curve, allowing tuning of the simplified shape CZM laws for particular conditions. For the property identification technique and the inverse method, a simplified parameterized shape is often selected (e.g. bilinear or trilinear) for the CZM, based on an assumption made by the user depending on the material behaviour to be simulated (Campilho et al. 2009a). On the other hand, in the direct method the precise shape of the CZM laws is easily defined, since it computes the CZM laws of an adhesive bond from the measured data of fracture characterization tests (Pandya and Williams 2000) by differentiation of the $G_{\mathrm{I}}\text{-}\delta_n$ or $G_{\mathrm{II}}\text{-}\delta_s$ curves. However, for subsequent use by FE strength prediction techniques, it is common practice to build a simplified approximation for easier implementation.

Notwithstanding the parameter identification method, deviations are expected to occur between the quantitative prediction of the cohesive parameters and the real behaviour of the adhesive bond (Leffler et al. 2007). Contrarily to the property identification method, the inverse and direct methods provide more accurate estimations as the adhesive can be characterized under identical adherend restraining conditions to real applications (Pardoen et al. 2005). However, the direct method is still considered to give the most accurate results, since it provides the precise shape of the CZM laws, while the inverse method parameters are based on previously assumed CZM shapes. Concerning the CZM parameters, G_c is usually the key parameter to be determined, because of the role that it plays on the overall results. For its estimation, the LEFM-based methods are usually easier to apply, although these can only be used for brittle or moderately ductile adhesives. For adhesives that endure extensive plasticization, LEFM techniques are rendered unfeasible, and the J-integral method emerges as a viable alternative. However, it is more complicated to apply, as generally additional data is required from the tests, such as the adherends rotation during loading.

Disregarding the parameter determination method, the CZM parameters invariably depend of t_A and t_P, which emphasizes the importance of the t_A and t_P consistency between the fracture tests and the structures to be simulated (Leffler et al. 2007).

7.4.3 Fracture Characterization Methods

The values of G_{Ic} and G_{IIc} are usually estimated by standardized test methods for all of the three methods. Additionally, the inverse and direct methods also take advantage of the P-δ results of the fracture tests for the accurate derivation of the CZM laws in each mode of loading. The DCB test for tension (Fig. 7.5a) and the ENF test for shear (Fig. 7.5b) are the most widespread test methods (B is the specimen width, a_0 the initial crack length and L the specimen length, DCB test, or half-length, ENF test).

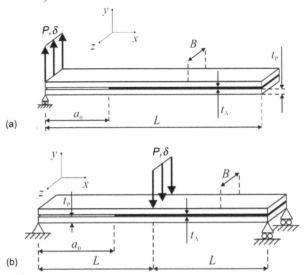

FIG. 7.5 DCB (a) and ENF (b) tests for fracture characterization of thin adhesive bonds.

Several works are available for fracture characterization in tension by the DCB test (Hojo et al. 2006). The main advantages of this test method include its simplicity and the possibility to obtain G_{Ic} mathematically using the beam theory (Yoshihara 2007). However, some issues must be taken into account for an accurate estimation. In fact, unstable crack propagation may occur, which hinders a clear crack length (a) monitoring during the test. In other cases, in the DCB test of adhesively bonded structures, the crack tip may not be clearly visible depending of the adhesive. This can induce non-negligible errors on the derivative of the compliance ($C = \delta/P$) used in the compliance calibration method (CCM). On the other hand, the energy dissipated at the FPZ can be large for ductile adhesives, implying that beam theory-based methods without any corrections will underestimate G_{Ic}. A few methods are available that allow the direct extraction of G_{Ic}, accounting for the FPZ effects and not requiring the measurement of a during propagation (de Moura et al. 2008). Blackman et al. (2003) considered Tapered Double-Cantilever Beam (TDCB) and peel tests instead to assess G_{Ic} of an adhesive layer and to study fracture of adhesively bonded joints.

Fracture characterization in shear is still not well addressed owing to some inherent features of the most popular tests: ENF, End-Loaded Split (ELS) and Four-Point End-Notched Flexure (4ENF). Actually, the ELS test involves clamping one of the specimen edges, constituting a source of variability and increasing the complexity of data reduction (Blackman et al. 2005). On the other hand, the 4ENF test requires a complex setup and presents some problems related to large friction effects (Schuecker and Davidson 2000). Therefore, the ENF test is the most suited for shear fracture characterization of adhesively bonded structures (Leffler et al. 2007). However, problems related to unstable crack growth and to crack monitoring during propagation are yet not fully solved. In addition, the classical data reduction schemes, based on beam theory analysis and compliance calibration, require the monitoring of a during propagation. Added to these issues, a quite extensive FPZ develops ahead of the crack tip for ductile adhesives, which affects the measurement of G_{IIc}. Consequently, its influence should be taken into account, which does not occur when the real value of a is used in the selected data reduction scheme. To overcome these limitations, data reduction schemes for the measurement of G_{IIc} that are only based on the specimen's compliance are also available (de Moura et al. 2009).

An important issue to account for when using the direct or inverse methods for the estimation of the CZM laws is to match as possible the values of t_P, lay-up (for stacked composite laminates) and t_A between the fracture tests and the bonded structures to be simulated, since the cohesive parameters of the adhesive bond largely depend of the adherends restraining to the adhesive bond. Pardoen et al. (2005) performed a parametric study on the effect of t_P on the values of G_{Ic}, using TDCB and compact tension (CT) fracture tests. It was shown that the increased bending of thin adherends promoted plasticity in the adhesive bond, greatly increasing G_{Ic}. Actually, the large transverse deflection of thin adherends is associated with large root rotation and substantial shear stresses, which tend to promote plastic yielding in the adhesive, increasing plastic strains as well. Additionally, extra plastic dissipation associated with the bending of the adhesive bond develops around the crack tip.

Particular attention should also be paid to match t_A between the fracture characterization tests and the real structures to be simulated, as previously mentioned. In fact, adhesives in the form of thin bonds behave differently than as a bulk material, because of the strain constraining effects of the adherends and the respective typical mixed-mode crack propagation (Leffler et al. 2007). Different studies reported on an obvious dependence of G_c of adhesive bonds with t_A (Lee et al. 2004). Typically, the value of G_c of a thin adhesive bond increases with t_A up to a peak value, bigger than the bulk quantity. After, G_c decreases with t_A to a steady-state value, corresponding to G_c of the bulk adhesive. This tendency is consistent with the work of Yan et al. (2001), which studied the influence of t_A on the fracture properties of DCB and CT joints with aluminium adherends and a rubber-modified epoxy adhesive. Using a large deformation FE technique and the peak loads measured in the experiments, the critical value of the J-integral was calculated for different values of t_A. Biel (2005) also emphasized on the bigger values of G_c near the optimal value of t_A than as a bulk. This was due to the predominantly state of prescribed deformation at the region where the crack could propagate, which enlarged the FPZ. Actually, with tough engineering adhesives, near the optimal value of t_A the FPZ typically extends several times larger than the value of t_A, and substantially longer than in bulk adhesives, resulting on bigger values of G_c.

7.4.4 Property Identification Method

In the property determination method, at least one of the CZM law parameters is approximated by consideration of bulk adhesive properties. Campilho et al. (2008) evaluated the tensile strength of bonded single-strap repairs on laminated composites as a function of the overlap length and t_P. A FE analysis including a trapezoidal CZM was used to simulate a thin ductile adhesive bond of Araldite® 420 (Hunstman) by the continuum approach. In the authors work, t_n^0 and δ_n^s were obtained from the σ-ε curve of the bulk adhesive, substantiated by previous evidence (Andersson and Stigh 2004) that t_n^0 is of the same order of magnitude of the tensile strength measured in bulk tests, and that δ_n^s and δ_s^s do not significantly influence the numerical results. On the other hand, t_s^0 was derived from t_n^0 by the von Mises yield criterion for bulk isotropic materials and, owing to its less influence on the results, δ_s^s was defined considering a similar softening slope to the tensile CZM law. The values of G_{Ic} and G_{IIc} were estimated from DCB and ENF tests, respectively. The interlaminar, intralaminar and fibre failure properties of the composite were obtained from previous works, considering triangular CZM laws (local approach).

Different fracture paths were equated (Fig. 7.6; A-A represents the line of symmetry for the FE simulations), in order to account for different failure modes. A reasonable agreement was found for the stiffness and failure load/displacement, despite the aforementioned CZM parameter approximations. Nonetheless, for an evaluation of the effect of the described approximations on the predictions, a sensitivity analysis was also performed by the authors, showing that δ_n^s and δ_s^s of the adhesive bond CZM laws do not change the failure mode of the repairs nor do they affect the failure load. Accounting for the observed failures, at lines P1 and P3 (interlaminar failure of the composite), the shear interlaminar CZM parameters showed a larger influence on the strength than the tension ones.

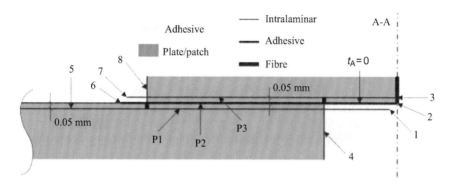

FIG. 7.6 Location of the CZM elements in the single-strap repairs (Campilho et al. 2008).

Pinto et al. (2009) tested a trapezoidal CZM with a continuum formulation to model a t_A = 0.2 mm adhesive bond (3M DP-8005®) for the estimation of the tensile strength of single-lap joints between adherends with different values of t_A and materials (polyethylene, polypropylene, glass-epoxy and carbon-epoxy composites). The FE analysis was carried out in Abaqus®, with user developed CZM elements implemented in a sub routine. For the tensile CZM law, t_n^0 and δ_n^s (Fig. 7.4b) were assumed to be equal to the corresponding bulk quantities. Although the authors did emphasize on possible miscalculations arising from this procedure, this course of action was substantiated by previous results (Andersson and Stigh 2004). G_{Ic} was estimated from DCB tests using the ASTM D3433-99 standard. The value of δ_n^s was approximated the product of the average failure strain obtained in the adhesive bulk tests with t_A, as this parameter does not significantly influence the FE results. Oppositely, the CZM parameters in shear loading were obtained with an inverse method (the basic principles of this technique are described in detail in Section 7.4.5), considering the block-shear test method (ASTM D4501-01). The block shear test was adopted since the adhesive layer is mainly loaded in shear, while normal stresses are minimized. The FE simulations captured fairly accurately the experimental behaviour of the joints, in terms of stiffness, and maximum load and the corresponding value of δ for all of the geometry and material configurations tested.

7.4.5 Inverse Method

The inverse method consists of an iterative curve fitting procedure between experimentally measured data and the respective FE predictions, considering a precise description of the experimental geometry and approximate cohesive laws, established based on the typical behaviour of the material to be simulated. The inverse characterization of adhesive bonds should be applied individually for each tested specimen to account for slight geometry variations between specimens (Campilho et al. 2009a). By this technique, the value of G_{Ic} or G_{IIc}, which corresponds to the steady-state value of G_I or G_{II} during crack propagation in the respective

R-curve built from the fracture characterization test data, is input in the FE model. To completely define the CZM law, approximate bulk values can be used for t_n^0 or t_s^0 (Fig. 7.3) for the initiation of the trial and error iterative process (Campilho 2009). Tuning of the cohesive parameters is performed by a few numerical iterations until an accurate prediction of the experimental data is achieved. Examples of reliable experimental data for the iterative fitting procedure are the R-curve (Flinn et al. 1993), the crack opening profile (Mello and Liechti 2004), and more commonly the P-δ curve (Li et al. 2005).

In the work of Campilho (2009), an inverse fitting methodology was considered for the definition of a shear CZM law by the ENF test (continuum approach) for a ductile epoxy adhesive bond (Hunstman Araldite® 2015) with t_A = 0.2 mm. Notwithstanding the crack measurement difficulties arising from crack growth without opening, the use of correction fluid along the crack growth path allowed a precise estimation of the current value of a by an imaging procedure, consisting of picture recording during the tests with 5 s intervals using a 10 MPixel digital camera. Estimation of G_{IIc} was made possible by correlation of P-δ-a by the time elapsed from the test initiation. The testing time of each P-δ data point was calculated from the current value of δ and the loading rate. The correspondence to the values of a was established knowing the testing time of each picture. G_{IIc} was estimated by three methods: the CCM, the corrected beam theory (CBT) and the compliance-based beam method (CBBM), this last not one requiring the measurement of a during crack propagation as it computes G_{IIc} only using the experimental compliance. Data reduction by the three methods for five tested specimens showed similar CCM and CBBM results, and smaller values of CBT. The FE analysis faithfully represented each specimen geometry and measured value of a_0. A user defined continuum-based trapezoidal CZM formulation (Campilho et al. 2008), coupled to Abaqus®, was considered to account for the adhesive ductility. Despite this fact, in terms of spatial modelling, the cohesive elements had zero thickness (Fig. 7.7).

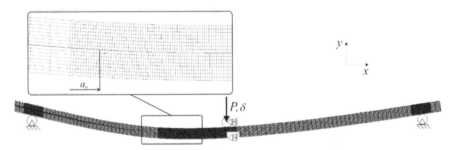

FIG. 7.7 Deformed shape of the ENF specimen during propagation, with boundary and loading conditions (Campilho 2009).

The CBBM values of G_{IIc}, corresponding to the steady-state value of the respective R-curves, were used as input in the FE models. The remaining cohesive parameters (t_s^0 and δ_s^s) were estimated by fitting the experimental and numerical P-δ curves of each specimen. Figure 7.8 shows the experimental and FE P-δ curves for one tested specimen after the fitting procedure.

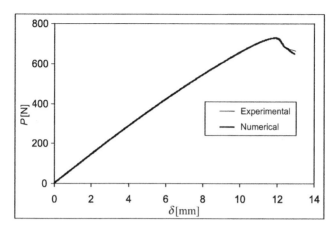

FIG. 7.8 Experimental and FE P-δ curves comparison for one tested specimen (Campilho 2009).

Figure 7.9 shows the average shear CZM law and respective values of G_{IIc} (J_{IIc}), t_s^0 ($\sigma_{\mathrm{u,II}}$), $\delta_s^s(\delta_{2,\mathrm{II}})$ and $\delta_s^f(\delta_{\mathrm{u,II}})$, and also the CZM laws range after individual application of the inverse fitting principles to five experimental P-δ curves.

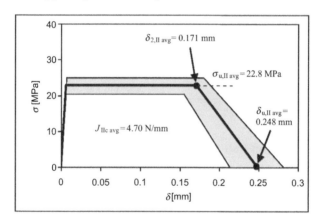

FIG. 7.9 Average shear CZM law and deviation after application of the inverse fitting principles to five specimens (Campilho 2009).

The manual fitting allowed a clear insight on the influence of the CZM parameters on the P-δ curves shape. G_{IIc}, which is the input value in the simulations, mainly influences the peak load. Higher values of t_s^0 increase the peak load and the specimen stiffness up this value, leading to a more abrupt post-peak load reduction. Finally, δ_s^s plays an important role on the roundness of the P-δ curve near the peak value. These findings indicated that a unique solution for the shear CZM law of the adhesive could be guaranteed by the inverse technique.

Lee et al. (2010) proposed a systematic procedure to estimate the local CZM parameters of an adhesive bond, using Single-Leg Bending (SLB) mixed-mode tests with tension or shear as the dominant modes, for the extrapolation of the pure mode laws, thus simplifying the inverse fitting technique. SLB specimens with co-cured adherends (upper carbon fibre reinforced composite adherend and lower steel adherend) were tested under different mode mixities, for validation of the obtained pure tension and shear data, since this parameter can be controlled by changing the specimens' geometry. All fractured specimens showed a cohesive failure of the bond, with evidence of adhesive and fibres on the bonding surface of the steel adherends. Measurement of G_c was carried out from the test data, and the mode mixity was defined from the classical beam theory. Linear extrapolation of the measured data for both tension and shear mode dominant tests allowed the definition G_{Ic} and G_{IIc}, as depicted in Fig. 7.10. The mode decompositions resulted on $G_{Ic} = 140$ N/m and $G_{IIc} = 280$ N/m for pure tension and shear, respectively, within the range of typical values for epoxy matrices.

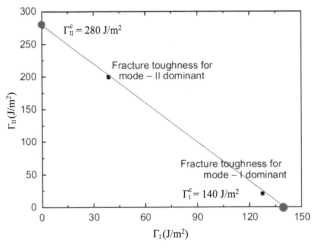

FIG. 7.10 Linear extrapolation of G_{Ic} and G_{IIc} from the mixed-mode test results (Lee et al. 2010).

A mixed-mode CZM was considered for the reproduction of the test results, built using triangular CZM laws for pure tension and shear, and whose criteria for stress softening and complete separation are consistent with those of equation (7.2) and equation (7.3), respectively. Definition of the missing cohesive parameters (K_{nn}, K_{ss}, t_n^0 and t_s^0) used the design of experiments (DoE) and kriging metamodel (KM) techniques. The DoE was built considering the aforementioned parameters to be determined as the DoE design variables, and setting the variable levels and required sampling rates for the experiments by the central composite method. As a result of the analysis, the quantity of 50 simulations was considered as minimum number of experiments to obtain the four parameters. The KM, by allowing a mathematical basis between the system inputs (sampling points of the cohesive parameters) and outputs (errors between the simulation results and experimental data), permits attaining an optimized solution by minimization of the error (*f*-*y* in Fig. 7.11).

An error function was defined as the load difference between the FE and experimental P-δ curves, at different pre-established values of δ within the test range (the vertical dashed line in Fig. 7.11 relates to the limit of the evaluation region).

The CZM parameters to be defined in pure tension and shear were then estimated by a nonlinear optimization algorithm, the sequential quadratic programming algorithm, for minimization of the experimental/FE deviation. The obtained values of CZM parameters in each pure mode of loading were considered to build the pure tension and shear CZM laws, and were regarded as the optimized values to describe the co-cured SLB tests under the mixed-mode tension or shear dominant modes, although applicable to pure modes or different mode mixities. Validation of the proposed CZM laws was carried out by consideration of two intermediate mode mixities (modification of the composite adherend thickness to 1.28 and 1.48 mm), by comparison of test results and FE simulations with the previously defined parameters for the dominant modes mixities. Comparison of the FE and test P-δ curves is provided in Fig. 7.12, giving maximum errors of $\approx 5\%$ for both modelling conditions.

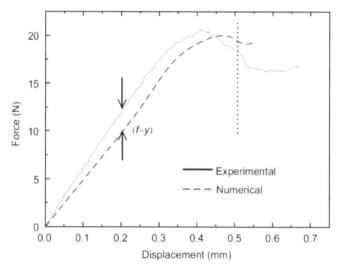

FIG. 7.11 Error defined by the KM, by comparison of the FE and experimental data (Lee et al. 2010).

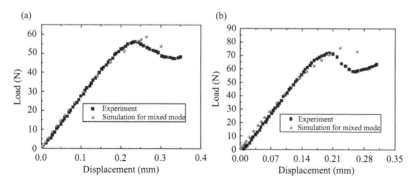

FIG. 7.12 P-δ curves for SLB specimens with varying mode mixities: composite adherend thickness of 1.28 (b) and 1.48 mm (b) (Lee et al. 2010).

The mode mixities for the two reported conditions were superimposed on the plot of Fig. 7.10, further reinforcing the linear extrapolation assumption of G_{Ic} and G_{IIc}, as the new data fitted perfectly on top of the proposed linear relationship. From the global results, the authors concluded that the proposed procedure accurately described the fracture behaviour of mixed-mode joints, showing the advantage of non-requirement of two separate tests as it is usually performed (e.g., DCB for tension and ENF for shear characterization).

7.4.6 Direct Method

By the direct method, the complete CZM law and respective shape for a given material strip or interface can be precisely estimated by the differentiation of the G_I–δ_n or G_{II}–δ_s curves. In this Section, a distinction is made between CZM law estimation in tension (DCB test) and in shear (ENF test).

7.4.6.1 Direct Method Applied to the DCB Specimen

For the DCB specimen, the path-independence of the J-integral can be used to extract relations between the specimen loads and the cohesive law of the crack path (Stigh et al. 2010). Based on the fundamental expression for J defined by Rice (1968), it is possible to derive an expression for the value of G_I applied to the DCB specimen from the concept of energetic force and also the beam theory for this particular geometry, as follows (the following formulae are developed assuming that the J-integral gives a measurement of G_I) (Banea et al. 2010)

$$G_I = 12\frac{(P_u a)^2}{E_a t_P^3} + P_u \theta_o \quad \text{or} \quad G_I = P_u \theta_p \qquad (7.4)$$

where P_u represents the applied load per unit width at the adherends edges, E_a the Young's modulus of the adherends, θ_o the relative rotation of the adherends at the crack tip and θ_p the relative rotation of the adherends at the loading line (Fig. 7.13). In previous works (Campilho et al. 2013a), the first expression of (7.4) was considered, using θ_o instead of θ_p, due to a simpler extraction of the parameter by the optical method. The J-integral can be calculated along an arbitrary path encircling the start of the adhesive layer, giving (Stigh et al. 2010)

$$G_I = \int_0^{\delta_n^f} t_n(\delta_n)\, d\delta_n \qquad (7.5)$$

where δ_n^f is measured at the initial crack tip. G_{Ic} can be considered the value of G_I at the beginning of crack growth. Thus, G_{Ic} is given by the steady-state value of G_I, at a δ_n value of δ_n^f (Carlberger and Stigh 2010). The $t_n(\delta_n)$ curve can be easily obtained by differentiation of equation with respect to δ_n

$$t_n(\delta_n) = \frac{\partial G_I}{\partial \delta_n} \qquad (7.6)$$

As a result, the procedure of an experiment is to measure the history of P, a, δ_n and θ_o. The cohesive law in tension can then be estimated by plotting G_I in equation

(7.4) as a function of δ_n, polynomial fitting of the obtained curve and differentiation (Stigh et al. 2010).

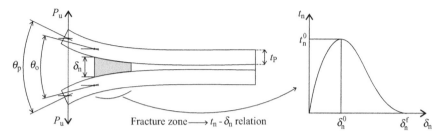

FIG. 7.13 DCB specimen under loading, with description of the analysis parameters.

7.4.6.2 Direct Method Applied to the ENF Specimen

This Section describes the direct method for G_{IIc} and cohesive law estimation by ENF experiments (Zhu et al. 2009, Carlberger and Stigh 2010, Leffler et al. 2007). This technique relies on the simultaneous measurement of the J-integral and δ_s. As discussed previously (Campilho et al. 2013a), the J-integral is suited to the non-linear elastic behaviour of materials, but it remains applicable for plastic monotonic loadings, as it is the case of the ENF test protocol. The proposed G_{II} evaluation expression results from using alternate integration paths to extract the J-integral (Rice 1968), resulting on the following closed-form expression for G_{II} (Leffler et al. 2007)

$$G_{II} = \frac{9}{16}\frac{(P_u a)^2}{E_a t_P^3} + \frac{3}{8}\frac{P_u \delta_s}{t_P} \qquad (7.7)$$

where P_u is measured at the loading cylinder (Fig. 7.14). The first term corresponds to the LEFM solution. The second term relates to the influence of a flexible adhesive layer, and it can give a large contribution to G_{II} (Leffler et al. 2007). The accuracy of this expression requires the linear elastic behaviour of the adherends, which needs to be checked during the design of the specimens' geometry and verified after each test is done. Similarly to the DCB specimen, by evaluating the J-integral around the damage region or FPZ of the adhesive layer, it is possible to write

$$G_{II} = \int_{0}^{\delta_s^f} t_s(\delta_s)\,d\delta_s \qquad (7.8)$$

Expression (7.8) gives a direct relation between the stress state at the crack tip and G_{II}. The evolution of G_{II} with δ_s is as follows: before crack propagation, G_{II} increases up to attaining G_{IIc}. At this stage, the crack begins to propagate. G_{IIc} is thus obtained by the steady-state value of G_{II} in the G_{II}–δ_s plot. This point in the plot corresponds to cancelling of the corresponding stress component and consequent formation of a crack. The $t_s(\delta_s)$ plot or shear cohesive law of the adhesive layer is thus estimated by fitting of the resulting G_{II}–δ_s curve and differentiation with respect to δ_s (Leffler et al. 2007)

$$t_s(\delta_s) = \frac{\partial G_{\text{II}}}{\partial \delta_s} \qquad (7.9)$$

Differentiation can be performed by polynomial functions or least square adaptions of Prony-series to the G_{II} data, this last technique to be recommended when the polynomial expressions reveal to be rough approximations of the experimental data (Carlberger and Stigh 2010). Because of compression and friction effects at the crack faces of the ENF specimen eventually compromising the cohesive law accuracy, some authors (Zhu et al. 2009) considered instead the Arcan test method to obtain the shear cohesive law of adhesive layers.

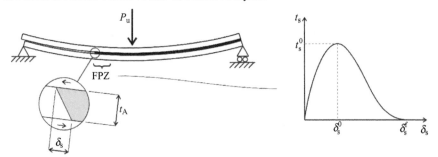

FIG. 7.14 ENF specimen under loading, with description of the analysis parameters.

7.4.6.3 Optical Algorithm for Parameter Extraction – DCB Specimen

This Section describes an optical algorithm developed by the authors (Campilho et al. 2014) for the extraction of δ_n and θ_o, required for the application of the direct method, as explained in Section 7.4.6.1. Other techniques exist to measure these parameters, such as coupling physical sensors to the specimens, and these will also be addressed further in this chapter. By this optical method, for calculating δ_n and θ_o for a given image, the identification of 8 points is required (Fig. 7.15): two points (p_3, p_4) to measure the current t_A value at the crack tip (t_A^{CT}) during loading in image units (pixels), two points (p_7, p_8) identifying a line segment in the image for which the length (d) is known in real world units (mm), and two points (p_1, p_5) on the top specimen and two points (p_2, p_6) on the bottom specimen to compute θ_o.

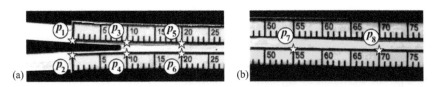

FIG. 7.15 Points taken by the optical method to measure δ_n and θ_o.

All eight points are manually identified in the first picture of a trial using an in-house software tool. The identification of the points is aided by the ruler attached to the specimens, which helps finding their correct locations. In addition, each point between p_1 and p_6 is printed with a distinct colour (although this is not perceptible in Fig. 7.15). Using the location of the points in the first picture, the points of the following pictures are automatically identified using a computer algorithm implemented in Matlab®. Basically, for each point p_i, a rectangular region centred in p_i is extracted from the first image forming a template (t). This template describes the image pattern that surrounds the point and is used for locating the point in the next image. This is done by finding the position (u, v) in the next image (I) that has the highest normalized cross-correlation with the template. The normalized cross-correlation is a measure of similarity between two images that is invariant to linear changes in the pixel intensities and that quantifies the correlation between two images/regions. This measure of similarity was chosen due to its low computational requirements, which is a critical factor given the high resolution of the images, and because changes in rotation and scale of the specimens are expected to be small between two consecutive acquisitions (gapped by 5 seconds). To take advantage of the colour information, the colour space of the images (and consequently, of the templates) was transformed to the CIELAB colour space. The CIELAB system represents the value of a pixel by three components, L, a and b, where L represents luminosity and a and b define colour. Since points p_1 to p_6 are differentiated by their colour, only the a and b components are used when detecting points. The normalized cross-correlation (γ) of template t with image I at the position (u, v) of image I for the colour component c is defined as

$$\gamma(u, v, c) = \frac{\sum_{x,y}\left[I(x, y, c) - \bar{I}_{u,v,c}\right] \cdot \left[t(x - u, y - v, c) - \bar{t}_c\right]}{\left\{\sum_{x,y}\left[I(x, y, c) - \bar{I}_{u,v,c}\right]^2 \cdot \sum_{x,y}\left[t(x - u, y - v, c) - \bar{t}_c\right]^2\right\}^{0.5}} \quad (7.10)$$

where $I(x, y, c)$ is the intensity of the colour component c of the pixel (x, y) of image I; $t(x, y, c)$ is the intensity of the colour component c of the pixel (x, y) of the template t; $\bar{I}_{u,v,c}$ is the average intensity of the colour component c of the region of image I centred at pixel (u, v) and with the same size as t, and $\bar{t}_c$ is the average intensity of the colour component c for the template t. Finally, the normalized cross-correlation for a single pixel taking into account the colour components a and b is defined as

$$\gamma(u, v) = \sqrt{\gamma(u, v, a)^2 + \gamma(u, v, b)^2} \quad (7.11)$$

Calculating γ for all the pixels of I results in a matrix where the maximum absolute value yields the location of the region in I that has the highest correlation with t and, thus, the most likely location of p_i in the next image. This is done for every one of the eight points identified in the first image. After successfully identifying all the points of the second image, new templates are computed from the second image to search for the eight points in the third image, and so on until processing all images.

Computation of δ_n

The value of t_A^{CT} in real world units (mm) is calculated as follows

$$t_A^{CT} = d\frac{|p_3 - p_4|}{|p_7 - p_8|} \quad (7.12)$$

assuming that the lens distortion is negligible, which is valid for the central area of pictures acquired with modern CCD cameras. A length of d = 15 mm was used for all trials (illustrated in Fig. 7.15). The pixel size was on average 0.021 mm and, thus, the estimated maximum error of the image acquisition process is ±0.011 mm. Finally, δ_n can be defined as

$$\delta_n = t_A^{CT} - t_A \qquad (7.13)$$

where t_A is the theoretical design value of t_A of the specimens. Since t_A can show small variations due to the fabrication process, an adjustment to δ_n can be applied to make δ_n = 0 at the beginning of the test. Figure 7.16 shows the δ_n-testing time plot for a specimen, more specifically the raw curve, the 6th degree fitting curve and the corrected polynomial and final curve (Campilho et al. 2014). Due to scaling difficulties, the raw curve in the figure is already translated such that δ_n (testing time = 0) = 0.

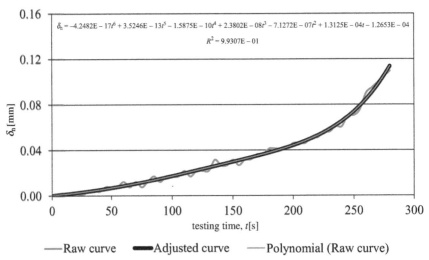

FIG. 7.16 Evolution of δ_n for one test specimen: raw curve obtained from the optical method, polynomial fitting curve and corrected polynomial curve.

Computation of θ_o

θ_o is calculated as the angle between the tangents to the horizontal curves of the 2 scales closest to the adhesive, measured at the crack tip (Fig. 7.17). The curvature of the top adherend is first computed by fitting a quadratic function to points p_1, p_3 and p_5. The first derivative of the quadratic function at p_3 yields the slope of the top curve (m_{top}) at the crack tip, which is then used to define a direction vector $\vec{v}_{top}$ = (1, m_{top}). The same process is repeated for points p_2, p_4 and p_6, yielding the slope of the tangent to the bottom curve at the crack tip (m_{bottom}) and its direction vector $\vec{v}_{bottom}$ = (1, m_{bottom}). Finally, θ_o is obtained by measuring the angle between the two vectors

$$\theta_0 = \arccos\left(\frac{\vec{v}_{top} \cdot \vec{v}_{bottom}}{|\vec{v}_{top}||\vec{v}_{bottom}|}\right) \qquad (7.14)$$

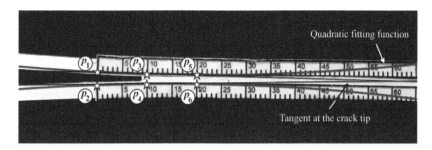

FIG. 7.17 Calculation of θ_o. Quadratic functions were fitted to points p_1, p_3, p_5 and p_2, p_4, p_6, representing the curvature of the top and bottom specimen, respectively, while the straight lines show the tangents to the curves at the crack tip (corresponding to 10 mm in the scales).

Figure 7.18 gives an example of the evolution of θ_o for a given test specimen (Campilho et al. 2014). Shown in the graphic are the raw curve, the 4th degree fitting curve and the corrected polynomial and final curve, adjusted to make θ_o (testing time = 0) = 0. This polynomial adjustment is required to smooth the raw data and remove the experimental measurement scatter, but also to cancel any eventual misalignment between the glued scales in both adherends.

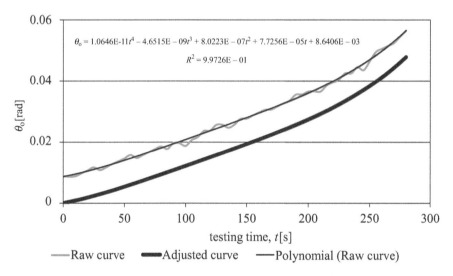

FIG. 7.18 Evolution of θ_o for one test specimen: raw curve obtained from the optical method, polynomial fitting curve and corrected polynomial curve.

7.4.6.4 Optical Algorithm for Parameter Extraction – ENF Specimen

A numerical algorithm was developed to measure δ_s based on image processing and tracking of a set of reference points throughout a sequence of images. The optical method requires the identification of 6 points, from p_1 to p_6 (Fig. 7.19), which define the curvatures of the top and bottom specimen. δ_s is given by the arc length between p_3' and p_4', which are the projections of p_3 and p_4 into the medial curve of the specimen.

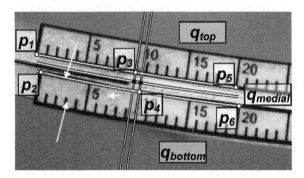

FIG. 7.19 Illustration of the points taken by the optical method (p_1 to p_6), the curves fitted to those points (q_{top} and q_{bottom}) and the medial curve (q_{medial}) where δ_s is measured.

The automatic point tracking algorithm has some similarities to that presented in Section 7.4.6.3 and, thus, it is explained here in a more condensed manner. The process starts by manually identifying the 6 points in the first picture of a test. Points p_1 to p_6 are printed with a distinct colour, which helps finding their correct locations. The manually identified locations only work as seeds since the locations of the points are automatically refined so that they lay at the centroid of the colour regions, making the algorithm invariant to the user input. Starting from the refined points in the first picture, the points of the following pictures are automatically tracked with an algorithm in Matlab®. For each point p_i, a rectangular region centred in p_i is extracted from the first image forming a template (t). This template describes the image pattern that surrounds the point and is used for locating the point in the next image. This is done by finding the position (u, v) in the next image (I) that has the highest normalized cross-correlation with the template (equation (7.10) defined in Section 7.4.6.3). The normalized cross-correlation for a single pixel, $\gamma(u, v)$, taking into account the colour components of the image is defined as depicted in equation (7.11). By calculating γ for all the pixels of I results in a matrix where the maximum absolute value yields the location of the region in I that has the highest correlation with t and, thus, the most likely location of p_i in the next image. This is done for every one of the 6 points identified in the first image. After successfully identifying all the points of the second image, new templates are computed from the second image to search for the 6 points in the third image, and so on until processing all images that comprise the sequence.

Computation of δ_s

The value of δ_s is obtained by measuring the arc length between p_3 and p_4 on the curve that lies between the two adherends, the medial curve. The curvatures of the adherends are described by quadratic polynomials

$$q_i(x) = a_i x^2 + b_i x + c_i \qquad (7.15)$$

where q_i represents the curvature of adherend i, and a_i, b_i and c_i are the coefficients of the polynomial. The coefficients are found by fitting the quadratic functions to the y coordinates of the points in the least squares sense, such as $q_{top}([x_1, x_3, x_5]^T)$ $= [y_1, y_3, y_5]^T$ and $q_{bottom}([x_2, x_4, x_6]^T) = [y_2, y_4, y_6]^T$. In theory, q_{top} and q_{bottom} would be strictly parallel and δ_s could be calculated by projecting p_3 to the bottom curve, or p_4 to the top curve. However, in practise, this is not verified and, thus, δ_s is calculated over the projections of p_3 and p_4 into the medial curve of the specimen. The medial curve, q_{medial}, is defined by averaging the coefficients of the bottom and top curvatures,

$$q_{medial}(x) = a_{medial}x^2 + b_{medial}x + c_{medial}$$

$$= \frac{a_{top} + a_{bottom}}{2}x^2 + \frac{b_{top} + b_{bottom}}{2}x + \frac{c_{top} + c_{bottom}}{2} \qquad (7.16)$$

The projection of a point $p_i = (x_i, y_i)$ into the medial curve is obtained by finding the line that is perpendicular to the medial curve and that passes by p_i. The perpendicular to the medial curve at a given point is the line that is perpendicular to the tangent of the curve at that point, which is given by the first derivative of q_{medial},

$$q'_{medial}(x) = 2a_{medial}x + b_{medial} \qquad (7.17)$$

The projection (x'_i, y'_i) of a given point (x_i, y_i) is found by solving the system

$$\begin{cases} a_{normal} = -\dfrac{1}{2a_{medial}x'_i + b_{medial}} \\ b_{normal} = y'_i - a_{normal}x'_i \\ y_i = a_{normal}x_i + b_{normal} \\ y'_i = a_{medial}x'^2_i + b_{medial}x'_i + c_{medial} \end{cases} \qquad (7.18)$$

where the first two equations define the coefficients a_{normal} and b_{normal} of the line that is perpendicular to q_{medial} and that passes by the projected point (x'_i, y'_i), the third equation forces the perpendicular to pass by the original point (x_i, y_i), and the forth equation forces the projected point to belong to the medial curve. This non-linear system of equations is numerically solved with Malab's *vpasolve* function, and always returns a single solution in the real numbers domain.

Having p'_3 and p'_4, δ_s^p (the value of δ_s in pixels) may be found by measuring the arc length of q_{medial} between these two points,

$$\delta_s^p = \int_{x'_3}^{x'_4} \sqrt{1 + \left(\frac{dy}{dx}\right)^2}\, dx = \int_{x'_3}^{x'_4} \sqrt{1 + (a_{medial}x + b_{medial})^2}\, dx \qquad (7.19)$$

δ_s^p is in image units (pixels) and needs to be converted to real world units (e.g. millimeters). Since the length of the arcs (p_1, p_5) and (p_2, p_6) is known to be $d = 20$ mm, this is used to find the pixel size and, thus, convert δ_s^p to millimeters,

$$\delta_s^t = \frac{\delta_s^p d}{2}\left(\int_{x_1}^{x_5} \sqrt{1 + (a_{top}x + b_{top})^2}\, dx + \int_{x_2}^{x_6} \sqrt{1 + (a_{bottom}x + b_{bottom})^2}\, dx\right)^{-1} \qquad (7.20)$$

The pixel size was on average 0.011 mm and, thus, the estimated maximum error of the image acquisition process is ±0.006 mm. Finally, δ_s can be defined as

$$\delta_s = \delta_s^t - \delta_s^0 \qquad (7.21)$$

where δ_s^0 is the initial value of δ_s^t.

Figure 7.20 gives a representative example of the variation of δ_s with the time elapsed since the beginning of the test for three different adhesives. These results pertain to the example of application that is going to be detailed further in Section 7.4.6.6. The raw curves from the point tracking algorithm and the adjusted polynomial laws, attained by making δ_s (testing time = 0) = 0, are presented in this figure. The curves were truncated at crack initiation since, from this point on, the data is no longer relevant for the direct method. It should be mentioned that, depending of the specimen under analysis, different degree polynomials were selected in order to attain the best correlation factor, R (which is also valid for the polynomial to fit the G_{II}–δ_s law, to be discussed further). While the polynomial approximation is necessary to remove the noise from the raw curve, the mentioned procedure to obtain the adjusted polynomial laws was required on account of eventual initial offsets while preparing the specimens that made the δ_s value not to be nil at the beginning of each test. The evolution of δ_s in Fig. 7.20 is exponential with the testing time, and this is consistent with previous works. Leffler et al. (2007) obtained an exponentially increasing evolution of the shear deformation rate in ENF tests by using an extensometer attached to the both adherends at the crack tip. Between adhesives, a clear difference can be found in the maximum δ_s values, which closely follow the adhesive ductility. Inclusively, δ_s at crack initiation for the Araldite® AV138 is under 1/10 of a millimetre. After having the δ_s-testing time plots, it was possible to estimate the G_{II}–δ_s relationship by direct application of equation (7.4). The G_{IIc} estimate is given by the steady-state value of G_{II} in the G_{II}–δ_s curve, which corresponds to the onset of crack propagation (Carlberger and Stigh 2010).

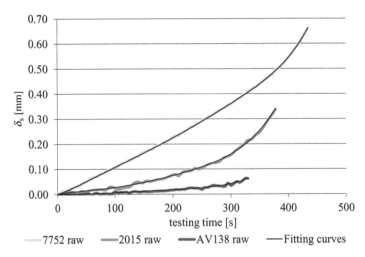

FIG. 7.20 Plot of δ_s – testing time for a specimen of each tested adhesive: raw curve and polynomial approximations.

7.4.6.5 Application of the Direct Method/ Optical Algorithm to the DCB Specimen

The following discussion regards previous works by the authors (Campilho et al. 2013a, 2014, 2015) that applied the previously described direct method/optical algorithm for G_{Ic} and CZM law estimation. Three case studies are presented: (1) adhesive bonding for adhesive joints with natural fibre composite as adherends, (2) adhesive bonding between aluminium adherends to study the effect of t_P on G_{Ic}, and finally (3) adhesive bonding between aluminium adherends considering varying values of t_A.

Characterization of the Materials

Three joint configurations were tested, presented in Table 7.1, considering the DCB test geometry. The joints for configuration 1 considered jute-epoxy composites as adherends, consisting of 8 stacked weave plies with a fibre volume fraction of $\approx$30%. For configuration 1, typical properties of jute are as follows: density of 1.3-1.4 g/cm^3, elongation at failure (ε_f) of 1.5-1.8%, tensile strength (σ_f) of 400-800 MPa and E of 15-30 GPa (Wambua et al. 2003, Ku et al. 2011). Epoxy was chosen for the matrix material on account of the good mechanical (strength and stiffness) and toughness properties, and also because of the superior wetting characteristics on natural fibres (Herrera-Franco and Valadez-González 2005). The epoxy resin type SR 1500 and SD 2505 hardener from Sicomin Epoxy Systems were used. The matrix properties, as specified by the manufacturer, are as follows: $E = 3.1$ GPa, $\sigma_f = 74$ MPa, strain at maximum load $\varepsilon_m = 4.4\%$ and $\varepsilon_f = 6.0\%$. The plates were fabricated by hand lay-up and cured at room temperature in a vacuum bag. The jute-epoxy composite was composed of 30% of jute fabric (by weight) and gave the following properties in tensile testing: $E = 5.7$ GPa and $\sigma_f = 124.3$ MPa. For configurations 2 and 3, the aluminium adherends were cut from a high strength aluminium alloy sheet (AA6082 T651). This material was characterized in bulk tension in previous works by the authors (Campilho et al. 2011b) using dogbone specimens and the following mechanical properties were obtained: $E = 70.07 \pm 0.83$ GPa, tensile yield stress (σ_y) of 261.67 ± 7.65 MPa, $\sigma_f = 324 \pm 0.16$ MPa and $\varepsilon_f = 21.70 \pm 4.24\%$.

Table 7.1 Configurations tested to measure G_{Ic}.

Configuration	Adherends	Adhesive
1	Jute-epoxy composite	SikaForce® 7888
2	Aluminium	SikaForce® 7888
3	Aluminium	SikaForce® 7752-L60

Configurations 1 and 2 used the polyurethane adhesive SikaForce® 7888, which was characterized in the work of Neto et al. (2012) by bulk tensile tests for the determination of E, σ_f and ε_f, and DCB and ENF tests to define the values of G_{Ic} and G_{IIc}, respectively. The bulk characterization was performed as specified in the ISO standard 527-2 (2012). The obtained results gave $E = 1.89 \pm 0.81$ GPa, $\sigma_f = 28.60 \pm 2.0$ MPa, $\varepsilon_f = 43.0 \pm 0.6\%$, $G_{Ic} = 0.7023 \pm 0.1233$ N/mm and $G_{IIc} = 8.721 \pm 0.792$ N/mm.

Configuration 3 used a novel polyurethane structural adhesive, SikaForce® 7752-L60. This is a two-part adhesive, and it consists of a filled polyol based resin and an isocyanate based hardener. It is characterized by a room temperature cure, high impact resistance and flexibility at low temperatures, with $\sigma_f \approx 10$ MPa and $\varepsilon_f \approx 25\%$ (manufacturer's values).

Joint Geometries

The geometry of the DCB specimens is shown in Fig. 7.5a. The dimensions for each joint configuration are presented in Table 7.2.

Table 7.2 Dimensions of the three joint configurations (in mm).

Configuration	L	a_0	t_P	B	t_A
1	160	50	5	15	1
2	160	40	1, 2, 3 and 4	25	1
3	160	55	3	25	0.1, 0.2, 0.5, 1.0 and 2.0

Some dimensions differ between configurations, but these do not affect the G_{Ic} measurement. For the three joint configurations, for a uniform value of t_A, calibrated spacers were inserted between the adherends. These spacers were inserted at both bonding edges between the adherends to control the value of t_A. For the calibrated spacer at the crack tip, 3 plies were stacked and glued together, composed of a 0.1 mm thick razor blade between steel spacers to achieve the desired value of thickness, to create a pre-crack. For all specimens, stainless steel piano hinges were glued to both faces of the specimens at the cracked edge with a ductile adhesive, to provide a loading means in the testing machine grips. Also, a metric scale was glued with cyanoacrylate in both adherends to measure a and of the input data for the extraction of the J-integral. Six specimens of each configuration were tested at room temperature ($\approx 20°C$), relative humidity of $\approx 40\%$ and 2 mm/min in an electro-mechanical testing machine (Shimadzu AG-X 100) with a load cell of 100 kN. Data recording was carried out at 5 Hz for the values of P and δ, registered during the test as a function of the time elapsed since its initiation. Pictures were recorded during the specimens testing with 5 s intervals using a 15 MPixel digital camera with no zoom and fixed focal distance to approximately 100 mm.

Results – Configuration 1

For the bonded specimens with jute-epoxy adherends, δ_n and θ_o were defined as specified previously. The values of G_{Ic} for the bonded joints were defined by plotting the $G_I-\delta_n$ curves, considering G_{Ic} as the steady-state value of G_I in the $G_I-\delta_n$ curve (Carlberger and Stigh 2010). Figure 7.21 plots the experimental $G_I-\delta_n$ law and the corresponding 6th degree polynomial fitting curve for a given specimen. At the beginning of the test, G_I slowly increases with δ_n, but the growth rate of G_I rapidly increases up to nearly $\delta_n = 0.02$-0.04 mm, and a steady-state value of G_I is attained at $\delta_n \approx 0.09$ mm. For this specimen, the measured value of G_{Ic} is 1.429 N/mm. For the six bonded specimens, the obtained data gave $G_{Ic} = 1.182 \pm 0.215$ N/mm.

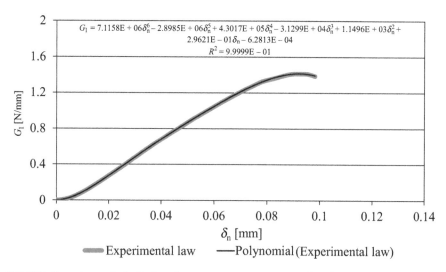

$$G_1 = 7.1158E + 06\delta_n^6 - 2.8985E + 06\delta_n^5 + 4.3017E + 05\delta_n^4 - 3.1299E + 04\delta_n^3 + 1.1496E + 03\delta_n^2 + 2.9621E - 01\delta_n - 6.2813E - 04$$
$$R^2 = 9.9999E - 01$$

Experimental law   **—Polynomial (Experimental law)**

FIG. 7.21 Experimental G_1–δ_n law for one test specimen and polynomial fitting curve (configuration 1).

Figure 7.22 shows the obtained experimental t_n–δ_n law, showing the ductile characteristics of the adhesive after the peak value of t_n is attained. For this specimen, the following values were found: $t_n^0 = 20.73$ MPa and $\delta_n^f = 0.0935$ mm. For the complete batch of tested specimens, average values and deviations were as follows: $t_n^0 = 23.18 \pm 3.57$ MPa and $\delta_n^f = 0.0843 \pm 0.156$ mm. Proposed triangular and trapezoidal simplified CZM laws are also presented, allowing concluding that for the adhesive SikaForce® 7888 a trapezoidal law is particularly suited, since it accounts the best for the adhesive ductility.

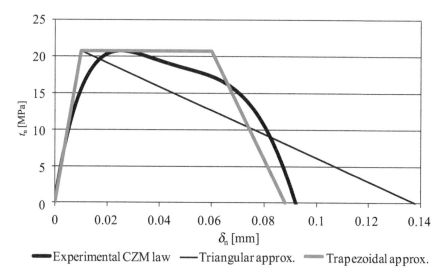

Experimental CZM law   —Triangular approx.   Trapezoidal approx.

FIG. 7.22 Experimental t_n–δ_n law for one test specimen.

Results – Configuration 2

G_{Ic} was calculated by equation (7.4). The experimental $G_{I}–\delta_{n}$ laws were identical in shape to Fig. 7.21, and an example for $t_{p} = 3$ mm is presented in Fig. 7.23.

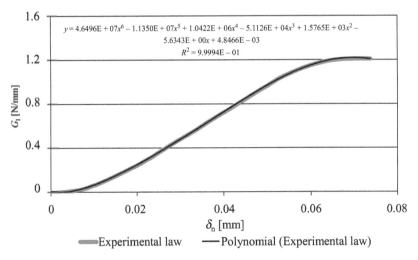

$$y = 4.6496E + 07x^6 - 1.1350E + 07x^5 + 1.0422E + 06x^4 - 5.1126E + 04x^3 + 1.5765E + 03x^2 - 5.6343E + 00x + 4.8466E - 03$$
$$R^2 = 9.9994E - 01$$

Experimental law   Polynomial (Experimental law)

FIG. 7.23 Experimental $G_{I}–\delta_{n}$ law for one test specimen with $t_{p} = 3$ mm and polynomial fitting curve (configuration 2).

The G_{Ic} results by applying this procedure for all tested specimens as a function of t_{p} are shown in Fig. 7.24. The deviation is somehow large, and whose justification lies on the experimental process to obtain G_{Ic}, which relies on a number of measured parameters and approximation functions, which are difficult to adjust to the experimental data (Campilho et al. 2013a). While for the specimens with $t_{p} = 1$ mm, a value of $G_{Ic} = 0.781 \pm 0.146$ N/mm was obtained, improvements of 12.6, 37.7 and 40.2% were attained by increasing h up to 4 mm. These results show the stabilization of G_{Ic} for a given value of t_{p} (in this case between $G_{Ic} = 1.075 \pm 0.226$ N/mm for $t_{p} = 3$ mm and $G_{Ic} = 1.095 \pm 0.195$ N/mm for $t_{p} = 4$ mm a stabilization of G_{Ic} was found).

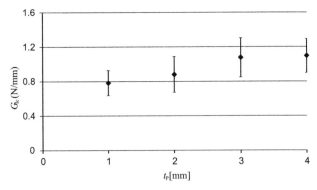

FIG. 7.24 Average values and deviation of G_{Ic} as a function of t_{p} by the J-integral.

This increase of G_{Ic} is reported in the literature because of the stress field variations ahead of the crack tip being dependent on the joint geometry, which highly influences the shape and size of the damage zone, and the local yield stress as well (Bell and Kinloch 1997). As it was discussed in previous works (Blackman et al. 2001), thicker adherends provide an elevation of peel stresses further within the joint, shifting the loading conditions from peeling to cleavage, and giving a larger length for the damage zone. These findings are corroborated in the work of Azari et al. (2012), regarding the adherend stiffness influence on the fatigue failure of bonded joints, which proved by FE that the plastic zone in adhesive joints between steel adherends was consistently higher than identical joints between aluminium adherends during the entire damage uptake process up to crack initiation. Mangalgiri et al. (1987) justified this tendency with the plastic zone and stress distributions ahead of the debond tip. Actually, the plastic zone was bigger in length across the adhesive layer with increasing number of composite plies (and thus, increasing t_p). Also, thicker adherends used a larger amount of the input energy to the specimen to develop a lengthier plastic zone, thus leaving less available energy for damage growth (Azari et al. 2013). On account of this, higher values of G_{Ic} can be expected for joints with higher degrees of restraint (i.e., stiffer or thicker adherends).

Results – Configuration 3

Following the method described in Section 7.4.6.1, G_{Ic} was calculated identically to the previous cases, which considered θ_o instead of θ_p to obtain G_I. The aforementioned method was applied to all tested specimens and the G_{Ic} results for each t_A value and respective deviation are presented in Fig. 7.25. For the specimens with $t_A = 0.1$ mm, the obtained results gave $G_{Ic} = 1.83 \pm 0.24$ N/mm. The increase of G_{Ic} from this point was of 14.5% ($t_A = 0.2$ mm), 57.8% ($t_A = 0.5$ mm), 105.6% ($t_A = 1.0$ mm) and 195.9% ($t_A = 2.0$ mm).

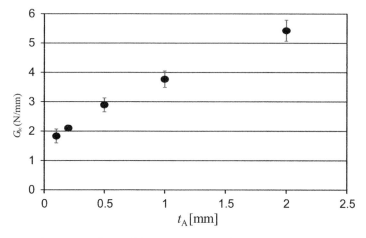

FIG. 7.25 Average values and deviation of G_{Ic} as a function of t_A by the J-integral.

Regarding the available studies (for epoxy adhesives), Yan et al. (2001) studied the influence of t_A on the fracture properties (G_{Ic}) of DCB and CT joints with aluminium adherends and a rubber-modified epoxy adhesive. Using a large deformation FE technique and the peak loads measured in the experiments, the critical value of the J-integral was calculated for different values of t_A. A G_{Ic} increase was found up to t_A = 1 mm and a decrease afterwards. An identical conclusion was found by Khoo and Kim (2011) for an epoxy adhesive between 0.2 < t_A < 1.5 mm, with the maximum G_{Ic} being found for t_A = 1 mm. The increasing trend of G_{Ic} with t_A obtained in the present work is linear up to t_A = 2.0 mm, and this result is consistent with previous studies in this matter, except from a reduction of G_{Ic} for big values of t_A that is common with less ductile epoxy adhesives. Another exception is the work of Marzi et al. (2011), which attained a maximum G_{Ic} between t_A = 1 and 2 mm for the polyurethane SikaPower® 498TM, a modern crash resistant epoxy adhesive, without a reduction tendency of G_{Ic} up to t_A = 2 mm, due to its large ductility. An identical trend to the presented results regarding the G_{Ic}–t_A law was found by Banea et al. (2014) with the high elongation polyurethane adhesive SikaForce® 7888, characterized with conventional fracture methods in the range of 0.2 ≤ t_A ≤ 2 mm. In both this and the present work, the peak value of G_{Ic} is attained for a t_A value bigger than 2 mm, but in this range of values the joints are more likely to have fabrication defects, be more difficult to fabricate, which justifies its limited industrial applicability.

Discussion of Results

The proposed technique, applied to the 3 joint configurations, showed that the proposed J-integral methodology can be a valuable tool to estimate G_{Ic} of adhesive joints. Moreover, with the measurement of δ_n, the cohesive law of the adhesive layer can be obtained as well. By analyzing the obtained results between the three tested configurations, a direct analogy cannot be formed between configuration 3 and configurations 1 and 2, because a different adhesive was considered (although both tested adhesives are ductile polyurethanes). In the comparison between configurations 1 and 2, it should be noted that, as depicted in Table 7.2, the value of t_p varied. This has a significant influence on the plastic zone size and, thus, also on the G_{Ic} measurements. The measured data for the bonded joint of configuration 1 gave G_{Ic} = 1.182 ± 0.215 N/mm (t_p = 5 mm), while for configuration 2 and t_p = 4 mm the value of G_{Ic} = 1.095 ± 0.195 N/mm was obtained. These values agree quite well, although the difference in bending stiffness of the adherends has to be considered: the values of t_p between these two configuration differ and, additionally, the value of E of the adherends for configuration 2 (aluminium) is much higher than that of configuration 1 (jute-epoxy composite). Since the results of configuration 2 show that, for aluminium adherends, for t_p values above 3 mm the plastic zone effect ceases to affect the results, the G_{Ic} measurement theoretically should be identical between configurations 1 and 2. In view of this discussion, the 7.4% different between these two configurations is attributed to experimental scatter and related issues.

7.4.6.6 Application of the Direct Method/Optical Algorithm to the ENF Specimen

A case study is presented regarding the estimation of G_{IIc} and shear CZM laws of bonded joints for three adhesives with distinct ductility, considering direct method and optical algorithm of Section 7.4.6.4.

Characterization of the Materials

The aluminium alloy AA6082 T651 was selected to prevent adherend plasticization, which would cancel the validity of the G_{IIc} results by plastic dissipations in the adherends. This aluminium alloy was previously characterized in Section 7.4.6.5. To evaluate the method's capability under different scenarios, either brittle/ductile failure or varying orders of magnitude in the measured quantities by the optical method, the following adhesives were used: the brittle epoxy Araldite® AV138, the ductile epoxy Araldite® 2015 and the ductile polyurethane SikaForce® 7752. A comprehensive mechanical and fracture characterization of these adhesives was recently undertaken in different studies (Campilho et al. 2011b, 2013a, 2013b, Faneco 2014). Bulk specimens loaded in tension enabled obtaining E, σ_y, σ_f and ε_f. The values of G_{Ic} and G_{IIc} were obtained by conventional data reduction schemes using the DCB and ENF tests, respectively. Table 7.3 presents the relevant mechanical and fracture data of the adhesives.

Table 7.3 Properties of the adhesives Araldite® AV138, Araldite® 2015 and SikaForce® 7752 (Campilho et al. 2011b, 2013a, 2013b, Faneco 2014).

Property	AV138	2015	7752
Young's modulus, E [GPa]	4.89 ± 0.81	1.85 ± 0.21	0.49 ± 0.09
Poisson's ratio, v	0.35^a	0.33^a	0.30^a
Tensile yield strength, σ_y [MPa]	36.49 ± 2.47	12.63 ± 0.61	3.24 ± 0.48
Tensile failure strength, σ_f [MPa]	39.45 ± 3.18	21.63 ± 1.61	11.48 ± 0.25
Tensile failure strain, ε_f [%]	1.21 ± 0.10	4.77 ± 0.15	19.18 ± 1.40
Shear modulus, G [GPa]	1.56 ± 0.01	0.56 ± 0.21	0.19 ± 0.01
Shear yield strength, τ_y [MPa]	25.1 ± 0.33	14.6 ± 1.3	5.16 ± 1.14
Shear failure strength, τ_f [MPa]	30.2 ± 0.40	17.9 ± 1.8	10.17 ± 0.64
Shear failure strain, γ_f [%]	7.8 ± 0.7	43.9 ± 3.4	54.82 ± 6.38
Toughness in tension, G_{Ic} [N/mm]	0.20^b	0.43 ± 0.02	2.36 ± 0.17
Toughness in shear, G_{IIc} [N/mm]	0.38^b	4.70 ± 0.34	5.41 ± 0.47

[a] manufacturer's data
[b] estimated in reference (Campilho et al. 2011b)

Joint Geometries

Figure 7.5b shows the characteristic geometry and dimensions of the ENF joints: $L = 100$ mm (in this case representing the mid-span), $a_0 \approx 60$ mm, $t_P = 3$ mm, $B =$

25 mm and $t_A = 0.2$ mm. The joints were assembled under controlled conditions of temperature and humidity. Before bonding, the adherends' faces to join were roughened by grit blasting followed by cleaning with acetone. Curing was carried out in a steel mould to assure the correct alignment between adherends and also the position of the calibrated spacers. Actually, calibrated steel spacers were inserted between the adherends to obtain a constant value of t_A in the adhesive bond, after applying demoulding agent to enable removal after curing. The crack tip spacer, which has to produce the pre-crack in the adhesive layer, was composed of a 0.1 mm thick razor blade between the steel spacers. Curing was performed at room temperature. Final set-up of the specimens was then undertaken with the removal of the steel spacers, painting the crack tip path with brittle white paint to better visualize a during the tests, and gluing a black numbered scale in both upper and lower adherends to aid the parameter extraction for the direct method. The tests comprised a total of twenty-four specimens (eight for each adhesive) and they were carried out at room temperature in a Shimadzu AG-X 100 testing machine equipped with a 100 kN load cell. For the required test documentation, an 18 MPixel digital camera was used, with no zoom and fixed focal distance to ≈100 mm. This procedure made possible obtaining a and δ_s, necessary for the application of the direct method. The correlation of the mentioned parameters with the P-δ data was done by the time elapsed since the beginning of each test.

Results

Figure 7.20 previously presented representative δ_s-testing time curves for the current case study. Figure 7.26 shows the G_{II}–δ_s curve for the same specimens of Fig. 7.20 and the selected polynomial approximations (the Araldite® AV138 curve is in secondary axis for clarity).

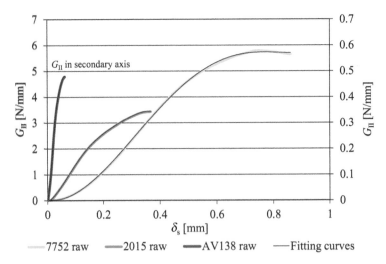

FIG. 7.26 Plot of G_{II}–δ_s for a specimen of each tested adhesive: raw curve and polynomial approximations.

The shape of these curves is consistent with published data in the literature (Ji et al. 2010, Campilho et al. 2013a). It can be found that, at the beginning of the test, G_{II} increases very slowly, but that the rate of improvement quickly increases and ultimately a steady-state value is attained. As previously mentioned, this last stage indicates the onset of crack growth and gives the G_{IIc} estimate. A clear difference is observed between the three adhesives regarding the range of G_{II} values up to attaining the steady-state value corresponding to G_{IIc}. Moreover, the horizontal span of the curves also shows the different in δ_s previously discussed for Fig. 7.20.

Table 7.4 Values of G_{IIc} [N/mm] for the three adhesives obtained by the J-integral.

Specimen	Araldite® AV138	Araldite® 2015	SikaForce® 7752
1	0.552	3.444	–
2	0.732	3.585	5.067
3	0.676	2.873	6.050
4	–	3.298	5.360
5	0.566	3.123	6.070
6	0.533	3.140	5.173
7	0.523	3.080	5.790
8	0.479	2.901	6.160
Average	0.580	3.181	5.667
Deviation	0.090	0.249	0.459

For the specimens depicted in the figures, the measured values of G_{IIc} [N/mm] are 0.479 (Araldite® AV138), 3.444 (Araldite® 2015) and 5.790 (SikaForce® 7752). The overall results for all specimens are presented in Table 7.4. These reveal a good repeatability, with percentile deviations under 10%, except for the Araldite® AV138. To apply the differentiation procedure represented by equation (7.9), polynomial functions were applied to the raw data of each specimen. Generally, it was possible to achieve a high degree of accuracy between 4th and 6th degree functions (R varied between 0.96 and 0.99). Figure 7.27 shows the full set of t_s–δ_s curves obtained by the direct method for the three adhesives.

The curves show identical t_s^0 values between the Araldite® AV138 and 2015, while the SikaForce® 7752 gives smaller values. The value of δ_s^f also differs in accordance with the previously mentioned difference discussed for Fig. 7.20. For all adhesives, a good agreement was found between curves regarding the initial stiffness of the curves, t_s^0, descending part of the curves and δ_s^f. The average and deviation of the cohesive parameters (with percentile deviation in parenthesis) were

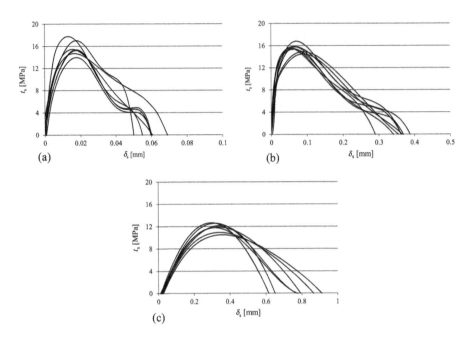

FIG. 7.27 Comparison of the full set of t_s–δ_s curves for each of the adhesives: Araldite® AV138 (a), Araldite® 2015 (b) and SikaForce® 7752 (c).

as follows. Araldite® AV138: $t_s^0 = 15.6 \pm 1.39$ MPa (8.9%), $\delta_s^0 = 0.0167 \pm 0.00163$ mm (9.80%) and $\delta_s^f = 0.0618 \pm 0.0214$ mm (34.6%), Araldite® 2015: $t_s^0 = 15.5 \pm 0.683$ MPa (4.4%), $\delta_s^0 = 0.0702 \pm 0.0122$ mm (17.4%) and $\delta_s^f = 0.372 \pm 0.0246$ mm (6.6%) and SikaForce® 7752: $t_s^0 = 11.8 \pm 0.807$ MPa (6.9%), $\delta_s^0 = 0.328 \pm 0.0182$ mm (5.5%) and $\delta_s^f = 0.781 \pm 0.107$ mm (13.7%). It can be considered that the scatter between specimens of the same adhesive is acceptable and shows a high correlation between specimens. The percentile deviation was generally under 10%, except a few cases in which δ_s^0 or δ_s^f exceeded this value. Figure 7.28 compares typical CZM laws for each adhesive and the parametrized CZM law that fits best each raw curve, either triangular or trapezoidal. The brittle Araldite® AV138 is best modelled by a triangular CZM law, while the other two adhesives are presented with more accuracy with a trapezoidal CZM law. However, the Araldite® 2015 could be modelled with a triangular CZM law as well with some level of accuracy to the raw curve. These results can be compared with the analysis of Constante et al. (2015) relating to the tensile CZM law estimation by the direct method considering the Araldite® AV138 and 2015, and a polyurethane with similar characteristics to the SikaForce® 7752 (the SikaForce® 7888). In fact, under tension, the Araldite® AV138 also reveals to be extremely brittle and a triangular law was proposed, while the ductility of the Araldite® 2015 and SikaForce® 7888 could be reproduced by a trapezoidal CZM law.

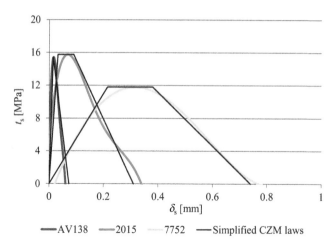

FIG. 7.28 Comparison of representative t_s–δ_s curves for each of the adhesives with simplified CZM laws.

7.4.6.7 Further Examples of Application

A few works currently exist on the direct parameter determination of adhesive bonds (Zhu et al. 2009) and fiber-reinforced composites (Sørensen and Jacobsen 2003). Andersson and Stigh (2004) used a direct method to determine the continuum CZM parameters in tension of a ductile adhesive bond of Dow Betamate® XW1044-3 in a DCB test configuration, after approximation of the G_1–δ_n data to a series of exponential functions to reduce errors in the measured data. The results showed that the t_n–δ_n relationship can be divided in three parts. Initially t_n increases proportionally to δ_n (linear elastic behaviour of the adhesive), until a limit stress is achieved. A steady-state region followed, corresponding to the development of plasticity in the adhesive. The curve ended with a parabolic softening region, giving an approximate trapezoidal shape.

Högberg and Stigh (2006) proposed a direct integration scheme to capture the CZM laws (continuum approach) of an adhesive bond by a mixed-mode DCB specimen, which consisted of the differentiation of the J-integral vs. δ_n (tension CZM law) or δ_s (shear CZM law). The J-integral was derived from the test data by a developed formulation that required the adherends rotation at the loading point, θ_p. Results showed that substantial differences exist between pure tensile and shear CZM laws, and that ten evenly distributed mode mixities can be used to build the mixed-mode constitutive behaviour of an adhesive system for any load and geometry conditions.

In the work of Carlberger and Stigh (2010), the continuum CZM laws of a thin bond of a ductile adhesive (Dow Betamate® XW1044-3) were determined in tension and shear using the DCB and ENF test configurations, respectively, considering $0.1 \leq t_A \leq 1.6$ mm. The values of G_I and G_{II} were derived by a J-integral formulation to accurately capture the large plastic straining effects present at the crack tip of the ductile adhesive. Actually, the J-integral is a viable means to capture the adhesive nonlinearity for a monotonic loading process, i.e., if no unloading occurs. The cohesive laws were derived by a direct method that used a least square adaption of a Prony-series to the G_I/G_{II} vs. δ_n/δ_s data, to avoid errors on the measured data resulting from direct differentiation of the experimental results. Although the J-integral solutions for this solution have their advantages, the respective formulae require additional data to the P-δ data available from the tests, compared to LEFM techniques. Actually, for G_I determination by the DCB test method, the values of P and θ_o or θ_p are necessary parameters to be defined as functions of δ_n. On the other hand, for shear characterization by the ENF test method, the simultaneous measurement of P and δ_s are required. With this data, by direct application of formulae derived in the authors work, the evolution of G_I and G_{II} (in the authors work addressed as J) with the deformation of the adhesive bond can be established (Fig. 7.29).

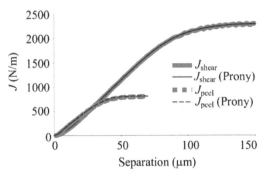

FIG. 7.29 G_I-δ_n and G_{II}-δ_s relations for the adhesive Dow Betamate® XW1044-3 and $t_A = 0.2$ mm (Carlberger and Stigh 2010).

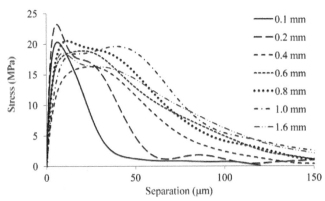

FIG. 7.30 CZM laws in tension for the adhesive Dow Betamate® XW1044-3 and $0.1 \le t_A \le 1.6$ mm (Carlberger and Stigh 2010).

Differentiation of the Prony-series adaptions of G_I or G_{II} relatively to the relative displacements, δ_n or δ_s, gives the full cohesive law up to failure, provided that the curves of Fig. 7.29 are obtained up to a steady-state value (Campilho et al. 2013a). Figure 7.30 shows the average CZM laws in tension for each value of t_A, considering 5 to 8 specimens for each condition. The obtained results clearly show the transition between approximate triangular CZM laws for small values of t_A to trapezoidal laws for bigger values. This is obviously related to an increase of G_{Ic}, because of larger FPZ developments in the adhesive bond (Duan et al. 2004, Choi et al. 2004).

Figure 7.31 represents the shear CZM laws for t_A values between 0.1 and 1.0 mm, averaged over at least two specimens. The reported results are qualitatively consistent with the tension data, i.e., a CZM shape dependence with t_A, with an approximate triangular shape for $t_A = 0.1$ mm and modification to a trapezoidal shape for bigger values of t_A. However, above $t_A = 0.2$ mm the value of G_{IIc} is virtually unaffected. The value of t_s^0 slightly diminished with t_A because of minor reductions of the effective strain rate with increasing values of t_A.

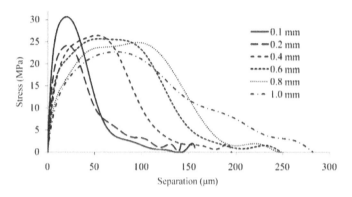

FIG. 7.31 CZM laws in shear for the adhesive Dow Betamate® XW1044-3 and $0.1 \leq t_A \leq 1.0$ mm (Carlberger and Stigh 2010).

It was thus concluded that the CZM shapes and respective parameters significantly vary with t_A, ranging from a rough triangular shape for the smaller values of t_A to a trapezoidal shape for bigger values of t_A.

The continuum CZM analysis of Ji et al. (2010) addressed the influence of t_A on t_n^0 and G_{Ic} for a brittle epoxy adhesive (Loctite® Hysol 9460), by using the DCB specimen and the direct method for parameter identification. The analysis methodology relied on the measurement of G_{Ic} by the analytical J-integral method proposed by Andersson and Stigh (2004), requiring the measurement of θ_p. Figure 7.32 shows the testing setup for the measurement of G_I.

FIG. 7.32 DCB specimen in the testing machine with inclinometers attached to the adherends and camera for measurement of θ_p and δ_n, respectively (Ji et al. 2010).

The testing machine was set to collect P, δ (with linear transducers placed at the crack tip) and, for the measurement of θ_p, two digital inclinometers with a 0.01° precision were attached at the free end of each adherend. A charge-coupled device (CCD) camera with a resolution of 3.7 × 3.7 μm/pixel was also used during the experiments, adjusted perpendicularly to one of the DCB specimen side edges. The camera focused at the crack tip and it collected images, which were input in an image processing toolkit for the estimation of δ_n at the crack tip, necessary for correlation with P and θ_p for the definition of G_I. The t_n–δ_n cohesive laws were quickly obtained by differentiation of the G_I–δ_n data. Figure 7.33 reports on a typical G_I–δ_n relationship for $t_A = 0.2$ mm and respective 6th degree polynomial function for differentiation and respective estimation of the CZM law.

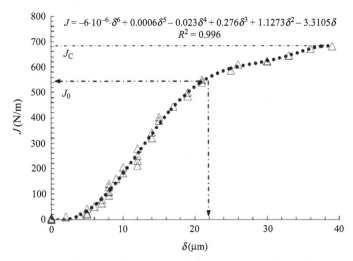

$$J = -6 \cdot 10^{-6} \cdot \delta^6 + 0.0006\delta^5 - 0.023\delta^4 + 0.276\delta^3 + 1.1273\delta^2 - 3.3105\delta$$
$$R^2 = 0.996$$

FIG. 7.33 Relationship between G_I and δ_n for $t_A = 0.2$ mm (Ji et al. 2010).

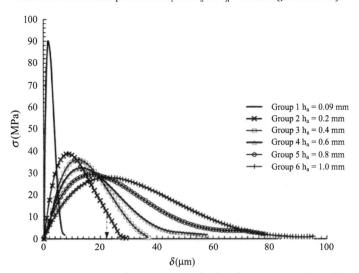

Group 1 $h_a = 0.09$ mm
Group 2 $h_a = 0.2$ mm
Group 3 $h_a = 0.4$ mm
Group 4 $h_a = 0.6$ mm
Group 5 $h_a = 0.8$ mm
Group 6 $h_a = 1.0$ mm

FIG. 7.34 Representative CZM law shapes in tension loading for $0.09 \leq t_A \leq 1.0$ mm (Ji et al. 2010).

Representative CZM laws for the adhesive layer are depicted in Fig. 7.34, showing some interesting variations, such as the previously mentioned increase of G_{Ic} with t_A, perceptible by the increase of area under the $t_n-\delta_n$ curve (Carlberger and Stigh 2010). A reduction of t_n^0 was also found with bigger values of t_A. Another important finding was related to the value of $t_n^0 = 88$ MPa found for $t_A = 0.09$ mm, compared to the bulk strength of 30.3 MPa. However, for increasing values of t_A the estimated t_n^0 values quickly approached the bulk strength of the adhesive (Fig. 7.34). The authors concluded that the J-integral method for the evaluation of G_{Ic} and the direct method for the CZM law calculation give accurate and calibrated CZM law parameters for specific geometry and material conditions.

7.5 CONCLUSIONS

This chapter performed an extensive review of the available parameter identification methods for cohesive parameters to be used in numerical simulations of bonded joints with selected examples of application, after a state-of-the-art review regarding strength prediction techniques for bonded joints. Different techniques are nowadays available for the definition of the cohesive parameters (G_{Ic}, G_{IIc}, t_n^0 and t_s^0), such as the property identification technique, the direct method and the inverse method. Notwithstanding the method, DCB and ENF tests are usually used to define some or all of the cohesive parameters. The property identification method revealed to be the most straightforward to apply, but it suffers from approximating some of the cohesive properties, usually t_n^0 and t_s^0, to the bulk values, which can lead to deviations to the real properties because of constraining effects by the adherends. The inverse method, although surpassing this limitation, relies on a fitting procedure, which can be difficult to perform. Both these methods assume parametrized CZM shapes, based on the previous knowledge of the adhesive behaviour. On the other hand, the direct method requires the experimental measurement of additional parameters when performing the fracture tests (typically DCB or ENF tests), but it has the major advantage of providing the exact shape of the CZM law that may or not be subsequently approximated to a parametrized shape for strength prediction purposes. Different case studies and related works were discussed for all three parameter identification methods, with emphasis on the direct method. This is undoubtedly the recommended method for strength prediction purposes on account of the more complete characterization of adhesive layers, accuracy and repeatability of the obtained results.

7.6 REFERENCES

527-2, ISO Standard. 2012. Plastics - Determination of tensile properties - Part 2: Test conditions for moulding and extrusion plastics, ISO.

Abaqus® 2009. Documentation. Vélizy-Villacoublay, Dassault Systèmes.

Adams, R. D. and Peppiatt, N. A. 1974. Stress analysis of adhesive-bonded lap joints. The Journal of Strain Analysis for Engineering Design 9: 185-196.

Adams, R. D., Comyn, J. and Wake, W. C. 1997. Structural Adhesive Joints in Engineering. Chapman & Hall, London.

Andersson, T. and Stigh, U. 2004. The stress–elongation relation for an adhesive layer loaded in peel using equilibrium of energetic forces. International Journal of Solids and Structures 41: 413-434.

Ashcroft, I. A. 2011. Fatigue load conditions. pp. 845-874. *In*: L. F. M. da Silva, A. Öchsner and R. D. Adams (eds.). Handbook of Adhesion Technology. Springer, Berlin Heidelberg.

Azari, S., Ameli, A., Datla, N. V., Papini, M. and Spelt, J. K. 2012. Effect of substrate modulus on the fatigue behavior of adhesively bonded joints. Materials Science and Engineering: A 534: 594-602.

Azari, S., Ameli, A., Papini, M. and Spelt, J. K. 2013. Adherend thickness influence on fatigue behavior and fatigue failure prediction of adhesively bonded joints. Composites Part A: Applied Science and Manufacturing 48: 181-191.

Banea, M. D. and da Silva, L. F. M. 2009. Adhesively bonded joints in composite materials: An overview. Proceedings of the Institution of Mechanical Engineers, Part L: Journal of Materials Design and Applications 223: 1-18.

Banea, M. D., da Silva, L. F. M. and Campilho, R. D. S. G. 2010. Temperature dependence of the fracture toughness of adhesively bonded joints. Journal of Adhesion Science and Technology 24: 2011-2026.

Banea, M. D., da Silva, L. F. M. and Campilho, R. D. S. G. 2014. The effect of adhesive thickness on the mechanical behavior of a structural polyurethane adhesive. The Journal of Adhesion 91: 331-346.

Barenblatt, G. I. 1959. The formation of equilibrium cracks during brittle fracture. General ideas and hypothesis. Axisymmetrical cracks. Journal of Applied Mathematics and Mechanics 23: 622-636.

Bell, A. J. and Kinloch, A. J. 1997. The effect of the substrate material on the value of the adhesive fracture energy, Gc. Journal of Materials Science Letters 16: 1450-1453.

Biel, A. 2005. Constitutive behaviour and fracture toughness of an adhesive layer. Lic. Eng. Dissertation, Chalmers University of Technology.

Blackman, B. R. K., Kinloch, A. J. and Paraschi, M. 2001. The effect of the substrate material on the value of the adhesive fracture energy, Gc: Further considerations. Journal of Materials Science Letters 20: 265-267.

Blackman, B. R. K., Hadavinia, H., Kinloch, A. J. and Williams, J. G. 2003. The use of a cohesive zone model to study the fracture of fibre composites and adhesively-bonded joints. International Journal of Fracture 119: 25-46.

Blackman, B. R. K., Kinloch, A. J. and Paraschi, M. 2005. The determination of the mode II adhesive fracture resistance, G_{IIC}, of structural adhesive joints: an effective crack length approach. Engineering Fracture Mechanics 72: 877-897.

Campilho, R. D. S. G., de Moura, M. F. S. F. and Domingues, J. J. M. S. 2005. Modelling single and double-lap repairs on composite materials. Composites Science and Technology 65: 1948-1958.

Campilho, R. D. S. G., de Moura, M. F. S. F. and Domingues, J. J. M. S. 2008. Using a cohesive damage model to predict the tensile behaviour of CFRP single-strap repairs. International Journal of Solids and Structures 45: 1497-1512.

Campilho, R. D. S. G. 2009. Repair of composite and wood structures. PhD Dissertation, Engineering Faculty of Porto University.

Campilho, R. D. S. G., de Moura, M. F. S. F., Pinto, A. M. G., Morais, J. J. L. and Domingues, J. J. M. S. 2009a. Modelling the tensile fracture behaviour of CFRP scarf repairs. Composites Part B: Engineering 40: 149-157.

Campilho, R. D. S. G., de Moura, M. F. S. F., Ramantani, D. A., Morais, J. J. L. and Domingues, J. J. M. S. 2009b. Tensile behaviour of three-dimensional carbon-epoxy adhesively bonded single- and double-strap repairs. International Journal of Adhesion and Adhesives 29: 678-686.

Campilho, R. D. S. G., Banea, M. D., Chaves, F. J. P. and Silva, L. F. M. d. 2011a. eXtended Finite Element Method for fracture characterization of adhesive joints in pure mode I. Computational Materials Science 50: 1543-1549.

Campilho, R. D. S. G., Banea, M. D., Pinto, A. M. G., da Silva, L. F. M. and de Jesus, A. M. P. 2011b. Strength prediction of single- and double-lap joints by standard and extended finite element modelling. International Journal of Adhesion and Adhesives 31: 363-372.

Campilho, R. D. S. G., Moura, D. C., Gonçalves, D. J. S., da Silva, J. F. M. G., Banea, M. D. and da Silva, L. F. M. 2013a. Fracture toughness determination of adhesive and co-cured joints in natural fibre composites. Composites Part B: Engineering 50: 120-126.

Campilho, R. D. S. G., Banea, M. D., Neto, J. A. B. P. and da Silva, L. F. M. 2013b. Modelling adhesive joints with cohesive zone models: Effect of the cohesive law shape of the adhesive layer. International Journal of Adhesion and Adhesives 44: 48-56.

Campilho, R. D. S. G., Moura, D. C., Banea, M. D. and da Silva, L. F. M. 2014. Adherend thickness effect on the tensile fracture toughness of a structural adhesive using an optical data acquisition method. International Journal of Adhesion and Adhesives 53: 15-22.

Campilho, R. D. S. G., Moura, D. C., Banea, M. D. and da Silva, L. F. M. 2015. Adhesive thickness effects of a ductile adhesive by optical measurement techniques. International Journal of Adhesion and Adhesives 57: 125-132.

Carlberger, T. and Stigh, U. 2010. Influence of layer thickness on cohesive properties of an epoxy-based adhesive—An experimental study. The Journal of Adhesion 86: 816-835.

Carver, D. R. and Wooley, G. R. 1971. Stress concentration factors for bonded lap joints. Journal of Aircraft 8: 817-820.

Chai, H. 1992. Experimental evaluation of mixed-mode fracture in adhesive bonds. Experimental Mechanics 32: 296-303.

Chandra, N., Li, H., Shet, C. and Ghonem, H. 2002. Some issues in the application of cohesive zone models for metal–ceramic interfaces. International Journal of Solids and Structures 39: 2827-2855.

Chen, Z., Adams, R. D. and da Silva, L. F. M. 2011. Prediction of crack initiation and propagation of adhesive lap joints using an energy failure criterion. Engineering Fracture Mechanics 78: 990-1007.

Choi, J. Y., Kim, H. J., Lim, J. K. and Mai, Y. W. 2004. Numerical analysis of adhesive thickness effect on fracture toughness in adhesive-bonded joints. Key Engineering Materials 270-273: 1200-1205.

Clough, R. W. 1960. The Finite Element Method in Plane Stress Analysis. Second ASCE Conference on Electronic Computation, Pittsburgh.

Constante, C. J., Campilho, R. D. S. G. and Moura, D. C. 2015. Tensile fracture characterization of adhesive joints by standard and optical techniques. Engineering Fracture Mechanics 136: 292-304.

Courant, R. 1943. Variational methods for the solution of problems of equilibrium and vibrations. Bulletin of the American Mathematical Society 49: 1-23.

da Silva, L. F. M. and Campilho, R. D. S. G. 2011. Advances in Numerical Modelling of Adhesive Joints. Springer, Heidelberg.

da Silva, L. F. M., Öchsner, A. and Adams, R. D. (Eds.). 2011. Handbook of Adhesion Technology. Springer, Heidelberg.

de Moura, M. F. S. F., Campilho, R. D. S. G. and Gonçalves, J. P. M. 2008. Crack equivalent concept applied to the fracture characterization of bonded joints under pure mode I loading. Composites Science and Technology 68: 2224-2230.

de Moura, M. F. S. F., Campilho, R. D. S. G. and Gonçalves, J. P. M. 2009. Pure mode II fracture characterization of composite bonded joints. International Journal of Solids and Structures 46: 1589-1595.

Duan, K., Hu, X. and Mai, Y.-W. 2004. Substrate constraint and adhesive thickness effects on fracture toughness of adhesive joints. Journal of Adhesion Science and Technology 18: 39-53.

Dugdale, D. S. 1960. Yielding of steel sheets containing slits. Journal of the Mechanics and Physics of Solids 8: 100-104.

Faneco, T. M. S. 2014. Caraterização das propriedades mecânicas de um adesivo estrutural de alta ductilidade. MsC Thesis, Instituto Superior de Engenharia do Porto.

Feraren, P. and Jensen, H. M. 2004. Cohesive zone modelling of interface fracture near flaws in adhesive joints. Engineering Fracture Mechanics 71: 2125-2142.

Flinn, B. D., Lo, C. S., Zok, F. W. and Evans, A. G. 1993. Fracture resistance characteristics of a metal-toughened ceramic. Journal of the American Ceramic Society 76: 369-375.

Herrera-Franco, P. J. and Valadez-González, A. 2005. A study of the mechanical properties of short natural-fiber reinforced composites. Composites Part B: Engineering 36: 597-608.

Högberg, J. L. and Stigh, U. 2006. Specimen proposals for mixed mode testing of adhesive layer. Engineering Fracture Mechanics 73: 2541-2556.

Hojo, M., Ando, T., Tanaka, M., Adachi, T., Ochiai, S. and Endo, Y. 2006. Modes I and II interlaminar fracture toughness and fatigue delamination of CF/epoxy laminates with self-same epoxy interleaf. International Journal of Fatigue 28: 1154-1165.

Ji, G., Ouyang, Z., Li, G., Ibekwe, S. and Pang, S.-S. 2010. Effects of adhesive thickness on global and local Mode-I interfacial fracture of bonded joints. International Journal of Solids and Structures 47: 2445-2458.

Kafkalidis, M. S. and Thouless, M. D. 2002. The effects of geometry and material properties on the fracture of single lap-shear joints. International Journal of Solids and Structures 39: 4367-4383.

Khoo, T. T. and Kim, H. 2011. Effect of bondline thickness on mixed-mode fracture of adhesively bonded joints. The Journal of Adhesion 87: 989-1019.

Kinloch, A. J. 1987. Adhesion and Adhesives: Science and Technology. Chapman & Hall, London.

Ku, H., Wang, H., Pattarachaiyakoop, N. and Trada, M. 2011. A review on the tensile properties of natural fiber reinforced polymer composites. Composites Part B: Engineering 42: 856-873.

Lee, D.-B., Ikeda, T., Miyazaki, N. and Choi, N.-S. 2004. Effect of bond thickness on the fracture toughness of adhesive joints. Journal of Engineering Materials and Technology 126: 14-18.

Lee, M. J., Cho, T. M., Kim, W. S., Lee, B. C. and Lee, J. J. 2010. Determination of cohesive parameters for a mixed-mode cohesive zone model. International Journal of Adhesion and Adhesives 30: 322-328.

Leffler, K., Alfredsson, K. S. and Stigh, U. 2007. Shear behaviour of adhesive layers. International Journal of Solids and Structures 44: 530-545.

Li, S., Thouless, M. D., Waas, A. M., Schroeder, J. A. and Zavattieri, P. D. 2005. Use of a cohesive-zone model to analyze the fracture of a fiber-reinforced polymer–matrix composite. Composites Science and Technology 65: 537-549.

Liljedahl, C. D. M., Crocombe, A. D., Wahab, M. A. and Ashcroft, I. A. 2006. Damage modelling of adhesively bonded joints. International Journal of Fracture 141: 147-161.

Mangalgiri, P. D., Johnson, W. S. and Everett, R. A. 1987. Effect of adherend thickness and mixed mode loading on debond growth in adhesively bonded composite joints. The Journal of Adhesion 23: 263-288.

Marzi, S., Biel, A. and Stigh, U. 2011. On experimental methods to investigate the effect of layer thickness on the fracture behavior of adhesively bonded joints. International Journal of Adhesion and Adhesives 31: 840-850.

Mello, A. W. and Liechti, K. M. 2004. The effect of self-assembled monolayers on interfacial fracture. Journal of Applied Mechanics 73: 860-870.

Neto, J. A. B. P., Campilho, R. D. S. G. and da Silva, L. F. M. 2012. Parametric study of adhesive joints with composites. International Journal of Adhesion and Adhesives 37: 96-101.

Pandya, K. C. and Williams, J. G. 2000. Measurement of cohesive zone parameters in tough polyethylene. Polymer Engineering & Science 40: 1765-1776.

Pardoen, T., Ferracin, T., Landis, C. M. and Delannay, F. 2005. Constraint effects in adhesive joint fracture. Journal of the Mechanics and Physics of Solids 53: 1951-1983.

Pereira, A. B. and de Morais, A. B. 2003. Strength of adhesively bonded stainless steel joints. International Journal of Adhesion and Adhesives 23: 315-322.

Pinto, A. M. G., Magalhães, A. G., Campilho, R. D. S. G., de Moura, M. F. S. F. and Baptista, A. P. M. 2009. Single-lap joints of similar and dissimilar adherends bonded with an acrylic adhesive. The Journal of Adhesion 85: 351-376.

Rice, J. R. 1968. A path independent integral and the approximate analysis of strain concentration by notches and cracks. Journal of Applied Mechanics 35: 379-386.

Schuecker, C. and Davidson, B. D. 2000. Effect of friction on the perceived mode II delamination toughness from three and four point bend end notched flexure tests. ASTM STP 1383: 334-344.

Shahin, K. and Taheri, F. 2008. The strain energy release rates in adhesively bonded balanced and unbalanced specimens and lap joints. International Journal of Solids and Structures 45: 6284-6300.

Song, S. H., Paulino, G. H. and Buttlar, W. G. 2006. A bilinear cohesive zone model tailored for fracture of asphalt concrete considering viscoelastic bulk material. Engineering Fracture Mechanics 73: 2829-2848.

Sørensen, B. F. and Jacobsen, T. K. 2003. Determination of cohesive laws by the J integral approach. Engineering Fracture Mechanics 70: 1841-1858.

Stigh, U., Alfredsson, K. S., Andersson, T., Biel, A., Carlberger, T. and Salomonsson, K. 2010. Some aspects of cohesive models and modelling with special application to strength of adhesive layers. International Journal of Fracture 165: 149-162.

Tsai, M. Y. and Morton, J. 1994. An evaluation of analytical and numerical solutions to the single-lap joint. International Journal of Solids and Structures 31: 2537-2563.

Turner, M. J., Clough, R. W., Martin, H. C. and Topp, J. L. 1956. Stiffness and deflection analysis of complex structures. Journal of the Aeronautical Sciences (Institute of the Aeronautical Sciences) 23: 805-823.

Turon, A., Dávila, C. G., Camanho, P. P. and Costa, J. 2007. An engineering solution for mesh size effects in the simulation of delamination using cohesive zone models. Engineering Fracture Mechanics 74: 1665-1682.

Wambua, P., Ivens, J. and Verpoest, I. 2003. Natural fibres: Can they replace glass in fibre reinforced plastics? Composites Science and Technology 63: 1259-1264.

Xie, D. and Waas, A. M. 2006. Discrete cohesive zone model for mixed-mode fracture using finite element analysis. Engineering Fracture Mechanics 73: 1783-1796.

Yan, C., Mai, Y.-W. and Ye, L. 2001. Effect of bond thickness on fracture behaviour in adhesive joints. The Journal of Adhesion 75: 27-44.

Yang, Q. D. and Thouless, M. D. 2001. Mixed-mode fracture analyses of plastically-deforming adhesive joints. International Journal of Fracture 110: 175-187.

Yang, Q. D., Thouless, M. D. and Ward, S. M. 2001. Elastic–plastic mode-II fracture of adhesive joints. International Journal of Solids and Structures 38: 3251-3262.

Yoshihara, H. 2007. Simple estimation of critical stress intensity factors of wood by tests with double cantilever beam and three-point end-notched flexure. Holzforschung 61: 182-189.

Zhao, X., Adams, R. D. and da Silva, L. F. M. 2011. Single lap joints with rounded adherend corners: Experimental results and strength prediction. Journal of Adhesion Science and Technology 25: 837-856.

Zhu, Y., Liechti, K. M. and Ravi-Chandar, K. 2009. Direct extraction of rate-dependent traction–separation laws for polyurea/steel interfaces. International Journal of Solids and Structures 46: 31-51.

Zienkiewicz, O. C. and Cheung, Y. K. 1967. The Finite Element Method in Structural and Continuum Mechanics. McGraw-Hill, London.

8

CHAPTER

Cohesive Zone Modelling for Fatigue Applications

A. Pirondi[1,*], G. Giuliese[1] and F. Moroni[2]

8.1 INTRODUCTION

Adhesively bonded metallic and/or composite material structures are nowadays present not only in the aerospace industry, but thanks to continuous performance improvement and cost reduction, also in many other industrial fields such as automotive, nautical and wind turbines. This in turn, requires extensive use of adhesive bonding and a more sophisticated capability to simulate and predict the strength of bonded connections where, for this purpose, analytical methods are being progressively integrated or replaced by finite element analysis (FEA). To ensure the safety of the resulting structures, it is imperative to understand their fatigue behavior. Thus, the rise of the application of adhesive bonding has gone hand in hand with the development of models capable to predict the fatigue life that is related to the initiation and propagation of defects, starting at the free edges of joining regions or other features such as through-thickness holes.

The cohesive zone model (CZM) has found wide acceptance as a tool for the simulation of debonding in adhesively bonded joints. This model is commonly used for the simulation of the quasi-static fracture problems, especially in the case of cracks in bonded joints and delamination in composites. The possibility to simulate the growth of a crack without any remeshing requirements and the relatively easy possibility to manipulate the constitutive law of the cohesive elements makes the cohesive zone model attractive also for the fatigue crack growth simulation.

The CZM approach is a finite element method used for crack propagation analysis. The idea for the cohesive model is based on the consideration that infinite

[1] Dipartimento di Ingegneria Industriale, Università di Parma, Parco Area delle Scienze 181/A, 43124 Parma, Italy.

[2] Centro Interdipartimentale SITEIA.PARMA, Università di Parma, Parco Area delle Scienze 181/A, 43124 Parma, Italy.

* Corresponding author: alessandro.pirondi@unipr.it

stresses at the crack tip are not realistic. Models to overcome this drawback have been introduced independently by Dugdale (1960) and Barenblatt (1962). Both authors divided the crack in two parts: one part of the crack surfaces, region I in Fig. 8.1, is stress free, the other part, region II, is loaded by cohesive stresses that represent the material resistance to fracture.

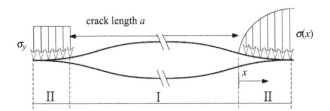

FIG. 8.1 Dugdale (left) and Barenblatt (right) crack models.

In Dugdale's model, the closure stress is the yield strength σ_y; in Barenblatt's model it represents the molecular force of cohesion and varies along the plastic zone (the stresses in the cohesive zone follow a prescribed distribution $\sigma(x)$, where x is the ligament coordinate). Most of the recently developed and proposed models are different from Barenblatt's model. In that, they define the traction acting on the ligament in dependency to the opening and not to the crack tip distance, as Barenblatt did. Needleman (1990) was the first, who used the model for crack propagation analyses of ductile materials. More than ten years earlier, Hillerborg et al. (1976) already applied the cohesive model to brittle fracture of concrete using the finite element method for the first time, followed in the 80's by Petersson (1981) and Carpinteri (1986), among others. In between, the cohesive model has been applied to almost any kind of materials, to many different size and time scales, from micro to macro and from impact to long term creep loading.

In Finite Element Method (FEM) models, the material separation and thus damage and failure of the structure is described by interface elements. Using this technique, the behavior of the material is split in two parts, the damage-free continuum with a given material law and the cohesive interfaces between the continuum elements, where material damage is specified. The interface elements open during loading and finally lose their stiffness such that the continuum elements are then disconnected. For this reason, the crack can propagate only along the path where cohesive elements are placed. If the crack propagation direction is not known in advance, the mesh generation has to make different crack paths possible.

The central point of all CZM is the function that describes the interaction force between the two interfaces (crack faces). This law represents a local material property that is independent of the external load. The so-called *cohesive law* or *separation law* is usually a relation between the boundary tractions σ and the separation $\delta_n = u_n^+ - u_n^-$ of the interfaces, i.e., the distance between the crack faces. Meanwhile, many cohesive laws exist in the literature, which differ according to various material and failure mechanism; for an overview see the review of Brocks et al. (2003). Some typical shapes are shown in Fig. 8.2.

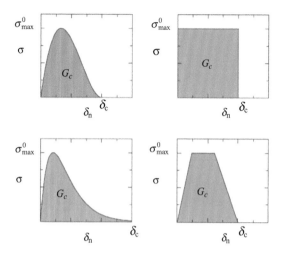

FIG. 8.2 Typical shapes of separation laws.

Initially, the stress increases with increasing opening up to a maximum called the *cohesive strength* σ_{max}^0 of the material. If the separation has reached a critical *decohesion opening* δ_c, then the material is completely separated and no stress can be transmitted. Notice that the area under the stress-separation curve corresponds to the dissipated work during a material's separation - the specific fracture energy per surface area G_C as introduced by Griffith (1920).

$$\Gamma = \int_0^{\delta_c^0} \sigma(\delta_n)d\delta_n = G_C \tag{8.1}$$

Thus, through Eq. (8.1) the relationship between CZM and classical fracture mechanics is established. The fracture process described above, implies that the part modelled with cohesive elements is loaded monotonically. In case of global unloading of the structure, it is necessary to define a behavior of the cohesive elements under decreasing separation which accounts for the irreversibility of the damage process. Since damage evolution is a nonlinear process like inelastic deformation, the cohesive models are established in analogy to the principles of plasticity, but allowing for strain softening. The terms "loading" and "unloading" will be used in the sense of increasing or decreasing separation, respectively, as the tractions decrease also under increasing separation beyond maximum stress, σ_{max}. More generally, "unloading" is any change of the deformation direction by which the stress state moves apart from the limiting traction-separation curve. This definition also applies for shear separation. Two principal mechanisms, depending on the material behavior, have to be distinguished:

- **Ductile unloading** The mechanical work for producing damage is totally dissipated, the void growth and hence the inelastic separation are irreversible and any reduction of separation occurs purely elastically with unchanged elastic stiffness, as shown in Fig. 8.3a.

- **Cleavage unloading** The elastic stiffness of the material is reduced by damage, but the separation vanishes when the stresses decrease to zero, as shown in Fig. 8.3b.

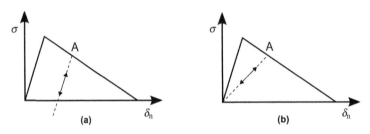

FIG. 8.3 "Ductile" (a) and "Cleavage" (b) unloading behavior.

Since the cohesive model is a phenomenological model, which can be used independently of the fracture mechanism, the two assumptions apply to different material behaviors. The first one is valid only for ductile crack propagation. Brittle fracture, which is characterized by micro cracks, would not be modelled correctly with a permanent separation on unloading, since the micro cracks close entirely (without getting back their stiffness). In this case the second unloading mechanism should be used.

CZM involves three important parameters: the critical separation δ_c, the cohesive strength σ_{max} and the cohesive energy Γ_0. It is usual to take Γ_0 and σ_{max} to be the constitutive parameters in the research. The cohesive parameters of the traction separation law in each fracture mode should be determined for application of this method. Both experimental and numerical methods are used for determining the cohesive parameters. The DCB test is typically used to obtain the opening Mode I fracture toughness G_{IC} of adhesives in bonded metal joints or fiber-reinforced composite materials. After determining the opening fracture toughness G_{IC}, σ_{max} can be obtained by fitting the experimental force-opening curve with FEA results.

8.2 FATIGUE CRACK PROPAGATION

8.2.1 Theoretical Framework

Fatigue fracture is one of the most occurring failure mode in engineering, where crack initiates, propagates, and finally results in the failure of components under cyclic loading. Fatigue of metallic materials develops as a result of alternating micro-plastic deformations (dislocations, slip bands), where microcracks are first formed on the surface or on microstructural in homogeneity (inclusion, grain boundaries) in the interior. Only above a crack length of about 7-10 grains diameter we call it a *crack*. From a macroscopic point of view, very small plastic deformations occur during fatigue crack growth (FCG). Therefore, this takes place in the K-controlled near-field, so Linear Elastic Fracture Mechanics (LEFM) is applicable (Fig. 8.4). If the crack initiation phase is negligible, lifetime prediction is based only on the crack propagation phase.

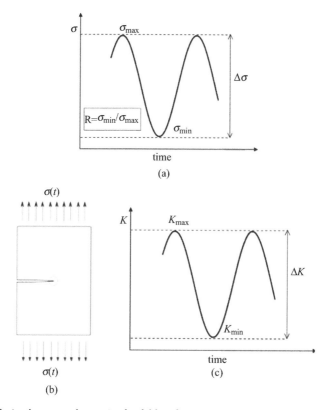

FIG. 8.4 Relation between alternating load (a) and stress intensity factors (c) for a component under fatigue loading (b).

An important parameter in FCG is the rate of fatigue crack propagation that is determined by subjecting fatigue-cracked specimens, like the compact specimen used in fracture toughness testing, to constant-amplitude cyclic loading. The increase in crack length a is recorded along with the corresponding number of elapsed load cycles N during the test (see Figs. 8.5 and 8.6).

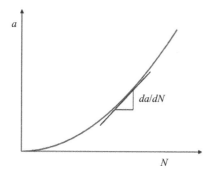

FIG. 8.5 a - N trend.

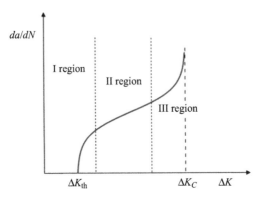

FIG. 8.6 *da/dN - ΔK* trend.

The *crack growth velocity* or *crack growth rate* is so defined as the ratio *da/dN* obtained by taking the derivative of the *a - N* curve (see Fig. 8.5). Plotting the experimentally determined crack velocity *da/dN* on a double logarithmic scale as a function of cyclic stress intensity factor ΔK, the typical behavior is shown in Fig. 8.6. The fatigue crack propagation curve, which has a sigmoidal shape, has three different regions: (I) threshold region defined by fatigue threshold ΔK_{th} below which no crack growth takes place, (II) linear or steady state crack growth region, which can be well described by the Paris' law (Paris and Erdogan 1963, Paris 1964), and (III) fast or unstable crack growth region where catastrophic failure takes place when the fracture toughness $\Delta K_C = K_C(1-R)$ is reached. The above mentioned Paris relation has been extensively used in the literature to relate crack growth rate to the fracture parameter ΔK:

$$\frac{da}{dN} = C(\Delta K)^n \qquad (8.2)$$

where C and n are empirical constants obtained by curve fitting of experimental data. The previous equation was modified and applied to fatigue delamination and debonding problems by Mostovoy and Ripling (1975), using the strain energy release rate ΔG as fracture parameter, to give:

$$\frac{da}{dN} = B(\Delta G)^d \qquad (8.3)$$

The exponent d and the coefficient B depend on the material, temperature, stress ratio $R = P_{min}/P_{max}$ of the cycle, and frequency (Russel and Street 1987, 1988). An accurate and efficient prediction of fatigue crack growth allows to adopt a "*damage-tolerant*" design philosophy, i.e. the component or structure may be safely operated even in the presence of some damage up to a limit value before a structure repair or replacement. In other words, a crack may grow in service, but it will not reach critical size before its detection. Essential ingredients of this approach are the knowledge of crack propagation as related with applied loading, and periodical inspections with a frequency ensuring that undetected damage in one inspection will not grow up to critical size before the next inspection. This approach leads to weight savings but also to increased maintenance costs, particularly those related with periodical inspections. Actually, a fail-safe design may not always possible and therefore the "*safe-life design*" approach must be used. This design philosophy is instead based on the intention of avoiding fatigue crack initiation during the entire lifetime.

8.2.2 CZM for Fatigue Crack Propagation

Numerical simulation offers nowadays outstanding possibilities for the prediction of the crack propagation process in debonding problems and has become an indispensable tool in performing this task. The mechanisms involved in both debonding and delamination are highly similar, and the prediction methods are therefore developed to deal with both of them.

CZM have been receiving increasing attention in this direction. In the CZM approach the interfaces, along which delaminations are expected to grow, are modelled using cohesive zone elements. These elements follow a prescribed traction-displacement relation. A damage parameter is used to progressively reduce the stiffness, simulating damage growth within the element. Thus the constitutive behavior of the cohesive element is generally defined as:

$$\sigma = K^0\delta \qquad \text{if} \quad 0 \leq \delta \leq \delta^0$$
$$\sigma = (1-D)K^0\delta \quad \text{if} \quad \delta^0 \leq \delta \leq \delta^c \qquad (8.4)$$
$$\sigma = 0 \qquad \text{if} \quad \delta^c \leq \delta$$

where δ is the current value of the relative displacement of the faces of the cohesive element, D is the damage parameter, δ^0 is the displacement at the onset of softening of the element, δ^c is the displacement at failure, K^0 is the stiffness and σ is the traction. Several attempts to establish the CZM approach to fatigue have been made.

According to Lemaitre et al. (1999), low-cycle fatigue occurs when the damage is localized in domains of stress concentrations but it can be measured and evaluated at the mesoscale. Low-cycle fatigue models account for fatigue damage evolution on a cycle-by-cycle basis defining an evolution of the damage variables during the unloading path. Within the context of CZM, there are several models that extend cohesive laws that were derived for monotonic loading into forms suitable for cyclic loading.

Foulk et al. (1998) simulated the interface failure under cyclic loading by adding an unloading condition to softening CZM. Unloading and subsequent reloading follow the same path, and consequently the traction-separation behavior stabilizes without further progress in material separation. de Andrés et al. (1999) took a similar approach, but added an unloading condition and a cycle dependent damage variable defined as:

$$D = \frac{\phi(\delta_{max})}{G_C} \qquad (8.5)$$

where δ_{max}, $\phi(\delta_{max})$, and G_C are the maximum separation attained during the cycle, corresponding dissipated energy to δ_{max}, and fracture energy, respectively. D ranges from 0 to 1, with these limits referring to an uncracked solid and a fully formed new fracture surface, respectively. As depicted in Fig. 8.7, the damage accumulation process is taken into account by a point on the cohesive zone at the crack tip.

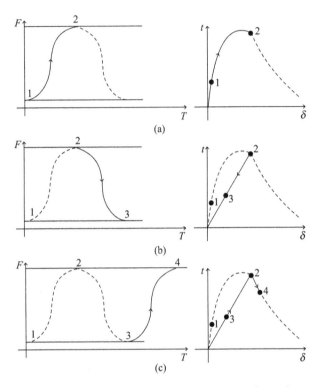

FIG. 8.7 Schematic representation of the process of damage accumulation during cyclic loading (de Andrés et al. 1999).

Suppose that loading leads to the increasing separation of cohesive surfaces. Upon unloading, the cohesive zone cannot close completely due to plastic deformation of the surrounded materials. Then the damage locus is reached again upon reloading and further damage accumulates. After sufficient loading cycles, the material in the cohesive zone will degenerate completely and form new fracture surfaces. The crack fronts of aluminum shafts subjected to axial loading have been predicted using this partial unloading-reloading configuration. However, in the work of de Andrés et al. (1999) the state of the specimen and thus the evolution of D was not an outcome of a cycle-by-cycle computation of FCG, while an extrapolation scheme was used to obtain estimates of D in dependence of elapsed number of cycles. This allows the damage length to be evaluated at a limited set of cycle numbers, without the need to find the intermediate behavior following in detail every loading cycle. A cohesive law with an unloading-loading hysteresis behavior was introduced by Nguyen et al. (2001) and Yang et al. (2001). Linear unloading combined with nonlinear reloading made it possible to take dissipative mechanisms into account, such as frictional interactions between asperities as well as crystallographic slip. Material degradation can accumulate below the limiting curve of the cohesive law for monotonic loading (the "damage locus") prior to failure due to the fact that unloading and reloading do not follow the same path, but show a hysteresis and thus dissipated energy (see Fig. 8.8).

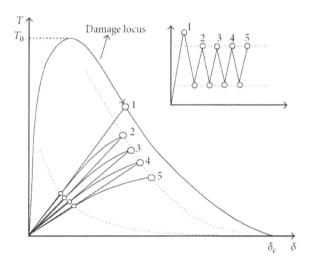

FIG. 8.8 Cohesive law with accumulating damage under cyclic loading (Liu et al. 2013).

The applied (global) loading range determines the upper and lower loading levels in the cohesive elements. Since damage evolution is a nonlinear process for inelastic deformation, CZM can be established in analogy to the principles of plasticity but allowing for strain softening. The well-known characteristics of typical elastic-plastic damage evolution laws include the following: (i) damage begins to accumulate once a deformation measure, accumulated or current, is greater than a critical magnitude; (ii) the increment of damage is related to the increment of deformation as weighted by the current load level; (iii) an endurance limit exists, which is a stress level below which cyclic loading can proceed infinitely without failure. Based on this consideration, Roe and Siegmund (2003) proposed the evolution equation for damage of the cohesive zone under cyclic sub-critical loads. Its increment form is written as

$$\Delta D_c = \frac{|\Delta\delta|}{\delta_\Sigma}\left[\frac{T}{\sigma_{\max}} - \frac{\sigma_f}{\sigma_{\max,0}}\right]H(\delta_n - \delta_n^0) \qquad (8.6)$$

with

$$\Delta D_c \geq 0 \qquad (8.7)$$

with H designating the Heaviside function. The present damage evolution law is formulated using effective cohesive zone quantities. $|\Delta\delta|$ and T are the incremental separation and the resultant traction, respectively. In the expression, two additional parameters are introduced, that is the cohesive zone endurance limit σ_f and the accumulated cohesive length δ_Σ, which determines the amount of accumulated effective separation necessary to fail the cohesive zone. δ_Σ is a multiple of the cohesive length δ_n^0, which is the material separation across the crack surfaces in the cohesive zone corresponding to the cohesive strength under normal loading. The magnitude of the incremental damage is then dependent on the two additional material parameters and proportional to the scaled and normalized incremental resultant separation, $|\Delta\delta|/\delta_\Sigma$, weighted by a measure of current traction reduced by the endurance limit. Then the damage is translated as the degradation of the cohesive properties in the CZM constitutive relation by

$$\sigma_{max} = \sigma_{max}^0 (1-D)$$
$$\tau_{max} = \tau_{max}^0 (1-D) \tag{8.8}$$

where σ_{max} and τ_{max} are the current cohesive normal and tangential strengths, respectively, used to substitute the initial ones. The unloading and reloading path in the investigation follow a linear relationship with a slope equal to that of the current traction-separation curve at zero separation. In the current model, the accumulated damage has been accounted for explicitly and incrementally. In almost all of these references, the fatigue damage accumulation is accounted for in a cycle-by-cycle analysis. For high-cycle fatigue, where the number of cycles is larger than 10^6, a cycle-by-cycle analysis would be computationally demanding.

According to Lemaitre et al. (1999), high-cycle fatigue occurs when the damage is localized at the microscale as a few micro-cracks. Since a cycle-by-cycle analysis is impractical, another strategy, called cycle-jump strategy, is usually adopted: damage evolution is extrapolated over a given number of cycles. In this way the computation of the whole load history is reduced to a certain number of cycles that are simulated blockwise. High-cycle fatigue models require the definition of the relation between the damage variable and the number of cycles as an input for the cycle-jump strategy (Fig. 8.9). This means also that numerically applied loads and numerically computed displacements are envelopes of the cyclic curves.

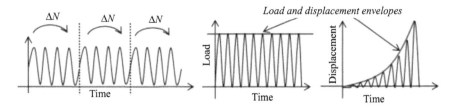

FIG. 8.9 Cycle Jump strategy (left), load and displacement envelopes (center and right).

The majority of the models relating the damage variable to the number of cycles use a phenomenological law established a priori and formulated as a function of the number of cycles. The damage evolution law is a function of several parameters that have to be adjusted to calibrate the numerical model with experimental results, usually by trial and error. An example of these models is the Peerlings' law (Peerlings et al. 2000) used to predict fatigue in metals. Peerlings' law has been adapted successfully to simulate high-cycle fatigue by means of an irreversible cohesive zone model (Robinson et al. 2005): it was proposed a damage parameter (D) that was split into two parts, a static (D_s) and a fatigue (D_f) part, respectively. Robinson et al. (2005) simulated the fatigue behavior by numerically applying a constant load equal to the maximum of the fatigue load, and treating the damage parameter and displacement as dependent on pseudo-time, as represented by the number of cycles. The static damage can be written in terms of the displacement as

$$D_s = \frac{\delta^c}{\delta_n} \frac{\delta - \delta^0}{\delta^c - \delta^0} \quad \text{for} \quad \delta^0 < \delta < \delta^c \tag{8.9}$$

with $D_s = 0$ when δ is below δ^0 and $D_s = 1$ otherwise. The change in static damage between N cycles and $N + \Delta N$ can then be computed from the respective displacements

$$\Delta D_s = \frac{\delta^0 \delta^c}{\delta^c - \delta^0} \left(\frac{1}{\delta(N)} - \frac{1}{\delta(N + \Delta N)} \right) \tag{8.10}$$

The fatigue damage is introduced in the following way, which was first proposed by Peerlings and then by Paas et al. (1993)

$$\dot{D}_f = \frac{\partial D_f}{\partial t} = A e^{\lambda D} \left(\frac{\delta}{\delta^c} \right)^\beta \frac{\dot{\delta}(t)}{\delta^c} \tag{8.11}$$

where β, λ and A are parameters which have to be determined so that the resulting crack growth is in agreement with the experimentally determined Paris law and δ/δ^c is a normalized displacement. Using this damage rate, fatigue damage can occur when the initial damage is zero and thus a crack can grow even in an initially undamaged interface. The fatigue damage after a number of cycles ΔN has elapsed can be found by integrating the fatigue damage rate over the respective number of cycles. This integration is performed numerically. A constant μ with $0 \leq \mu \leq 1$ is found, so that

$$\Delta D_f = \int_N^{N + \Delta N} \frac{A}{1 + \beta} e^{\lambda D(N)} \left(\frac{\delta_n(N)}{\delta_c} \right) dN = \Delta N \frac{\partial D_f(D_\mu, \delta_\mu)}{\partial N} \tag{8.12}$$

$$D_\mu = (1 - \mu) D(N) + \mu D(N + \Delta N)$$
$$\delta_\mu = (1 - \mu) \delta(N) + \mu \delta(N + \Delta N) \tag{8.13}$$

A value for μ of 0.7 was used in Eq. (8.13). Combining the previous equations, the damage evolution with respect to a cycle jump ΔN can thus be expressed as

$$D(N + \Delta N) = D(N) + \Delta D_s + \Delta D_f \tag{8.14}$$

This implicit formulation for $D(N + \Delta N)$ is approximated using a Newton-Raphson algorithm. In the model proposed by Robinson et al. (2005), a new set of parameters has to be determined for each mode-mix ratio in a manner similar to the way Blanco et al. (2004) related the Paris parameters B and d to the mode-mix. Tumino and Cappello (2007) modified this model by relating the model parameters A and β with the mode-mix. Muñoz et al. (2006), and Tumino and Cappello (2007) all presented comparisons between predictions produced with models derived from the one proposed by Robinson et al. (2005) and experimental data. In each case the same dataset is used and good agreement is shown. However this dataset was also used to find the required input parameters of the model. Thus the demonstrated agreement between the model and the experimental data is tautological. Until the model is compared to a new dataset it cannot be considered to be validated.

A different definition of damage parameter was proposed by Harper and Hallett (2008). In this model the static component of the damage parameter is defined as

$$D_s(\delta) = \frac{\delta - \delta^0}{\delta^c - \delta^0} \tag{8.15}$$

For the fatigue component, the model of Blanco et al. (2004) is used to predict the delamination growth rate. From this growth rate the matching fatigue damage zone size and fatigue damage growth rate are calculated. The fatigue component of the damage parameter follows from the fatigue damage growth rate. Such a linkage between CZM and fracture mechanics was originally proposed by Turon et al. (2007). The strain energy release rate (SERR) required as input for the Blanco model is calculated by integrating the traction–displacement curve of the cohesive element. Essentially, the model proposed by Harper and Hallett (2008) provides a more complex method of calculating the SERR, but is otherwise not much different from the model proposed by Blanco et al. (2004). Harper and Hallett (2008, 2010) present a comparison between their model and experimental data reported by Asp et al. (2001). However this data was also used to find the empirical parameters in the Blanco model (and thus in the Harper-Hallett model). Therefore this comparison cannot be regarded as a validation of the Harper-Hallett model, and one must conclude that a comparison with new experimental data remains necessary.

Khoramishad et al. (2010) proposed a damage parameter based on the maximum strain, according to:

$$\frac{\Delta D}{\Delta N} = \begin{cases} \zeta(\varepsilon_{max} - \varepsilon_{th})^\kappa & \varepsilon_{max} > \varepsilon_{th} \\ 0 & \varepsilon_{max} \leq \varepsilon_{th} \end{cases} \tag{8.16}$$

$$\varepsilon_{max} = \frac{\varepsilon_n}{2} + \sqrt{\left(\frac{\varepsilon_n}{2}\right)^2 + \left(\frac{\varepsilon_s}{2}\right)^2} \tag{8.17}$$

where ε_{max} is the maximum principal strain in the cohesive element, ε_{th} is the threshold strain, below which no fatigue damage occurs, ε_n is the normal component of the strain, ε_s is the shear component of the strain, and ζ and κ are material constants. The parameters ζ, κ and ε_{th} need to be calibrated against the experimental tests. The fatigue damage was modelled by degrading the bi-linear traction-separation response. As shown in Fig. 8.10, the fatigue damage variable was used to determine the degraded traction-separation response whereas the static damage parameter was utilized to define the material status within that traction-separation response.

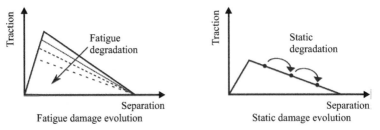

FIG. 8.10 Fatigue and static degradation of cohesive element properties (Khoramishad et al. 2010).

Khoramishad et al. (2010) produced a load-life diagram and predicted the back-face strain at the point where a strain gauge was applied during the experiments. The experimental and numerical results showed good agreement.

8.3 CZM FOR 2-D FATIGUE DEBONDING

8.3.1 Cohesive Law Accounting for Fatigue Damage

The model was developed by some of the authors (Pirondi and Moroni 2010, Moroni and Pirondi 2011, 2012) starting from the framework proposed by Turon et al. (2007). In the approach presented by Turon, the evolution of the damage variable associated with cyclic loading is derived from a Damage Mechanics description of the fatigue crack growth rate. Therefore, the proposed model links damage mechanics to fracture mechanics. Considering a representative surface element (RSE) (represented in the simulation by a cohesive element section) with a nominal surface equal to A_e, the accumulated damage can be related to the damage area due to micro voids or crack (A_d) (see Fig. 8.11) according to Lemaitre (1985)

$$D = \frac{A_d}{A_e} \tag{8.18}$$

In Turon et al. (2007), D is related to the ratio between the energy dissipated during the damage process and the critical energy release rate, G_c. Instead, in this work, D acts directly on stiffness, like in (Lemaitre 1985).

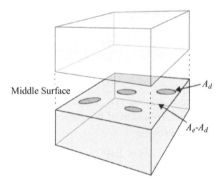

FIG. 8.11 Nominal and damaged area in a representative surface element.

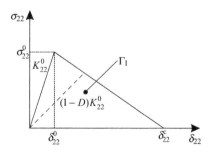

FIG. 8.12 Traction separation law in Mode I loading.

Referring to a mode I loading case, when the opening is relatively small the cohesive element behaves linearly; this happens until a given value of displacement, δ_{22}^0 (or equivalently until a certain value of stress σ_{22}^0). This initial step is characterized by a stiffness K_{22}^0 that remains constant until δ_{22}^0. Beyond this limit the stiffness is progressively reduced by D, until the final fracture in δ_{22}^c where the two surfaces are completely separated. Between δ_{22}^0 and δ_{22}^c the stiffness K_{22} can be computed as

$$K_{22} = K_{22,0}(1-D) \tag{8.19}$$

As mentioned earlier, the area Γ_1 underlying the cohesive law is the energy to make the defect grow of an area equal to the element cross-section and it is therefore representative of the fracture toughness, G_{IC}

$$\Gamma_I = \int_0^{\delta_c} \sigma_{22}\, d\delta_{22} \tag{8.20}$$

In the monotonic case, the damage variable D can be written as a function of the opening (δ_{22}) and of the damage initiation and critical opening (respectively δ_{22}^0 and δ_{22}^c)

$$D = \frac{\delta_{22,c}(\delta_{22} - \delta_{22,0})}{\delta_{22}(\delta_{22,c} - \delta_{22,0})} \tag{8.21}$$

When the element is unloaded, the damage cannot be healed, therefore, looking at Fig. 8.12, the unloading and subsequent loadings will follow the dashed line, until a further damage is attained. This simple model is able to describe the monotonic damage in case of mode I loading.

Considering the entire cohesive layer, the crack extension (A) can be computed as the sum of damaged areas of all the cohesive elements (A_d)

$$A = \Sigma A_d \tag{8.22}$$

When the fatigue damage is considered, from the previous equation, the crack growth (dA) can be written as a function of the increment of the damage area of all the cohesive elements (dA_d), therefore

$$dA = \Sigma dA_d \tag{8.23}$$

However, the damage increment would not concern the whole cohesive layer, but it will be concentrated in a relatively small process zone close to the crack tip, A_{CZ}. In order to estimate the size of A_{CZ}, analytical relationships can be found in the literature (Harper and Hallett 2008), where the size per unit thickness is defined as the distance from the crack tip to the point where $\sigma_{22\,max}^0$ is attained. In this work, different definition and evaluation methods are proposed: A_{CZ} corresponds to the sum of the nominal sections of the cohesive elements where the difference in opening between the maximum and minimum load of the fatigue cycle, $\Delta\delta_{22} = \delta_{22,max} - \delta_{22,min}$, is higher than a threshold value $\Delta\delta_{22}^{th}$. The value $\Delta\delta_{22}^{th}$ is supposed to be the highest value of $\Delta\delta_{22}$ in the cohesive layer when ΔG in the simulation equals ΔG_{th} experimentally obtained by FCG tests. It has to be underlined that in this way FCG may take place even at $\delta_{22,max} \leq \delta_{22,0}$, which is a condition that should be accounted for since $\delta_{22,0}$ results from the calibration of cohesive zone on fracture tests and may not be representative of a threshold for FCG. The process zone size (A_{CZ}) has therefore to be evaluated by FEA while performing the FCG simulation but, on the other hand, does not need to be assumed from a theoretical model. Eq. (8.23) can be therefore rewritten as

$$dA = \sum_{i \in A_{CZ}} dA_d^i \tag{8.24}$$

where only the elements lying in the process zone, A_{CZ}, are considered. In order to represent the crack growth due to fatigue (dA/dN), the local damage of the cohesive elements (D) has to be related to the number of cycles (N). This is done using the equation

$$\frac{dD}{dN} = \frac{dD}{dA_d} \frac{dA_d}{dN} \tag{8.25}$$

The first part of Eq. (1.25) can be easily obtained deriving Eq. (1.18). Therefore

$$\frac{dD}{dA_d} = \frac{1}{A_e} \tag{8.26}$$

The process to obtain the second part is quite more complicated: the derivative of Eq. (8.24) with respect to the number of cycles is

$$\frac{dA}{dN} = \sum_{i \in A_{CZ}} dA_d^i \tag{8.27}$$

At this point an assumption is introduced: the increment of damage per cycle is supposed to be the same for all the elements lying in the process zone. Therefore the value dA_d/dN is assumed to be the average value of the damaged area growth rate dA_d^i/dN for all of the elements in the process zone. Hence the crack growth rate can be rewritten as

$$\frac{dA}{dN} = \sum_{i \in A_{CZ}} \frac{dA_d^i}{dN} = n_{CZ} \frac{dA_d}{dN} \tag{8.28}$$

where n_{CZ} is the number of integration points (IP) lying on the process A_{CZ}. In FEA, n_{CZ} can be written as the ratio between the process zone extension (A_{CZ}) and the nominal cross section area (A_e) leading to

$$\frac{dA}{dN} = \frac{A_{CZ}}{A_e} \frac{dA_d}{dN} \tag{8.29}$$

The second part of Eq. (8.25) can be therefore written as

$$\frac{dA_d}{dN} = \frac{dA}{dN} \frac{A_e}{A_{CZ}} \tag{8.30}$$

The damage evolution rate can be finally expressed as a function of the applied strain energy release rate (ΔG), in the simplest version using Eq. (8.3)

$$\frac{dD}{dN} = \frac{1}{A_{CZ}} B\Delta G^d \tag{8.31}$$

8.3.2 Strain Energy Release Rate Computation

In the previous section, a relationship between the applied SERR and the increase of damage in the cohesive zone was defined. In order to simulate the fatigue crack growth, it is therefore required a general method to calculate the value of the SERR as a function of crack length. The most common methods for the SERR evaluation using FEA are the contour integral J, (Rice and Rosengren 1968, Rice 1968) and the Virtual Crack Closure Technique (VCCT) (Rybicki and Kanninen 1977). These two methods are usually available in finite element softwares, but VCCT is intended in general as alternative to using cohesive elements and, additionally, the software used in this work (Abaqus®) did not output the contour integral for an integration path including cohesive element. Given an arbitrary counter-clockwise path (Π) surrounding the tip of a crack (see Fig. 8.13), the is J-integral given by

$$J = \int_{\Pi} \left(W dy - \vec{T} \frac{\partial \vec{u}}{\partial x} ds \right) \tag{8.32}$$

where W is the strain energy density defined as

$$W = \int \sigma_{ij} d\varepsilon_{ij} \tag{8.33}$$

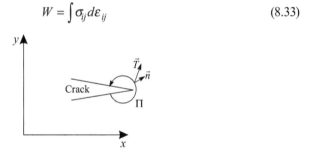

FIG. 8.13 The J contour integral.

Π is an arbitrary contour around the tip of the crack, $\vec{n}$ is the unit vector normal to Γ, $\vec{u}$ is the displacement field, $\vec{T}$; is the stress acting on the contour: $\vec{T} = \sigma \vec{n}$. Because of its path independence, the integral can be calculated in the remote field and characterizes also the near-tip situation.

In CZM the path Π is represented by the dashed line in Fig. 8.14 and it includes all the top and bottom nodes of the cohesive elements.

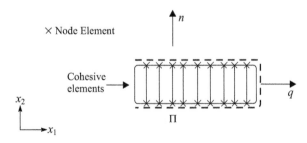

FIG. 8.14 J-integral.

According to Eq. (8.32), the strain energy density, W, the traction vector, $\vec{T}$, and the derivative of the displacement field with respect to x_1 are needed to compute J-integral. Eq. (8.32) in a two-dimensional (2D) plane stress case become

$$J = \int_{\Pi} \left(W dx_2 - \vec{T} \frac{\partial \vec{u}}{\partial x} ds \right) = \int_{\Pi} (\sigma_{12}\varepsilon_{12} + \sigma_{22}d\varepsilon_{22}) dx_2 - \int_{\Pi} (\sigma_{22}n_1 + \sigma_{12}n_2) \frac{\partial u_1}{\partial x_1} ds$$

(8.34)

where n_k is the outward unit vector normal to integration path and s is the path coordinate. If geometrical non linearity can be neglected, following a rectangular path around the cohesive element and considering that its vertical part (along x_2 direction) is far away from the crack tip, the Eq. (8.34) can be simplified in this way

$$J = \int_{\Pi} \left(-\sigma_{12} \frac{\partial u_1}{\partial x_1} - \sigma_{22} \frac{\partial u_2}{\partial x_1} \right) ds$$

(8.35)

Extracting the opening/sliding and the stresses in the cohesive elements at the beginning of the increment, the strain energy release rate is then computed. An interesting feature of this approach is that the mode I and the mode II component of the J-integral can be obtained by integrating separately the second or the first components of the integral in Eq. (8.35), respectively. This method can be easily implemented for a 2D problem, since there is only one possible path. In the case of a three dimensional (3D) problem, the implementation is more difficult since several paths can be identified along the crack width, and moreover their definition is rather troublesome, especially when dealing with irregular meshes.

8.3.3 Implementation in Abaqus®

The theoretical framework described in the previous section and the SERR calculation procedure are implemented using suitable Fortran subroutines in the finite element software Abaqus® (Abaqus® 6.11 Documentation). User-defined FORTRAN routines allow users to adapt Abaqus® to their particular analysis requirements. In particular the USDFLD subroutine is used here to modify the cohesive element stiffness by means of a field variable, which is the damage D. The URDFIL subroutine is used instead to get the result in terms of stresses, displacements and energies. Figure 8.15 shows the flow chart of the analysis from the start of an Abaqus®/Standard analysis to the end of an analysis step.

The values of stresses, displacements and energies are provided at the beginning of the increment; subroutine USDFLD must use the Abaqus® utility routine GETVRM to access this material point data. The flow diagram in Fig. 8.15 shows that the USDFLD is called at the start of the increment to define the properties of the cohesive elements. Subroutine URDFIL is used to read the results file at the end of an increment. Subroutine URDFIL must call the utility routine DBFILE to read records from the results file. Results are extracted from the results file, stored in COMMON blocks, and passed on to USDFLD.

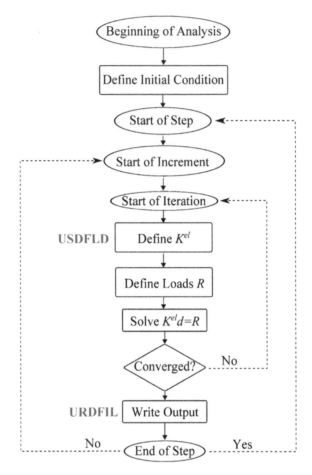

FIG. 8.15 Flow chart of the analysis in Abaqus®/Standard.

The fatigue analysis is carried out as a simple static analysis, divided in a number of increments. Each increment corresponds to a given number of cycles. Assuming that the fatigue cycle varies from a maximum value P_{max} to a minimum value P_{min}, the analysis is carried out applying to the model the maximum load P_{max}. The load ratio is defined as the ratio between the minimum and maximum load applied

$$R = \frac{P_{max}}{P_{min}} \tag{8.36}$$

The strain energy release rate amplitude is therefore

$$\Delta G = (1 - R^2) G_{max} \tag{8.37}$$

This latter is compared with the strain energy release rate threshold ΔG_{th}. If $\Delta G > \Delta G_{th}$ the analysis starts (or it continues is the increment is not the first); otherwise the analysis is stopped. The flow diagram in Fig. 8.16 shows the operations done within each increment.

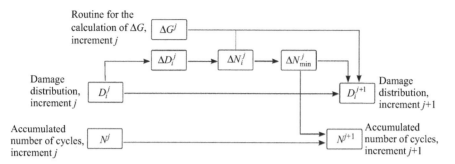

FIG. 8.16 Flow diagram of the procedure for the crack growth rate simulation.

At the beginning of the increment j the number of cycles (N_j) and the damage variable for each of the i-th element (D_i^j) are known. Now for each element the maximum possible damage change within the increment (ΔD_i^j) is computed. If ΔD_{max} is the maximum allowable variation in a single increment (it is a user defined value and it is used in order to ensure a smooth enough crack growth), ΔD_i^j is calculated as follows

$$\Delta D_i^j = \Delta D_{max} \quad \text{if} \quad 1 - D_i^j > \Delta D_{max}$$
$$\Delta D_i^j = 1 - D_i^j \quad \text{if} \quad 1 - D_i^j \leq \Delta D_{max}$$

$$(8.38)$$

In other words, ΔD_i^j is the minimum between ΔD_{max} and the amount needed for D to reach the unity. Therefore, for each element, the amount of number of cycles ΔN_i^j to produce ΔD_i^j is calculated by integrating Eq. (8.31) using ΔG evaluated within the URDFIL. The SERR is computed through Eq. (8.35) along a rectangular path extracting stresses and displacement of cohesive elements at the end of the current increment. After that, the routine searches for the minimum value among the calculated ΔN_i^j within the cohesive zone. This value, ΔN_{min}^j, is assumed to be the number of cycles of the increment, ΔN^j. Finally, the number of cycles is updated (N^{j+1}), and using Eq. (8.31) this time to calculate the ΔD_i^j corresponding to ΔN^j the new damage distribution ΔD_i^{j+1} is determined for all the elements belonging to the process zone. It is worth to underline that the procedure is fully automated, i.e. the simulation is performed in a unique run without stops.

8.3.4 FE Modelling

To simulate crack propagation in a bonded joint with the model described previously, the FEM model must contain the adherends and the interface layer where the crack has to increase its length. In Fig. 8.17 a typical specimen geometry is shown; depending on how loads and boundary conditions are set, different loading condition can be obtained.

For this type of simulation in 2D, plane stress elements are used for the adherends, while four-node 2D cohesive elements are used to model the adhesive layer (see Fig.

8.18). Plane stress elements can be used when the thickness of a body or domain is small relatively to its lateral (in-plane) dimensions. The stresses are functions of planar coordinates alone, and the out-of-plane normal and shear stresses are equal to zero. Plane stress elements must be defined in the X–Y plane, and all loading and deformation are also restricted to this plane. Instead, plane strain elements can be used when it can be assumed that the out-of-plane normal and shear strains are equal to zero. Plane strain elements must be defined in the X–Y plane, and all loading and deformation are also restricted to this plane. This modeling method is generally used for bodies that are very thick relatively to their lateral dimensions.

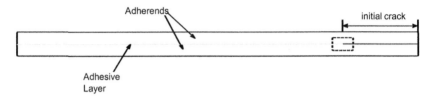

FIG. 8.17 Schematic representation of the adhesively bonded joint used in simulations.

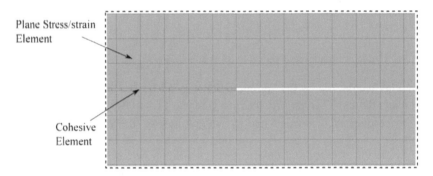

FIG. 8.18 Magnification of the initial crack tip of Fig. 8.17.

The cohesive zone must be discretized with a single layer of cohesive elements through the thickness. Since the cohesive zone represents an adhesive material with a finite thickness, the continuum macroscopic properties of this material are used directly for modeling the constitutive response of the cohesive zone.

Generally, for describing the constitutive behavior of the cohesive elements, the following features are needed:

- The initial stiffness (E_c) that relates the nominal stresses to the nominal strains across the interface. For taking into account the traction separation law that relates stress and displacements, the stiffness of the interface must be considered ($K_c = E_c/t_a$ with t_a the initial thickness of the layer);
- Damage initiation condition that refers to the beginning of degradation of the response of a material point. The process of degradation begins when the stresses and/or strains satisfy certain damage initiation criteria that one specifies. Several damage initiation criteria are available in Abaqus®;

- Damage evolution law that describes the rate at which the material stiffness is degraded once the corresponding initiation criterion is reached. Normally damage evolution can be defined based on the energy that is dissipated as a result of the damage process, also called the fracture energy. The user specifies the fracture energy as a material property and chooses either a linear or an exponential softening behavior. Abaqus® ensures that the area under the linear or the exponential damaged response is equal to the fracture energy. The dependence of the fracture energy on the mode mix can be specified either directly in tabular form or by using analytical forms provided in Abaqus®.

The damage initiation and its evolution are in this case directly implemented in USDFLD and URDFIL subroutines. The damage values (D_i^j), evaluated at the end of increment j, are used in the following increment as field variable in USDFLD routine to reduce the stiffness of the cohesive elements lying in the process zone.

8.3.5 Mixed Mode Loading

In order to extend the model to mixed-mode I/II conditions, a mixed mode cohesive law has to be defined. This is allowed by the knowledge of the pure mode I and pure mode II cohesive laws (index 22, refers to opening or mode I direction, index 12 refers to sliding or mode II direction).

First of all the mixed mode equivalent opening has to be defined. This is done using the relationship

$$\delta_{eq} = \sqrt{\left(\frac{\delta_{22} + |\delta_{22}|}{2}\right)^2 + (\delta_{12})^2} \tag{8.39}$$

In case of pure mode I this equation gives as δ_{eq}, the value of δ_{22} in case of positive δ_{22}, while it gives 0 in case of negative δ_{22}. This is done since it is supposed that compression stresses do not lead to damage of the adhesive layer. Of course δ_{22} assumes only positive values if crack surface penetration is properly prevented in the model. Moreover the mixed mode cohesive law is defined in terms of the initial stiffness (K_{eq}^0), damage initiation equivalent opening (δ_m^0) and critical equivalent opening (δ_c^0).

The equivalent initial stiffness is obtained by equating the equivalent strain energy (U_{EQ}) to the total strain energy (U_{TOT}), which in turn is equal to the sum of the strain energy in mode I (U_{22}) and in mode II (U_{12})

$$U_{EQ} = U_{TOT} = U_{22} + U_{12} = 1/2(\delta_{22} + |\delta_{22}|)^2 K_{22}^0 + 1/2\delta_{12}^2 K_{12}^0 \tag{8.40}$$

where K_{22}^0 and K_{12}^0 represent the initial stiffnesses of the mode I and mode II cohesive laws, respectively.

A further relationship is needed to define damage initiation: this is done using a quadratic failure criterion (Ungsuwarungsru and Knauss 1987)

$$\left(\frac{\sigma_{22}}{\sigma_{22\,max}^0}\right)^2 + \left(\frac{\sigma_{12}}{\sigma_{12\,max}^0}\right)^2 = 1 \tag{8.41}$$

The last relationship needed, regards the definition of the critical equivalent opening. Since the area underlying the cohesive law is representative of the critical strain energy release rate, using the KB theory (Kenane and Benzeggagh 1996) the area underlying the mixed mode equivalent cohesive law (Γ_{eq}) can be computed as

$$\Gamma_{eq} = \Gamma_I + (\Gamma_{II} - \Gamma_I) \cdot MM^\eta \tag{8.42}$$

where Γ_I and Γ_{II} are the areas underling the mode I and mode II cohesive laws, respectively, η is a mixed mode coefficient depending on the adhesive and MM is the mixed mode ratio defined as a function of the mode I and mode II SERR as follows:

$$MM = \frac{G_{II}}{G_{II} + G_I} \tag{8.43}$$

The KB mixed mode fatigue crack propagation model is the first considered, since it is the most general law that can be found in the literature. The fatigue crack growth rate is given by Eq. (8.3) where this time B and d are functions of the mixed mode ratio (MM)

$$d = d_I + (d_{II} - d_I) \cdot MM^{n_d} \tag{8.44}$$
$$\ln B = \ln B_{II} + (\ln B_I - \ln B_{II}) \cdot (1 - MM^{n_B}) \tag{8.45}$$

and d_I, B_I and d_{II}, B_{II} are, respectively, the parameters of the Paris law in mode I and mode II and n_d, n_B are material parameters.

8.3.6 FCG Simulations

The 2D model has been tested on various joint geometries characterized by varying mixed mode ratios, in order to verify accuracy, robustness and performance in terms of computational time. In particular, pure mode I, mode II and mixed-mode loaded cracks in woven composite joints have been simulated. The results obtained have been compared with VCCT fatigue simulations provided by Andrea Bernasconi and Azhar Jamil from the "Politecnico di Milano", Milano, Italy. These results were jointly published in Pirondi et al. (2014).

In particular the following geometries have been simulated with both methods:

- Double Cantilever Beam (DCB) geometry to test pure mode I loading;
- End Loaded Split (ELS) geometry to test pure mode II loading;
- Mixed Mode End Loaded Split (MMELS) geometry to test mixed mode I/II loading.

Additionally, a Single-Lap Joint (SLJ) has been modelled as a representative case of a real joint geometry. The propagation of the crack in the SLJ was allowed only on one side to simplify the comparison of the models results. All the specimen geometries are schematically described in Fig. 8.19, while in Table 8.1 the applied load and the specimens' dimensions are summarized.

The elastic properties of composite laminate are taken from Bernasconi et al. (2013) (see Table 8.2) while the cohesive law and FCG behavior is taken from Turon et al. (2007) (see Table 8.3). In all simulations a load ratio $R = 0$ is assumed.

The element type and mesh size are reported in Table 8.4, which represent a good balance between convergence on strain release rate and computational cost. Maximum damage increment, ΔD_{max} has been set to 0.2 value.

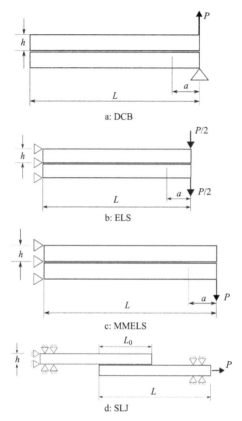

FIG. 8.19 Specimen geometries.

Table 8.1 Specimens dimension and applied load.

	DCB	**ELS**	**MMELS**	**SLJ**
P [N/mm]	10	20	15	200
a_0 [mm]	20	20	20	/
h [mm]	5	5	5	10.56
L [mm]	175	175	175	285.8
L_0 [mm]	/	/	/	110.8

Table 8.2 Engineering constants of the woven composite plies.

Parameter	**Value**
E_{11} [MPa]	54000
E_{22} [MPa]	8000
v	0.25
G_{12} [MPa]	285.8

Table 8.3 Cohesive zone parameters and FCG behavior for Mode I, Mode II, and Mixed-Mode I/II.

Parameter	Mode I	Mode II
Γ_0[N/mm]	0.266	1.002
σ_{max}[MPa]	30	30
δ^0[mm]	0.003	0.003
δ^c[mm]	0.0173	0.066
B	0.0616	4.23
d	5.4	4.5
Parameter		**Value**
η		2.6
n_d		1.85
n_B		0.35

Table 8.4 Element types and mesh sizes.

	Composite Laminate		Cohesive Zone	
	Element type	Size	Element type	Size
DCB	4-node bilinear plane stress quadrilateral, reduced integration	0.5 mm	4-node 2D cohesive element	0.2 mm
ELS	4-node bilinear plane stress quadrilateral, reduced integration	0.5 mm	4-node 2D cohesive element	0.5 mm
MMELS	4-node bilinear plane stress quadrilateral, reduced integration	0.5 mm	4-node 2D cohesive element	0.2 mm
SLJ	4-node bilinear plane stress quadrilateral	0.1 mm (next to cohesive elements)	4-node 2D cohesive element	0.1 mm

An initial crack length of 0.1 mm (1 element) has been specified for the SLJ when simulated using VCCT, while no initial crack length was needed in the case of CZM. The increment in crack length is fixed in the case of VCCT, i.e. equal to element size along the delamination/debonding interface (0.1 mm for the SLJ, 0.5 mm elsewhere), while in the case of CZM it comes as a result of the increment in damage ΔD. Therefore the increment in crack length is not generally constant as ΔD may vary from increment to increment according to Eq. (8.38). However, the average increment in crack length in the case of CZM ranged from 0.1 to 0.5 mm in the various cases simulated in this work.

The two methods are compared with respect to:

1. Agreement with each other;
2. Agreement with numerical integration of Eq. (8.3);
3. Calculation time.

Concerning the second point, the numerical integration was done using ΔG as a function of crack length coming from the FEA simulations. As ΔG is known by

FEA analysis, the trapezoidal rule (i.e. using the mean ΔG over the increment) was used. In this way, a closer estimate of the number of cycles at failure should be obtained with respect to both the CZM and VCCT where, for numerical reasons, ΔG at the beginning of the increment is used. As Eq. (8.3) represents the best fit of experimental data, the level of agreement between the number of cycles output by the models and the numerical integration of Eq. (8.3) represents also the level of agreement between experimental data and the simulations. Regarding the third point, it is the time the analyst has to wait for the crack to reach the knee of the *a-N* diagram, that is close to fracture. In the cases studied here this means a crack length of 40 mm for all the geometries except SLJ, for which the analysis have stopped at 40 mm of crack length even though still far from fracture. Only the outputs strictly necessary for each model were required, in order to minimize time spent in storing data. The PC used for calculations is an Athlon X2 Dual Core 2 GHz CPU, with 2 Gb RAM and 200 Gb HD (7200 rpm, 8 Mb cache).

8.3.7 Results

The crack growth in a cracked DCB specimen is simulated with CZM. Every increment correspond to a ΔN_j as described in Eq. 8.33. The joint deformed shape and damage distribution of the cohesive elements near the crack tip is reported in Figs. 8.20 and 8.21 for two different crack lengths.

a: Deformed shape of the DCB joint

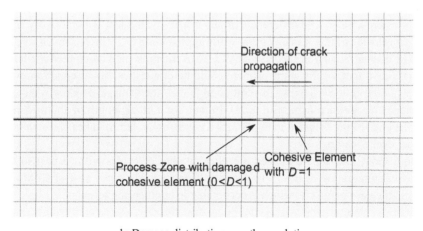

b: Damage distribution near the crack tip

FIG. 8.20 Fatigue simulation at *increment* 25 (1.28E+06 cycles).

a: Deformed shape of the DCB joint

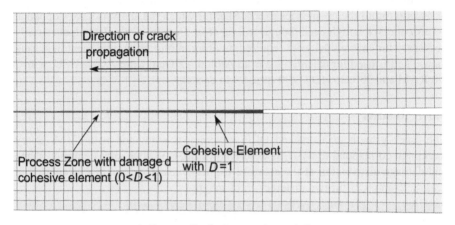

b: Damage distribution near the crack tip

FIG. 8.21 Fatigue simulation at *increment* 88 (1.85E+06 cycles).

The three sets show a very good correspondence with each other as expected, with only some small oscillation in the SERR calculated using the subroutine in the case of CZM (Fig. 8.22a).

The main result in terms of crack length versus number of cycles is shown in Fig. 8.22b, where a very little difference, of about 2.5%, is evident. Another very little difference is the gradient in a-N trend while approaching G_{Ic}, which is much steeper (almost discontinuous) in the case of VCCT. In the presented automated model, it is considered that locally some element could be in a critical stress condition with respect to the cohesive law; so in addition to the fatigue damage, eventually a static damage is determined using Eq. (8.21). In fact, for a crack length of about 30 mm (see the knee of the diagram in Fig. 8.22b) a catastrophic failure takes place without further cyclic loadings.

Both CZM and VCCT yielded a higher number of cycles with respect to the numerical integration of Eq. (8.3), with a difference 2.3% in the case of CZM and 1.8% in the case of VCCT, which is acceptable in engineering terms.

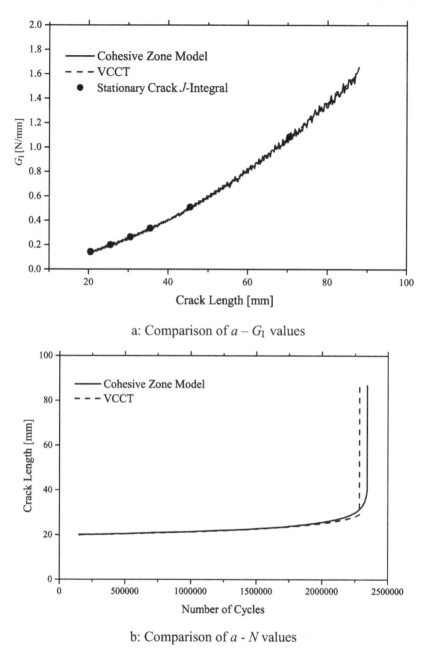

a: Comparison of $a - G_I$ values

b: Comparison of a - N values

FIG. 8.22 CZM 2D results compared with VCCT 2D ones (DCB case).

Concerning the ELS case, Fig. 8.23a shows the values of G_{II} obtained by CZM, VCCT and J-integral (stationary crack). The three sets show a very good correspondence with each other especially until 80 mm of crack length, while for longer cracks the CZM - G_{II} is slightly lower than the VCCT one and the J-integral lies in between. The main result in terms of crack length vs. number of cycles is shown in Fig. 8.23b, where a difference of about 11% comes out. The number of cycles at failure was obtained also by integrating Eq. (8.4) using the trapezoidal rule and ΔG as a function of crack length coming from the FEA simulations. Both CZM and VCCT yielded a higher number of cycles with respect to the numerical integration of Eq. (8.3), with a negligible difference both in the case of CZM and VCCT.

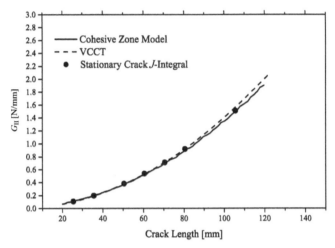

a: Comparison of $a - G_{II}$ values

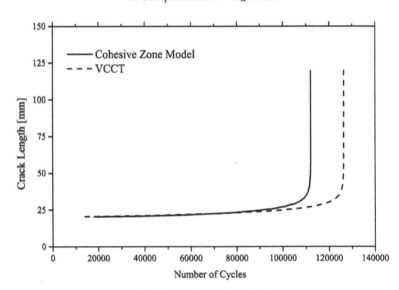

b: Comparison of a - N values

FIG. 8.23 CZM 2D results compared with VCCT 2D (ELS case).

Concerning the MMELS case, Fig. 8.24a shows the values of G_I and G_{II} obtained by CZM and VCCT. The values obtained with the two methods show a very good correspondence with each other in the case of the Mode I component, as for the DCB geometry. Under Mode II the agreement is good especially until 50 mm of crack length alike the ELS, while for longer cracks the CZM- G_{II} is lower than the VCCT one. The main result in terms of crack length vs. number of cycles is shown in Fig. 8.24b, where a very little difference of about 4% comes out. Both in the case of CZM and VCCT the numerical integration of Eq. (8.3) yielded a lower number of cycles, with a difference of 5.5% in the case of CZM and 8.3% in the case of VCCT, which may be still acceptable in engineering terms.

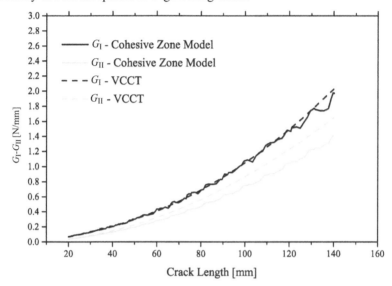

a: Comparison of $a - G_I/G_{II}$ values

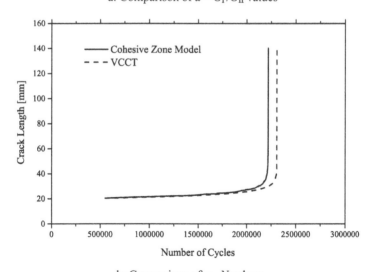

b: Comparison of a - N values

FIG. 8.24 CZM 2D results compared with VCCT 2D (MMELS case).

Regarding the SLJ simulation, Fig. 8.25a shows the values of G_I and G_{II} obtained by CZM and VCCT. The values obtained with the two methods show a good correspondence with each other in the case of both mode components in the first millimeters of propagation, while at longer cracks the CZM values are lower than the VCCT ones. A higher difference is noticed in the case of the Mode II component, someway similarly to Mode II and Mixed-Mode I/II loading. However, in those cases the difference in the number of cycles between the two models to failure was affected to a limited extent, while in the case of SLJ the discrepancy is much higher (see Fig. 8.25b). This discrepancy, however, is especially due to the fact that VCCT at present does not allow to modify the coefficient (B) and exponent (d) of Eq. (8.3) according to the mixed-mode ratio MM (see Eq. (8.44)-(8.45)) as CZM instead does. In the case of SLJ, the MM ratio increases steeply in the first 5 mm of propagation and then becomes almost stationary (ranges between 0.55 and 0.56, Fig. 8.26), and the VCCT simulation has been performed in this case using the stationary MM value. If the mixed-mode ratio MM versus the crack length is plotted for both approaches, a similar trend is obtained. In the VCCT case, the actual mixed-mode ratio is evaluated considering the SERR in Mode I and Mode II obtained from the simulations. Both in the case of CZM and VCCT the numerical integration of Eq. (8.3) yielded a lower number of cycles, with a difference (after 48 mm of crack propagation) of 2.6% in the case of CZM and 1.1% in the case of VCCT, which is absolutely acceptable in engineering terms.

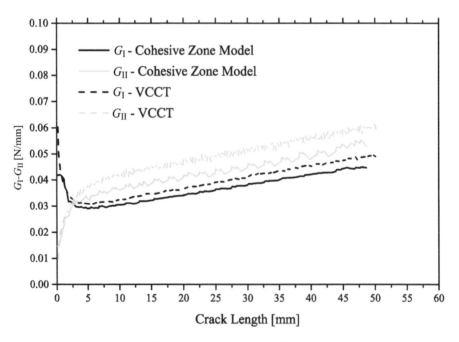

a: Comparison of $a - G_I/G_{II}$ values

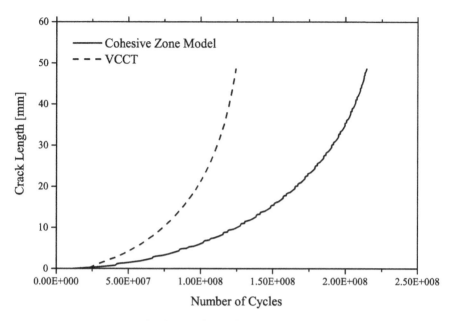

b: Comparison of a - N values

FIG. 8.25 CZM 2D results compared with VCCT 2D ones.

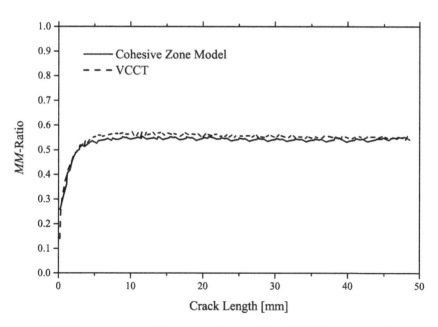

FIG. 8.26 Comparison of MM ratio obtained by CZM and VCCT in the case of SLJ.

8.3.8 Conclusions

The comparison of the performances of the CZM and the VCCT, embedded in the software Abaqus®, on mode I, mode II and mixed-mode I/II loaded cracks in composite assemblies has been carried out. The two models agree with each other to within 4% except in the case of SLJ, where VCCT at present does not allow to modify the coefficient (B) and exponent (d) of Eq. (8.3) according to the mixed-mode ratio MM as CZM does. Therefore, the rapid increase of MM in the first millimeters of propagation generates a large discrepancy between the two models. In this sense the CZM model offers an additional feature with respect to Abaqus® VCCT. Although the modeling effort is a bit higher (need of introducing a layer of cohesive elements), the CZM results in easier use. At the same time, it results more efficient as the computation is lower of about two orders of magnitude, even though the origin of this large difference in performance can be at least partly found in the Direct Cyclic procedure that is associated with VCCT in Abaqus®. The calculation times are reported in Table 8.5. In the case of SLJ, the increase in calculation time is related to the finer mesh, but the time required by VCCT is becoming so important that high performance computing may be needed if the model complexity would increase further.

Table 8.5 Computational time for both methods (time is reported in minutes).

	DCB	ELS	MMELS	SLJ
CZ	9.1	4.6	4.5	21.4
VCCT	676.2	688.6	727.5	2796.8

It is interesting to plot the stress of a cohesive element (mode I, DCB model) during the fatigue simulation in terms of stress and opening (see Figs. 8.27a, b). The approach described and its implementation allow cohesive elements to have stresses and opening within the traction separation law defined as input; but sometimes, during the fatigue simulation, the cohesive elements can be in stress-opening condition that does not respect the cohesive law. This is due to the fact that static and fatigue damage were assigned within the subroutine at the beginning of an increment. But at the end of the increment, the stresses coming out from the convergence iteration of the whole model can be a little higher than the damage locus. In this case, it means that the analysis would have required smaller maximum damage increments, ΔD_{max}.

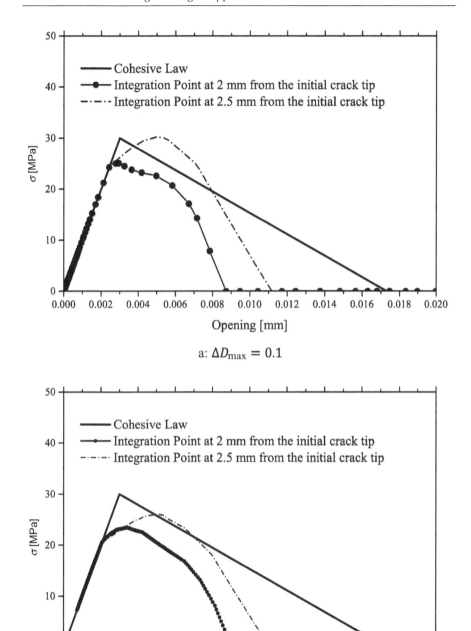

a: $\Delta D_{\max} = 0.1$

b: $\Delta D_{\max} = 0.01$

FIG. 8.27 *Stress-Opening* evolution during the fatigue simulation for two different IP.

8.4 CZM FOR 3-D FATIGUE DEBONDING (PLANAR CRACKS)

8.4.1 Model Implementation in 3D Problems

The approach presented for a 2D analysis can be easily extended to a 3D version for the case of planar crack geometries and regular cohesive mesh, for which Eq. (8.35) is evaluated on several parallel contours in order to obtain the J-integral along the crack front (Moroni et al. 2013, Giuliese et al. 2013).

Namely, the model is separated in slices (Fig. 8.28) where every slice corresponds to a row of cohesive elements in the direction of crack propagation. The COH3D8 element, used to mesh the cohesive zone, has eight nodes and four IP, as shown in Fig. 8.29. Thus, the routine, for each cohesive element lying on the same path, extracts stresses and openings at all the IP; then for evaluating the J-integral, stresses and displacements averaged value at the centroid are used.

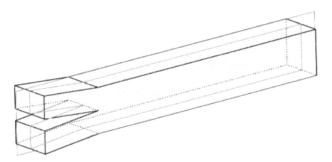

FIG. 8.28 Contour evaluation ("path") in a 3D specimen.

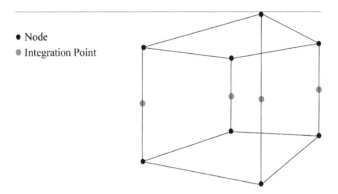

● Node
● Integration Point

FIG. 8.29 COH3D8 element.

The flow diagram, shown in Fig. 8.16, is still representative for this 3D modelling approach. The routine calculates for each IP the number of cycles ΔN_i^j to produce ΔD_i^j integrating Eq. (8.31), and searches for the minimum value among the calculated ΔN_i^j within all the path of the cohesive zone. This value, ΔN_{min}^j, is assumed to be the number of cycles of the increment, ΔN_i^j. The updated damage distribution is determined similarly to the procedure described for the 2D case.

8.4.2 Finite Elements Models

The CZM fatigue debonding model was tested on various joint geometries characterized by different mixed mode ratios, in order to verify the accuracy, robustness and performance in terms of computational time. In particular, pure mode I loading was simulated with a DCB geometry, pure mode II loading with an ELS geometry and mixed mode I/II loading with a MMELS geometry and a SLJ (see Fig. 8.19). The applied force and the specimens' dimensions are given in Table 8.6.

Table 8.6 Specimens dimension and applied load.

	Aluminum Joints			Composite Joints		
	DCB	ELS	MMELS	DCB	ELS	MMELS
P [N/mm]	10	25	20	10	20	15
a_0 [mm]	20	20	20	20	20	20
H [mm]	5	5	5	5	5	5
L [mm]	175	175	175	175	175	175
b^* [mm]	20	20	20	20	20	20

b^*: specimens width

In order to investigate the model sensitivity to material behavior, two kind of materials were simulated, one representing an elastic, isotropic aluminum alloy (E = 70 GPa; v = 0.3) and another one an elastic, orthotropic composite laminate (see Table 8.2). In all the simulations a load ratio R = 0 is assumed. The adherends were meshed with 3D *Continuum Shell* elements. From a modeling point of view continuum shell elements look like 3D continuum solids, but their kinematic and constitutive behavior is similar to conventional shell elements. For example, conventional shell elements have displacement and rotational degrees of freedom, while continuum solid elements and continuum shell elements have only displacement degrees of freedom. Continuum shell elements allow for thickness tapering, a more accurate contact modeling than conventional shells (they take into account two-sided contact and thickness changes) and stacking (they capture more accurately the through-thickness response for composite laminate structures). Thus a structured sweep mesh of continuum shell elements SC8R with an element size of 1 mm was introduced with a single element in the thickness.

The cohesive elements were kinematically tied to the two delaminating halves. The cohesive element size was 0.5 mm (CZM). The continuum shell and cohesive element dimensions have been chosen in order to keep computational time within a reasonable value. The maximum damage increment was taken ΔD_{max} = 0.2, based on the sensitivity analysis done in Pirondi and Moroni (2010).

The 3D model performance is compared with the 2D version concerning the value of G as a function of crack length. Due to crack front bowing (Fig 8.30), the average G (or G_I, G_{II}) and crack length along the crack front were considered for the comparison with the 2D model. In the case of elastic, isotropic material parts, analytical solutions for G (Mode I, Krenk 1992, Mode II, Wang and Williams 1992 and Mixed/Mode I/II, Blanco et al. 2006) were also introduced in the comparison; while in the case of the elastic, orthotropic composite only the value of G obtained by VCCT simulations, done at Politecnico di Milano, were considered. Additionally, Eq. (8.3) is integrated numerically using the G versus crack length coming from the analysis instead of taking directly the output number of cycles. The reason is that, as the CZM process zone needs some time to get to a steady state while VCCT does not, the G calculated by CZM may be rather different in the first millimeters of propagation yielding a different number of cycles. For this reason, a comparison with experiments is foreseen as a further validation step, while at the moment the authors focus on the comparison of numerical results of the two models after the transient phase of process zone formation.

8.4.3 Results

In Fig. 8.30 one of the two adherends is hidden to see the damage distribution of damage at a generic increment of a fatigue simulation in the case of Mode I loading. Both fully damaged element ($D = 1$) and partially damaged element ($0 < D < 1$) that forms the process zone, A_{CZ}, can be noticed.

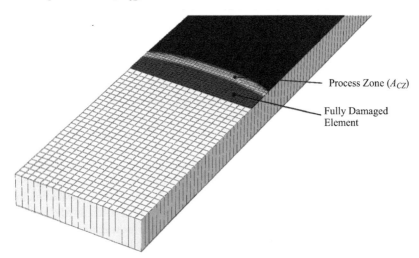

Process Zone (A_{CZ})

Fully Damaged Element

FIG. 8.30 Crack front during a fatigue simulation.

Regarding the calculation time, i.e., the time the analyst has to wait for the crack to reach the point where FCG rate increases steeply, that is close to fracture, was monitored. In the cases examined, this means a crack length of 30 mm for all the geometries except SLJ, for which the analyses were stopped at 10 mm of crack length even though still far from fracture. Only the outputs strictly necessary for each model were required, in order to minimize time spent in storing data. The PC used for calculations for CZM was an Intel1 CoreTM I, 2630 QM 2 GHz CPU, with 6 GB RAM and 579 Gb HD (7200 rpm, 6 MB cache), while for VCCT it was an

Intel1 XeonTM E5645 (Nehalem, 1 core) 2.4 GHz CPU, with 48 GB RAM and 900 Gb HD (10 k rpm, 12.3 MB cache).

In the case of aluminum joints, the four models show the same overall trend and an overall good correspondence with each other of the SERR plot (see Fig. 8.31a). In general, the VCCT yields slightly lower values than the two CZM and the analytical model. Moreover, both the CZM show a quite different trend in the first millimeter, where the process zone is under development. The average difference between the VCCT and both the CZM is about 5% on the 21-29 mm crack length span, while the difference is in the order of 2% whit respect to the analytical solution. As a consequence, the crack growth predictions given by both the CZM are rather similar to that given by the analytical model (Fig. 8.31b). Due to the differences shown in the SERR plot, the VCCT model gives a slight under-prediction of the crack growth rate with respect to both the CZM and the analytical model. It is believed that further mesh refinement can get the two models closer to each other.

Moving to the analysis of composite joints, the presence of an anisotropic behavior of the adherends increases the difference between the three models. In particular, for the SERR evaluation, the trends of both the CZM are more scattered (Fig. 8.32a): this in due to the combination of adherends anisotropy and stepwise element deletion.

However, a good agreement can be noticed for the three plots, although the CZM - 3D curve is slightly lower that the other (differences in the order of 12%). These differences produce a significant under prediction of the crack growth rate in the case of CZM - 3D with respect to the CZM - 2D and VCCT (Fig. 8.32b).

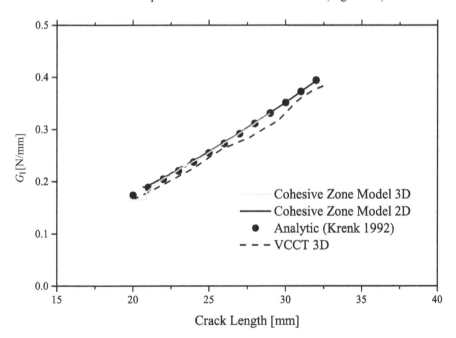

a: Comparison of $a - G_I$ values

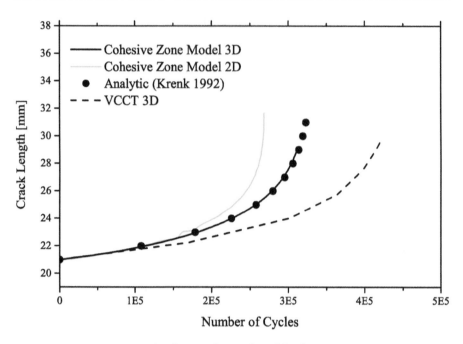

b: Comparison of a - N values

FIG. 8.31 CZM 3D results compared with CZM 2D, VCCT 3D and analytical ones (DCB, isotropic material).

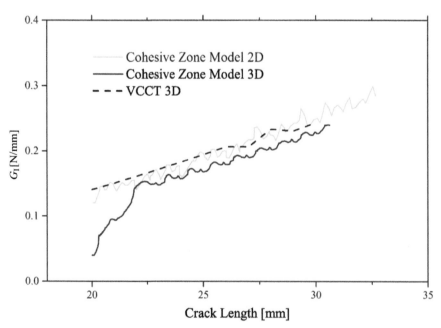

a: Comparison of $a – G_I$ values

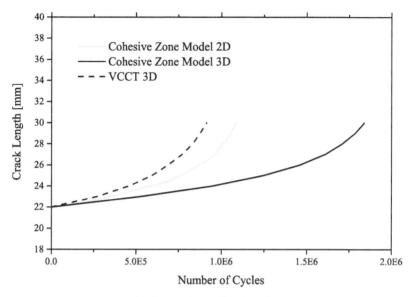

b: Comparison of a - N values

FIG. 8.32 CZM 3D results compared with CZM 2D, VCCT 3D and analytical ones (DCB, ortho-tropic material).

Concerning simulations in pure mode II, Fig. 8.33a shows the values of G_{II} obtained by both the CZM, VCCT and the analytical mode in the case of aluminum substrates. Again as in the case of DCB, the four sets show a good overall correspondence with each other.

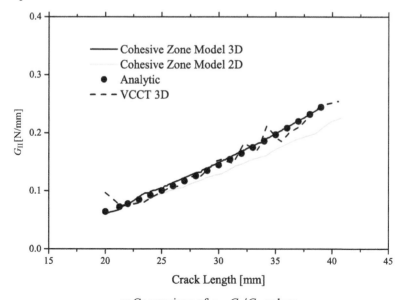

a: Comparison of $a - G_I/G_{II}$ values

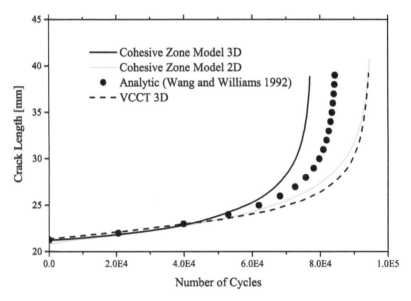

b: Comparison of $a - N$ values

FIG. 8.33 CZM 3D results compared with CZM 2D, VCCT 3D and analytic ones (ELS, isotropic material).

Some differences concern the CZM - 2D, that gives a slightly lower trend than the other models, and the VCCT, whose trend is rather jagged: probably this is related to a non-uniform (one row of elements per time of increment) crack front propagation that has been recorded in the simulation.

The main result in terms of crack length vs. number of cycles is shown instead in Fig. 8.33b, where a difference comes out lower than mode I (18% at 39 mm crack length), and also in this case a further mesh refinement can get the two models closer to each other.

In Fig. 8.34 the crack front during a ELS fatigue simulation is shown. It can be seen how the front appears straighter than the one of a DCB fatigue simulation.

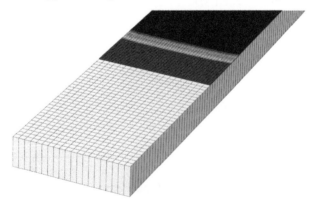

FIG. 8.34 Crack front during an ELS fatigue simulation.

In the case of composite adherends the three modes analyzed are again in good agreement between each other. Concerning the SERR plot (Fig. 8.35a), the CZM - 3D trend is quite lower than the other; however the differences are, in the average, lower than 7%. In terms of crack length vs. number of cycles prediction (Fig. 8.35b), this produces similar results for CZM - 2D and VCCT, while the CZM - 3D yields

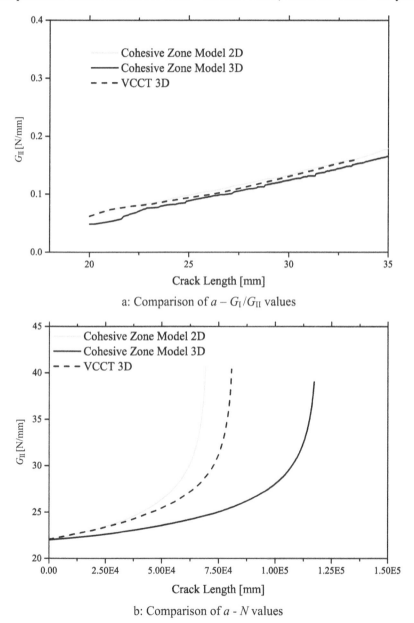

a: Comparison of $a - G_I/G_{II}$ values

b: Comparison of $a - N$ values

FIG. 8.35 CZM 3D results compared with CZM 2D, VCCT 3D and analytical ones (ELS, orthotropic material).

a higher fatigue life with respect to the others (about 55% for a propagation from 22 to 36 mm). The SERR plot of mode II composite joints appears less scattered with respect to the mode I composite joints: this is due to the lower stress concentrations in the ELS joints (respect to the DCB), which produce smaller changes of the stress fields when one or more elements change from "undamaged" to "damaged".

Regarding the model behavior in mixed-mode loadings, crack propagation in MMELS and SLJ has been simulated. Figure 8.36a shows the values of G_{TOT} ($G_{TOT} = G_I + G_{II}$) obtained again by both the CZMs, VCCT and analytical model

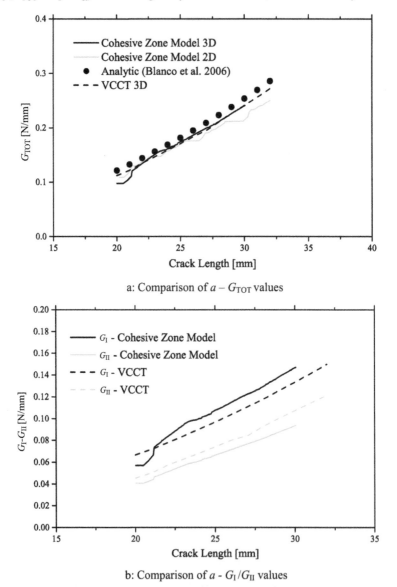

a: Comparison of $a - G_{TOT}$ values

b: Comparison of $a - G_I/G_{II}$ values

FIG. 8.36 Comparison of G_I, G_{II} and G_{TOT} trends (MMELS, isotropic material).

for the aluminum joint. The CZM - 3D trend is almost superimposed to the VCCT trend (1% average difference from 21 to 30 mm crack length), while the analytical solution gives a slight overestimation and CZM - 2D a slight underestimation with respect to the previous ones. The mode separation (G_I and G_{II}) is shown in Fig. 8.36b only for CZM - 3D and VCCT - 3D. The values obtained with the two methods in this case highlight a little bit the differences found for the single modes (see previous paragraphs).

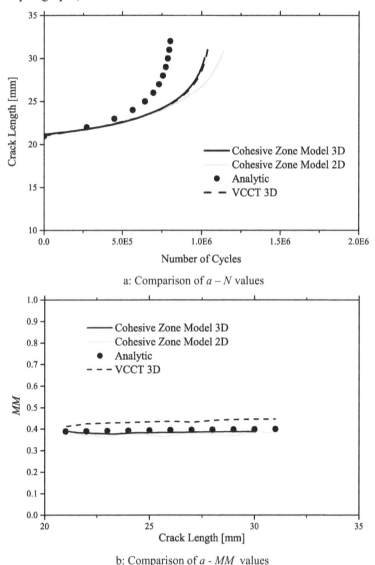

a: Comparison of $a - N$ values

b: Comparison of a - MM values

FIG. 8.37 Comparison of N and MM trends (MMELS, isotropic material).

These differences become larger in the crack length vs. number of cycles plot (Fig. 8.37a), where again the CZM - 3D and VCCT give similar results, while the analytical solution and the CZM - 2D give, respectively, higher and lower crack growth rate. Figure 8.37b shows the variation of *MM* during the propagation.

In the case of composite joints the three models give quite different results: although the trends of the total SERR ($G_{TOT} = G_I + G_{II}$) are similar (Fig. 8.38a), the mode separation produces differences of the mixed mode ratio in order of 30% (Fig. 8.38b) and this, in turn, produces significant differences in terms of crack length vs. number of cycles prediction (Fig. 8.39a). This phenomenon, still under investigation, is thought to be produced by the material anisotropy.

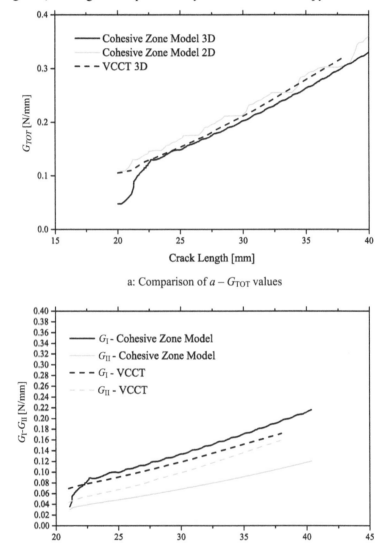

a: Comparison of $a - G_{TOT}$ values

b: Comparison of $a - G_I/G_{II}$ values

FIG. 8.38 Comparison of G_I, G_{II} and G_{TOT} trends (MMELS, orthotropic material).

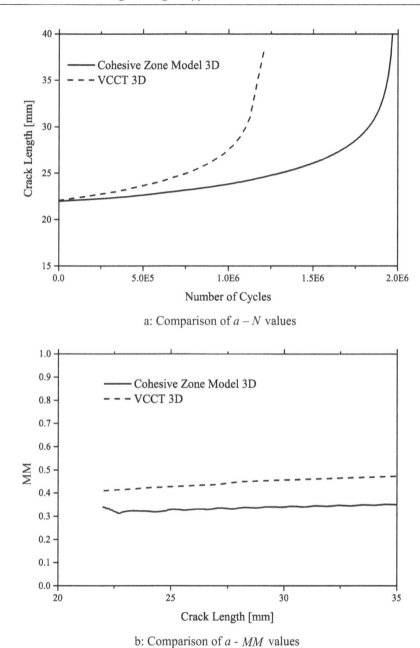

a: Comparison of $a - N$ values

b: Comparison of a - MM values

FIG. 8.39 Comparison of N and MM trends (MMELS, orthotropic material).

Figure 8.40b show the values of G_I and G_{II} obtained by CZM and VCCT for the SLJ. Firstly it will be discussed the case of adherends made of isotropic material. Concerning the comparison of the single modes, the values are closer to each other than in the case of MMELS, except the first 3 mm, necessary to establish a steady state process zone in CZM. After this, G_{TOT} (see Fig. 8.40a) is the same and so also the mixed-mode ratio (2.5% average difference from 3 to 10 mm crack length). With these premises, the crack length vs. number of cycles in Fig. 8.41a, skipping the first millimeter of propagation, the values are practically coincident (4.5% difference at 10 mm crack length). Figure 8.41b shows the variation of the mixed mode ratio (*MM*) during the propagation.

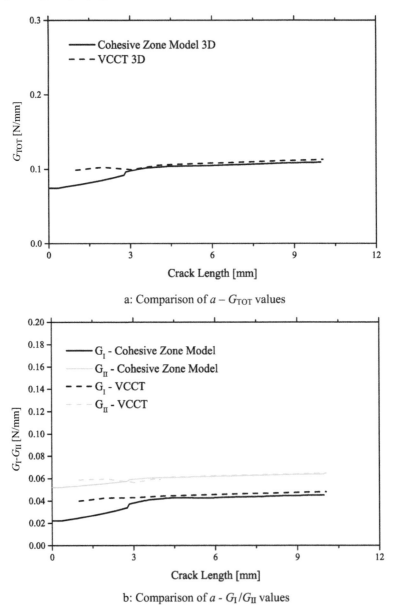

a: Comparison of $a - G_{TOT}$ values

b: Comparison of $a - G_I/G_{II}$ values

FIG. 8.40 Comparison of G_I, G_{II} and G_{TOT} trends (SLJ, isotropic material).

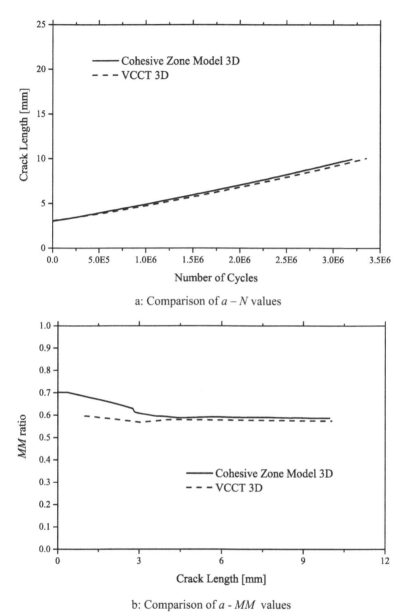

a: Comparison of $a - N$ values

b: Comparison of a - MM values

FIG. 8.41 Comparison of N and MM ratio trends (SLJ, isotropic material).

Regarding the composite assemblies, Fig. 8.42 shows the distribution of G_I, G_{II} and G_{III} along crack front in a VCCT simulation. It is clear that values are unevenly distributed probably due to the development of an asymmetrical crack front. It is also evident that where G_I and G_{II} values drop, a comparable G_{III} shows up. Since Mode III should be concentrated in correspondence of the surfaces, the occurrence of non-negligible G_{III} at some points in the interior is peculiar. Regarding the crack front in the CZM simulation, it appears symmetric at any increment; since SERR is an output unavailable in CZM model. In Fig. 8.43 the damage distribution and stress condition along the front are shown.

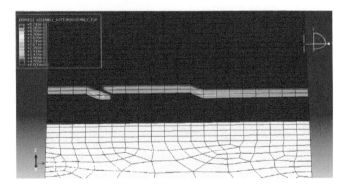

a: G_I distribution

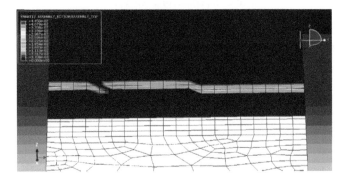

b: G_{II} distribution

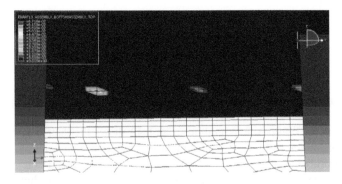

c: G_{III} distribution

FIG. 8.42 Comparison of N and MM ratio trends (SLJ, isotropic material).

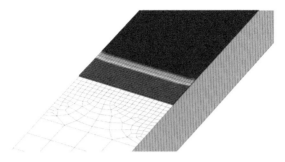

a: Damage distribution in the Cohesive Zone

b: Opening stress (σ_{33}) distribution in the Cohesive Zone

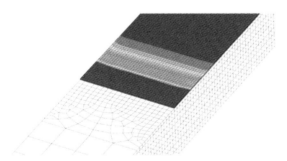

c: In-plane shear stress (σ_{13}) distribution in the Cohesive Zone

FIG. 8.43 Crack front in CZM - 3D fatigue simulations.

Keeping in mind that these results may affect the comparison with CZM, the values of G_I and G_{II} obtained by CZM and VCCT are reported in Fig. 8.44b. Even if the two sets show a very good correspondence with each other concerning G_{TOT} (see Fig. 8.44a), the single modes present some differences (Fig. 8.44b). The very low values in the case of CZM at the beginning are related instead to the absence of an initial crack and, therefore, the necessity of some millimeters of crack growth for the process zone to develop. Oscillations visible afterwards are instead related to the relatively coarse mesh used to keep calculation time within affordable limits. As already said in the case of MMELS, a general statement whether CZM under- or over predict the SERR out coming from VCCT cannot be drawn, unless further investigations, especially concerning mesh size effects, are done. Skipping the first five millimeters where the cohesive process zone develops, the crack length vs. number of cycles is shown in Fig. 8.45a. Given the differences shown in Fig. 8.44b in the values of G_I and G_{II} between the two models, sensible differences in a-N and in a-MM (see Fig. 8.45b) trends obviously turn out.

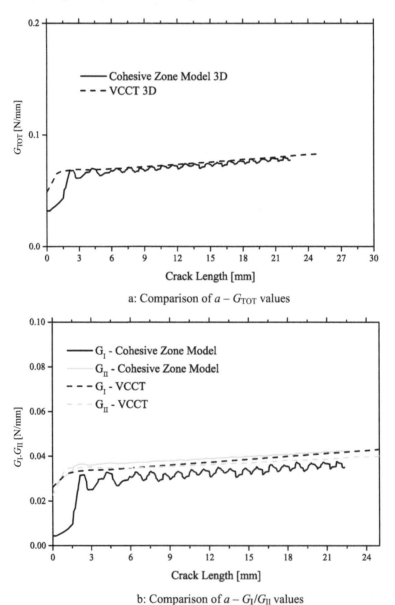

a: Comparison of $a - G_{TOT}$ values

b: Comparison of $a - G_I/G_{II}$ values

FIG. 8.44 Comparison of G_I, G_{II} and G_{TOT} trends (SLJ, orthotropic material).

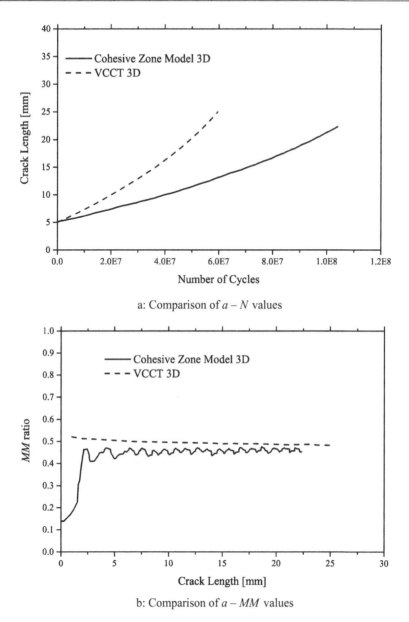

a: Comparison of $a - N$ values

b: Comparison of $a - MM$ values

FIG. 8.45 Comparison of N and MM ratio trends (SLJ, orthotropic material).

8.4.4 Conclusions

The comparison of the performances of the CZM presented in Moroni et al. (2013) and the VCCT embedded in the software Abaqus® on mode I, mode II and mixed-mode I/II loaded cracks in composite assemblies yielded that the two models are in good agreement concerning Mode I and Mode II conditions and also under

Mixed-Mode I/II loading concerning the total $G_{TOT} = G_I + G_{II}$. However, as the crack growth rate (Eq. 8.3) is strongly dependent on ΔG, even a limited difference causes a sensible difference in the elapsed number of cycles for a given crack growth. Since the element size is a compromise between the computation time and a simulation with sufficient detail, by just decreasing it a better agreement would be found. Hence, a general statement whether CZM under- or over predict the strain energy release rate out coming from VCCT cannot be drawn, unless further investigations, especially concerning mesh size effects, are done.

The calculation times are reported in Table 8.7. The CZM results are on average two-order of magnitude quicker than VCCT, with calculation times of the order of minutes instead of hours. In the case of SLJ, the increase in calculation time is related to the finer mesh, but the time required by the VCCT is becoming so important that high performance computing may be needed if the model complexity would increase further.

Table 8.7 Calculation time of VCCT and CZM (time is reported in minutes).

	DCB	ELS	MMELS	SLJ
CZ	186	54	90	57
VCCT	1012	86	229	1015

It must be also recalled that CZM simulation have been run on a less powerful PC. As mentioned above, Direct Cyclic procedure is computationally more expensive than the in-house CZM subroutine.

8.5 CZM FOR 3-D FATIGUE DEBONDING (GENERAL SHAPE CRACKS)

8.5.1 Crack Front Identification

In Section 8.3 a model of the cohesive zone able to correctly simulate the propagation of fatigue defects in 2D geometry was described. The procedure has been extended in Section 8.4 to 3D planar crack geometries, where G could be evaluated by the contour-integral on parallel slices along the crack front.

However the models in Sections 8.3 and 8.4 are able to simulate and predict the joint lifetime only in case the crack front keeps almost straight during all the fatigue simulation. Therefore a CZM approach able to handle propagation of cracks with a general crack front shape is still needed. In order to generalize the model, it is necessary that the Abaqus® subroutine first identifies the crack front (see Fig. 8.46) and the direction perpendicular to the front at each node, at every increment. The flow diagram shown in Fig. 8.16 is still valid to evaluate the damage distribution, but it has to be integrated with a crack front detection algorithm.

Inspired by Xie and Biggers Jr. (2006) who proposed a procedure to allow the determination of the actual shape of the delamination front for the VCCT, the routine looks for IP having at the same time:

- $D < 1$;
- at least one surrounding point with $D = 1$.

The segments that separate the points with $D < 1$ from the points with $D = 1$ define the crack front (see Fig. 8.47). Through the perpendicular lines n_1 and n_2 respectively to the segments, 1 and 2, the direction of the local crack propagation is identified (the vector n in Fig. 8.47). Once the front is defined, at each point the SERR is evaluated through the J-integral (where also Mode III can be taken into account) along the direction of propagation. However, further clarifications will be given in the reminder.

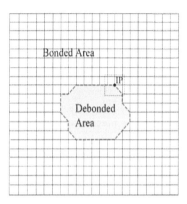

FIG. 8.46 Crack front of generic shape.

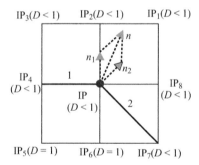

FIG. 8.47 Identification of local crack front at IP.

In the 3D CZM presented in the previous section, the J-integral was evaluated using stresses and strains extracted at IP; this was possible because the path perpendicular to the front that was assumed straight at each increment. In this model a condition as the one described in the picture at right of Fig. 8.48 has now to be managed. In fact, the J-integral has to be evaluated using stresses and opening at P path r points.

Therefore, using the bi-linear interpolation, for any variable in P path r, q, it can be written

$$q = (1 - v)(1 - u)q_{j,k} + v(1 - u)q_{j+1,k} + vuq_{j+1,k+1} + (1 - v)uq_{j,k+1} \qquad (8.46)$$

where $v = a/a_T$, $u = b/b_T$, j and k are subscripts referred to the four surrounding IP (see Fig 8.48).

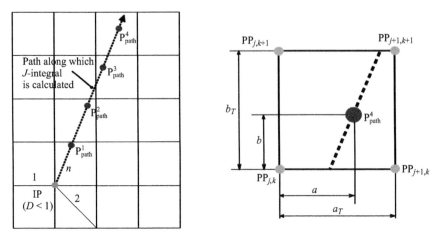

FIG. 8.48 Path for evaluating the J-integral (left) and the bi-linear interpolation.

With regard to the J-integral computation, Eq. (8.36) should be modified to consider the Mode III loading (Chiarelli and Frediani 1993)

$$J = \int_{\Pi} \left(-\sigma_{12}\frac{\partial u_1}{\partial x_1} - \sigma_{22}\frac{\partial u_2}{\partial x_1} - \sigma_{32}\frac{\partial u_3}{\partial x_1} \right) ds \qquad (8.47)$$

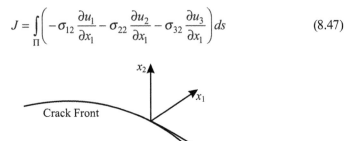

FIG. 8.49 Definition of the local coordinate systems for a crack front point.

Figure 8.49 can help explaining that Eq. (8.47) must be evaluated in a local coordinate system for each crack front point. For this reason the output from Abaqus®, which is referred in a global coordinate system, must be converted to local ones.

The strain energy release rate will be known for each crack front point and, therefore, it is possible to calculate ΔD_i^j and $\Delta N_{\min}^j$ using the same algorithm described in Fig. 8.16. The minimum number of cycles can this time be used to update the damage distribution of the crack front point only.

8.5.2 Mixed-Mode loading (I/II/III)

In order to obtain a comprehensive generalization of the new approach, also Mode III loading needs to be considered. Therefore, a general formulation for decohesion elements dealing with mixed-mode delamination onset and propagation is required. As mentioned before, under pure Mode I, II or III loading, the onset of damage at the interface can be determined simply by comparing the traction components with their respective allowables. However, under mixed-mode loading damage onset and the corresponding softening behavior may occur before any of the traction components involved reaches their respective allowable. It is assumed that mode III cohesive behavior is identical to mode II.

Cohesive Law

Concerning the cohesive law, the mode II and mode III displacements are combined in order to obtain a global sliding

$$\delta_s = \sqrt{\delta_{12}^2 + \delta_{23}^2} \tag{8.48}$$

This value is then combined with the mode I opening in order to obtain the equivalent opening

$$\delta_{eq} = \sqrt{\left(\frac{\delta_{22} + |\delta_{22}|}{2}\right)^2 + (\delta_s)^2} \tag{8.49}$$

Equivalent initial stiffness ($K_{eq,0}$), damage initiation equivalent opening ($\delta_{eq,0}$) and critical equivalent opening ($\delta_{eq,c}$) are also defined. The equivalent initial stiffness is obtained by equating the equivalent strain energy (U_{EQ}) to the total strain energy (U_{TOT}), which in turn is equal to the sum of the mode I (U_{22}) and in-plane shear (U_s) strain energies. Assuming that mode II and mode III undamaged stiffness is the same ($K_{12}^0 = K_{23}^0 = K_s^0$)

$$U_{EQ} = U_{TOT} = U_{22} + U_{12} = \frac{1}{2} \cdot \delta_{eq}^2 \cdot K_{eq}^0 = \frac{1}{2} \cdot (\delta_{22} + |\delta_{22}|)^2 \cdot K_{22}^0 + \frac{1}{2} \cdot \delta_s^2 \cdot K_s^0 \tag{8.50}$$

A further relationship is needed to define damage initiation: this is done using a quadratic damage initiation criterion (Ungsuwarungsru and Knauss 1987)

$$\left(\frac{\sigma_{22}}{\sigma_{22\,max}}\right)^2 + \left(\frac{\sigma_{12}}{\sigma_{12\,max}}\right)^2 + \left(\frac{\sigma_{23}}{\sigma_{23\,max}}\right)^2 = 1 \tag{8.51}$$

Assuming $\sigma_{23\,max} = \sigma_{12\,max}$ it follows that $\delta_{22}^0 = \delta_{12}^0$. The onset of damage is therefore ($\beta = \delta_s / \delta_{22}$)

$$\delta_{eq}^0 = \delta_{22}^0 \delta_{12}^0 \sqrt{\frac{1 + \beta^2}{\delta_{12}^{02} + (\beta \delta_{22}^0)^2}} \tag{8.52}$$

The last relationship needed, concerns the definition of the critical equivalent opening. Since the area underlying the cohesive law is representative of the critical strain energy release rate, using the Kenane and Benzeggagh (1996) theory the area underlying the mixed mode equivalent cohesive law (Γ_{eq}) can be computed as

$$\Gamma_{eq} = \Gamma_{22} + (\Gamma_s - \Gamma_{22}) \cdot MM^{m_m} \tag{8.53}$$

where (Γ_{22}) and (Γ_s) are the areas underlying the mode I and in-plane shear cohesive laws, respectively, m_m is a mixed mode coefficient depending on the adhesive and MM is the mixed mode ratio defined as a function of the mode I and in-plane SERR ($G_s = G_{II} + G_{III}$) as follows

$$MM = \frac{G_s}{G_I + G_s} \tag{8.54}$$

The underlying assumption is of course that $\Gamma_s = \Gamma_{12} = \Gamma_{23}$. Finally, the critical equivalent opening, δ_{eq}^c, is

$$\delta_{eq}^c = \frac{2}{K_{eq}^0 \delta_{eq}^0}\left[\Gamma_{22} - (\Gamma_s - \Gamma_{22})\left(\frac{\beta^2}{1+\beta^2}\right)^{m_m}\right] \tag{8.55}$$

FCG Rate

Concerning the fatigue crack growth rate, the KB mixed mode presented in 8.3.5 is used again. The fatigue crack growth rate is given again by Eq. (8.3) where B and d are functions of the mixed mode ratio MM. The latter in this case is defined as

$$MM = \frac{G_{II} + G_{III}}{G_{III} + G_{II} + G_{I}} \tag{8.56}$$

8.5.3 Results

The procedure described in this section has been validated so far only for aluminum adhesive joints under Mode I, Mode II and Mixed-Mode I-II. The specimens' dimensions and the applied loads are the same of Table 8.6. The cohesive law parameters and Paris law equation coefficients are given in Table 8.3. A structured sweep mesh of continuum shell elements SC8R with an element size of 1 mm was introduced with a single element in thickness, while the cohesive zone was modelled with COH3D8 element of the size 0.2 mm.

The first ten millimeters of crack propagation in a DCB test have been simulated. In order to verify the accuracy of this approach, the SERR has been compared with the analytical solution (Krenk 1992). The values of G_I obtained by 3D VCCT and planar cracks 3D CZM are plotted on the same diagram. As the 3D crack front is slightly bowed, the G_I and crack length are average values. All the sets show quite a good match (see Fig. 8.50).

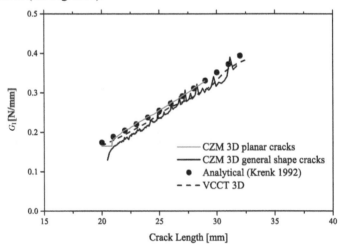

FIG. 8.50 Comparison of $a - G_I$ values, DCB joint.

The crack front during the propagation is shown in Fig. 8.51. Figure 8.52 shows the crack front as it appears if fully damaged element are removed.

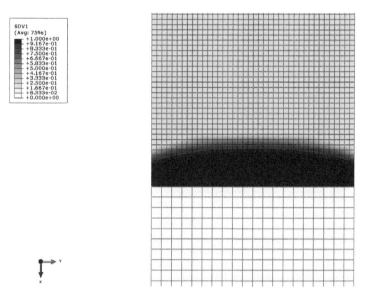

FIG. 8.51 Crack front during the fatigue simulation, DCB joint.

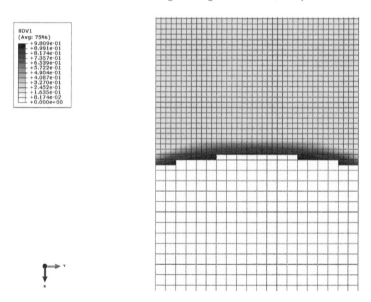

FIG. 8.52 Crack front during the fatigue simulation (fully damaged element are deleted), DCB joint.

Concerning the Mode II, ELS test is simulated with the generalized approach. Figure 8.53 shows results in terms of G_{II} against the VCCT and the planar cracks 3D CZM. Also in this case all sets show a good match.

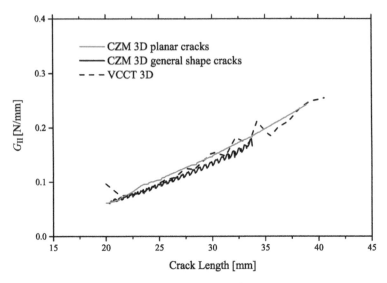

FIG. 8.53 Comparison of a – G_{II} values, ELS joint.

In the case of mixed-mode I/II MMELS adhesive joints instead (Fig. 8.54), the 3D general shape crack CZM is comparable to VCCT regarding the mode I component and with the 3D planar cracks CZM regarding mode II. The overall G_{TOT} ($G_{TOT} = G_I + G_{II}$) results therefore lower than that output by the other techniques and the analytical one; however the trend is the same for all models within the range of crack length simulated. On the other hand, the mixed-mode ratio $MM = G_{II}/G_{TOT}$ is perfectly aligned with the CZM for planar cracks and the analytical solution (Fig. 8.55).

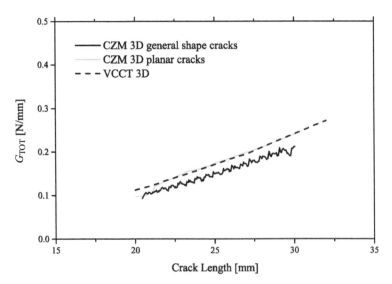

FIG. 8.54 Comparison of a – G_{TOT} values, MMELS joint.

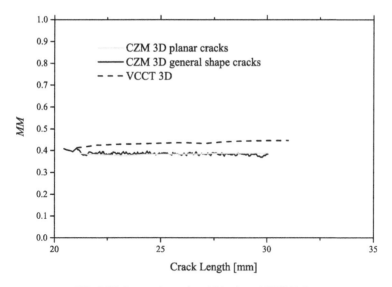

FIG. 8.55 Comparison of *a – MM* values, MMELS joint.

Figure 8.56 shows the stress-opening values in various IP, located at different crack length, during all the fatigue simulation. As one can note, the area underling the stress-opening values grows with increasing distance of IP from the initial crack tip. Obviously, this depends on the fact that the SERR increases with the length of the crack. It is also worth to notice that stress-opening data are much closer to the prescribed cohesive law than in the 2D simulation (Fig. 8.26). This is due to the higher number of increments needed to advance the crack in the 3D simulation.

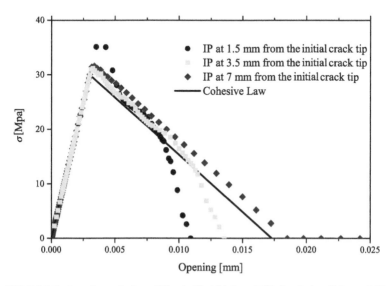

FIG. 8.56 Fatigue Degradation of IP at half width for a DCB simulation ($\Delta D_{max} = 0.2$).

8.6 CONCLUSIONS

A new procedure for simulating fatigue debonding/delamination with cohesive zone of arbitrarily shaped cracks has been developed. So far this approach has been tested on simple geometries (DCB, ELS and MMELS) for adhesively bonded joints with adherends made by isotropic material. The results have shown a good agreement with FEA analyses done using VCCT and the CZM for planar cracks of Section 8.4. Calculation time is increased in the worst case of about 2%, compared to the CZM 3D for planar cracks. A restriction of the model is that algorithms requires plane, square-shaped grids of cohesive elements. However, it has been shown that also the VCCT implemented in Abaqus® is highly sensitive to the mesh size and shape (see Section 8.4), restricting the possibility of practical use.

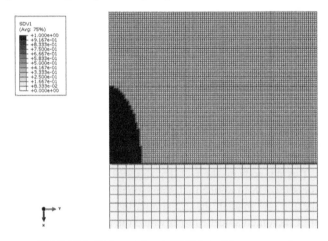

FIG. 8.57 Initial crack front, DCB joint with corner crack.

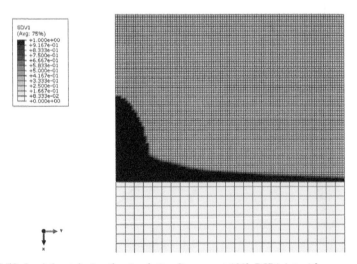

FIG. 8.58 Crack front during the simulation (increment 120), DCB joint with corner crack.

Finally, to show the potential of the model, the next figures show a qualitative example of a fatigue simulation in a DCB test where the initial defect shaped like one quarter of ellipse (Figs. 8.57-8.61).

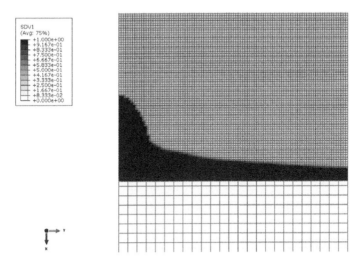

FIG. 8.59 Crack front during the simulation (increment 450), DCB joint with corner crack.

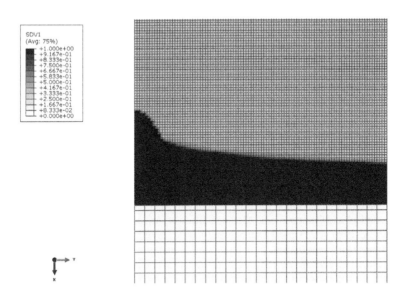

FIG. 8.60 Crack front during the simulation (increment 1045), DCB joint with corner crack.

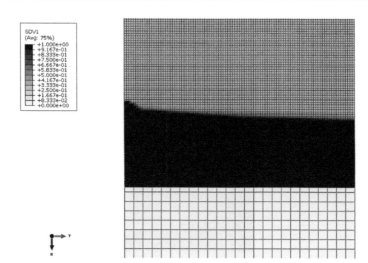

FIG. 8.61 Crack front during the simulation (increment 1400), DCB joint with corner crack.

8.7 REFERENCES

Abaqus® 6.11 Documentation, Analysis User's Manual, Dassault Systèmes.

Asp, L. E., Sjögren, A. and Greenhalgh, E. 2001. Delamination growth and thresholds in a carbon/epoxy composite under fatigue loading. Journal of Composites Technology and Research 23: 55–68.

Barenblatt, G. 1962. The mathematical theory of equilibrium cracks in brittle fracture. Advances in Applied Mechanics 7: 55-129.

Bernasconi, A., Jamil, A., Moroni, F. and Pirondi, A. 2013. A study on fatigue crack propagation in thick composite adhesively bonded. International Journal of Fatigue 50: 18-25.

Blanco, N., Gamstedt, E. K., Asp, L. E. and Costa, J. 2004. Mixed-mode delamination growth in carbon–fibre composite laminates under cyclic loading. International Journal of Solids and Structures 41: 4219-4235.

Blanco, N., Gamstedt, E. K., Costa, J. and Trias, D. 2006. Analysis of the mixed-mode end load split delamination test. Composite Structures 76: 14-20.

Brocks, W., Cornec, A. and Scheider, I. 2003. Computational aspects of nonlinear fracture mechanics. pp. 129-203. In: Milne, I., Ritchie, R.O. and Karihaloo, B. (eds.). Comprehensive Structural Integrity. Fracture of Materials from Nano to Macro, Vol. 3: Numerical and Computational Methods. Elsevier, Amsterdam.

Carpinteri, A. 1986. Mechanical Damage and Crack Growth in Concrete. Martinus Nijhoff Kluwer, Dordrecht Boston.

Chiarelli, M. and Frediani, A. 1993. A computation of the three-dimensional J-integral for elastic materials with a view to applications in fracture mechanics. Engineering Fracture Mechanics 44: 763-788.

de Andrés, A., Pérez, J. L. and Ortiz, M. 1999. Elastoplastic finite element analysis of three-dimensional fatigue crack growth in aluminum shafts subjected to axial loading. International Journal of Solids and Structures 36: 2231-2258.

Dugdale, D. S. 1960. Yielding of steel sheets containing slits. Journal of the Mechanics and Physics of Solids 8: 100-104.

Foulk III, J. W., Allen, D. H. and Helms, K. L. E. 1998. A model for predicting the damage and environmental degradation dependent life of SCS-6/Timetal 21S [0]4 metal matrix composite. Mechanics of Materials 29: 53-68.

Giuliese, G., Pirondi, A., Moroni, F., Bernasconi, A., Jamil, A. and Nikbakht, A. Fatigue delamination: A comparison between virtual crack closure and cohesive zone simulation techniques, presented at the 19th ICCM Toronto, Canada, July 2013.

Griffith, A. A. 1920. The phenomena of rupture and flows in solids. Transactions, Royal Society of London, Series A, 221: 163-198.

Harper, W. P. and Hallett, S. R. 2008. Cohesive zone length in numerical simulations of composite delamination. Engineering Fracture Mechanics 75: 4774-4792.

Harper, W. P. and Hallett, S. R. 2010. A fatigue degradation law for cohesive interface elements – development and application to composite materials. International Journal of Fatigue 32: 1774-1787.

Hillerborg, A., Modeer, M. and Petersson, P. E. 1976. Analysis of crack formation and crack growth in concrete by means of fracture mechanics and finite elements. Cement Concrete Research 6: 773-782.

Kenane, M. and Benzeggagh, M. L. 1996. Measurement of mixed mode delamination fracture toughness of unidirectional glass/epoxy composites with mixed mode bending apparatus. Composites Science and Technology 56: 439-449.

Khoramishad, H., Crocombe, A. D., Katnam, K. B. and Ashcroft, I. A. 2010. A generalized damage model for constant amplitude fatigue loading of adhesively bonded joints. International Journal of Adhesion & Adhesives 30: 513-521.

Krenk, K. 1992. Energy release rate of symmetric adhesive joints, Engineering Fracture Mechanics 43: 549-559.

Lemaitre, J. 1985. A continuous damage mechanics model for ductile fracture. Transactions of the ASME, Journal of Engineering Materials and Technology 107: 83-89.

Lemaitre, J., Desmorat, R. and Sauzay, M. 1999. Anisotropy damage law of evolution. European Journal of Mechanics - A/Solids 19: 187-208.

Liu, J., Li, J. and Wu, B. 2013. The cohesive zone model for fatigue crack growth. Advances in Mechanical Engineering 2013: Article ID 737392, 16 pages.

Moroni, F. and Pirondi, A. 2011. A procedure for the simulation of fatigue crack growth in adhesively bonded joints based on the cohesive zone model and different mixed-mode propagation criteria. Engineering Fracture Mechanics 78: 1808-1816.

Moroni, F. and Pirondi, A. 2012. A procedure for the simulation of fatigue crack growth in adhesively bonded joints based on a cohesive zone model and various mixed-mode propagation criteria. Engineering Fracture Mechanics 89: 129-138.

Moroni, F., Pirondi, A. and Giuliese, G. 2013. Simulation of fatigue debonding in adhesively bonded three dimensional joint geometries using a cohesive zone model approach, IGF XXII – 22nd Congress of the Italian Group on Fracture, Rome, Italy, 1-3 July.

Mostovoy, S. and Ripling, E. J. 1975. Flaw tolerance of a number of commercial and experimental adhesives. pp. 513–562 *In*: L. H. Lee (ed.). Adhesion Science and Technology. 9. B, Plenum Press, NY.

Muñoz, J. J., Galvanetto, U. and Robinson, P. 2006. On the numerical simulation of fatigue driven delamination with interface elements. International Journal of Fatigue 28: 1136-1146.

Needleman, A. 1990. An analysis of decohesion along an imperfect interface, International Journal of Fracture 42: 21-40.

Nguyen, O., Repetto, E. A., Ortiz, M. and Radovitzky, R. A. 2001. A cohesive model of fatigue crack growth. International Journal of Fracture 110: 351-369.

Paas, M. H. J. W., Schreurs, P. J. G. and Brekelmans, W. A. M. 1993. A continuum approach to brittle and fatigue damage: Theory and numerical procedures. International Journal of Solids and Structures 30: 579-599.

Paris, P. and Erdogan, F. 1963. A critical analysis of crack propagation laws. Journal of Basic Engineering 85: 528-534.

Paris, P. 1964. The fracture mechanics approach to fatigure. *In*: 10th Sagamore Army Material Research Conference, Syracuse University Press. pp. 107-132.

Peerlings, R. H. J., Brekelmans,W. A. M., de Borst, R. and Geers, M. G. D. 2000. Gradient-enhanced damage modelling of high-cycle fatigue. International Journal for Numerical Methods in Engineering 49: 1547-1569.

Petersson, P. E. 1981. Crack growth and development of fracture zones in plain concrete and similar materials. Technical Report, Division of Building Materials 1006, Lund Institute of Technology.

Pirondi, A. and Moroni, F. 2010. A progressive damage model for the prediction of fatigue crack growth in bonded joints. The Journal of Adhesion 86: 501-521.

Pirondi, A., Giuliese, G., Moroni, F., Bernasconi, A. and Jamil, A. 2014. Simulation of fatigue delamination/debonding using cohesive zone and virtual crack closure. pp. 369-400. *In*: A. P. Vassilopoulos (ed.). Fatigue and Fracture of Adhesively Bonded Composite Joints: Behaviour, Simulation and Modeling, 1st Edition. Woodhead Publishing, Elsevier, Cambridge, UK.

Rice, J. 1968. A path independent integral and the approximate analysis of strain concentration by notched and cracks. Journal of Applied Mechanics 35: 379-386.

Rice, J. and Rosengren, G. 1968. Plane strain deformation near a crack tip in a power-law hardening material. Journal of the Mechanics and Physics of Solids 16: 1-12.

Robinson, P., Galvanetto, U., Tumino, D., Bellucci, G. and Violeau, D. 2005. Numerical simulation of fatigue-driven delamination using interface elements. International Journal for Numerical Methods in Engineering 63: 1824-1848.

Roe, K. L. and Siegmund, T. 2003. An irreversible cohesive zone model for interface fatigue crack growth simulation. Engineering Fracture Mechanics 70: 209-232.

Russel, A. J. and Street, K. N. 1987. The effect of matrix toughness on delamination: Static and fatigue fracture under mode II shear loading of graphite fiber composites. Toughened Composites. ASTM STP 937. American Society for Testing and Materials. Philadelphia, PA. pp. 275-294.

Russel, A. J. and Street, K. N. 1988. A constant ΔG test for measuring mode I interlaminar fatigue crack growth rates. ASTM STP 972. American Society for Testing and Materials. Philadelphia, PA. pp. 259-277.

Rybicki, E. F. and Kanninen, M. F. 1977. A Finite element calculation of stress intensity factors by a modified crack closure integral. Engineering Fracture Mechanics 9: 931-938.

Tumino, D. and Cappello, F. 2007. Simulation of fatigue delamination growth in composites with different mode mixtures. Journal of Composite Materials 41: 2415-2441.

Turon, A., Costa, J., Camanho, P. P. and Dávila, C. G. 2007. Simulation of delamination in composites under high-cycle fatigue. Composites, Part A. 38: 2270-2282.

Ungsuwarungsru, T. and Knauss, W. G. 1987. The role of damage-softened material behaviour in the fracture of composites and adhesives. International Journal of Fracture 35: 221-241.

Wang, Y. and Williams, J. G. 1992. Corrections for mode II fracture toughness specimens of composite materials. Composites Science and Technology 43: 251-256.

Xie, D. and Biggers Jr., S. B. 2006. Strain Energy Release Rate calculation for a moving delamination front of arbitrary shape based on the virtual crack closure technique. Part I: Formulation and validation. Engineering Fracture Mechanics 73: 771-785.

Yang, B., Mall, S. and Ravi-Chandar, K. 2001. A cohesive zone model for fatigue crack growth in quasibrittle materials. International Journal of Solids and Structures 38: 3927-3944.

9

CHAPTER

Damage Mechanics for Static and Fatigue Applications

Ian A. Ashcroft[1,*] and Aamir Mubashar[2]

9.1 INTRODUCTION

Damage mechanics is an approach to predicting failure in a material or structure by representing load induced damage through the deterioration of material properties in a constitutive equation. This can be seen in Fig. 9.1, where the material follows the abcd path, with deterioration of properties along path cd, as damage response of the material is included in the model. This can be compared with the conventional elastic-plastic response in the figure, illustrated by path abcd′ where there is no deterioration of the properties at point c, instead, strain hardening continues until failure at point d′.

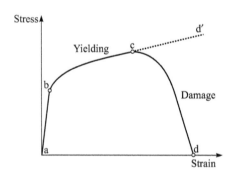

FIG. 9.1 Typical uniaxial stress-strain response of a ductile material comparing conventional elastic-plastic response (path abcd′) with a damage mechanics response (path abcd).

[1] Faculty of Engineering, University of Nottingham, University Park, Nottingham NG7 2RD, UK.
[2] Department of Mechanical Engineering, School of Mechanical and Manufacturing Engineering, National University of Sciences and Technology, H-12, Islamabad, Pakistan.
   E-mail: aamirmub@gmail.com
* Corresponding author: ian.ashcroft@nottingham.ac.uk

Damage mechanics can be considered as an alternative to the better known strength of materials and fracture mechanics approaches for predicting failure. In order to predict failure, a failure criterion relevant to the mechanism of failure needs to be identified. The simplest form of failure analysis is then to define failure, as occurring when the value of this failure criterion has exceeded a permissible limit, e.g. when the equivalent von Mises stress has exceeded the yield stress of a material. However, this type of analysis can be problematic when one has theoretical stress singularities, such as in adhesive joints, as the definition of failure becomes mesh dependent in a finite element analysis. Also, this approach doesn't predict the sequence of events leading to failure or behaviour post failure, which is useful for many reasons, such as in designing in-service monitoring or modelling energy absorption in an impact. In order to predict this, a progressive damage modelling approach is required. In ductile materials, a progressive damage approach can be based on plastic yielding, with failure defined when a complete path of yielded material has developed between load points. This may be defined by plastic collapse of the structure and may play a part in a limit state design approach. In a brittle material, a fracture mechanics approach may be more applicable where the propagation of a macro-crack through otherwise (assumed) undamaged material is predicted by modelling the conditions for crack growth. However, with many modern engineering materials, particularly those exhibiting a variety of concurrent micro-mechanical failure mechanisms, neither of these methods can fully capture the sequence of progressive damage under loading in a mechanistically accurate way. For example, a modern polymeric adhesive is a complex multi-phase material, typically comprised of a viscoelastic matrix with rubber particles for toughening, some form of filler particle and a carrier mat. Failure in such a material is complex and can involve the initiation and propagation of a macro-crack, accompanied by a region ahead and/or around the crack exhibiting numerous forms of micro-damage such as particle debonding and cavitation, carrier mat debonding and micro-cracking and yielding of the matrix material. In such a system, the state of the material after undergoing loading induced damage may be better represented by some form of damage mechanics. Damage mechanics models are useful because of their capability to simulate the response of both ductile and brittle materials under complex loading, including the repeated loading and unloading of structures under conditions of increasing damage, which is often represented by reduced stiffness. This is in contrast to traditional elastic-plastic models, which assume unloading at the initial stiffness and fracture mechanics, in which material ahead of a crack is assumed to be undamaged, regardless of the load history.

The distinctive feature of damage mechanics, then, is the representation of micro-damage, in a damage or process zone, as a reduction in material properties in that zone. This will affect the response of the structure to further loading and hence damage mechanics is often applied in a progressive damage modelling context. This approach is able to predict not just the final failure load, as is commonly the case when using the strength of materials approach, but also to predict the state of damage and response to loading at any point in the load-time history. This is extremely useful for a number of reasons, including the development of appropriate in-service health monitoring procedures and to aid in understanding the mechanisms of failure and their dependence on factors such as geometry, materials and environmental conditions. A better understanding of the response of a structure to loads can, in turn, lead to better designs and reduction in structural redundancy.

Two different forms of progressive damage modelling have generally been used with adhesive joints: cohesive zone modelling (CZM), where the failure is localised along a plane, and continuum damage modelling (CDM), where the failure can occur throughout the material. The potential sites for damage modelling in an adhesively bonded joint include the adhesive, the interface (or interfacial region) and the adherend. The CZM approach is more directly applicable to interfacial failure and certain forms of adherend failure, e.g. delamination in fibre reinforced composite adherends, whilst the CDM approach is more relevant to cohesive failure within the adhesive layer, particularly where an extended damage zone develops, such as in toughened adhesives. However, the CZM approach can also be used to model the cohesive failure of the adhesive and this has certain advantages, particularly in large structural models. CDM approaches can also be applied to typical adherend materials, such as aluminium or carbon-fibre reinforced plastics (CFRP). However, this is beyond the scope of this chapter and it will not be discussed. As CZM is discussed in Chapters 6-8, this chapter will be limited in scope to the application of CDM to the prediction of failure in adhesively bonded joints.

A relatively recent enhancement to the finite element method is the extended finite element method (XFEM). XFEM can be used to predict fracture in a material without prior consideration of the crack path in the finite element mesh. The crack can propagate within finite elements without adhering to the element boundaries and thus this alleviates the need to define crack paths, as required by CZM. The formulation of XFEM is based on the addition of enrichment functions to the finite element formulation. A damage initiation, propagation and failure framework, similar to CZM, can be implemented using XFEM. Damage initiation may be based on a stress or strain criterion and then it follows a damage curve, which may be linear or exponential. Damage propagation is based on a scalar damage variable, D, whose value increases from 0 to 1, where $D = 0$ represents no damage in the material and $D = 1$ represents fully damaged material. Details of XFEM and its applications for adhesive joints are discussed in Chapter 10.

The focus of this chapter, then, is to discuss the several approaches to CDM that have been applied to the prediction of failure in adhesive bonded joints. The underlying principles of the thermodynamic and micromechanical approaches to CDM are presented in the first part of the chapter, providing the general frameworks used for the implementation of CDM. This is followed by a review of the application of CDM to adhesive joints under static and fatigue loading. Environmental degradation is an important aspect when considering the durability of adhesive joints and this is discussed next in the context of a CDM framework. This is followed by a summary and conclusions.

9.2 CONTINUUM DAMAGE MECHANICS

9.2.1 Introduction

CDM has been the subject of extensive research over the last few decades and models have been developed for fatigue, creep, brittle and ductile damage. Several works and reviews of continuum damage mechanics have been carried out, such as Kachanov (1986), Kattan and Voyiadjis (2002) and Lemaitre and Desmorat (2005), and these should be referred to for a more complete coverage of the field. Damage variables are used in CDM models to determine the degradation in material properties, such as stiffness. The damage variables used are of two main types; the first type can predict the value of damage but does not characterise the damage itself, such as damage equivalent stress. The second type of damage variable is linked to some physical definition of damage such as porosity or relative area of micro-cavities. This type of variable is based on macroscopic material properties and its evolution is governed by a state equation. If several damage mechanisms occur in a material, each of them may be represented by an independent damage variable. The choice of damage variable is important as it characterises the type of failure and its practicability for engineering applications. Ideally, the selected damage criterion should be capable of an accurate continuum representation of the micro-damage features seen in the material and still be simple enough to enable the model to be applied to engineering applications.

Damage in a material can be considered in relation to a representative volume element (RVE), which is large enough compared to the damage induced material defects to enable material properties in the element to be represented by homogenized variables, but small enough to enable the representation of spatial variation in damage in a loaded structure. Kachanov (1958) proposed a scalar damage variable, D, in such an element for the case of isotropic damage, given by

$$D = \frac{\delta S_D}{\delta S} \tag{9.1}$$

where δS and δS_D are the cross section of a plane cutting the RVE and the area of microcracks and microvoids on the plane, respectively. $D = 0$ in the undamaged state, and $D = 1$ when failure has occurred. If several mechanisms of damage occur they may be represented by a number of different damage variables and, if the damage is non-isotropic, the variable D in equation 9.1 will be dependent on the direction of the normal of the plane of intersection, in which case a tensorial damage variable may be defined. Damage within the RVE will decrease the load bearing capacity of the element, which was described by Rabotnov (1968) in terms of an effective stress. For the case of isotropic damage and uniaxial loading an effective stress ($\tilde{\sigma}$) may be defined as

$$\tilde{\sigma} = \frac{\sigma}{1 - D} \tag{9.2}$$

where σ is the stress for undamaged material. In practice, determination of the density of micro defects is often difficult and it is more convenient to represent damage in terms of the effect of the damage on a measurable parameter, such as effective area, elastic modulus or void volume fraction. Hence, for the case of isotropic damage and uniaxial loading, damage may be related to an effective elastic modulus ($\tilde{E}$) by

$$D = 1 - \frac{\tilde{E}}{E} \tag{9.3}$$

where E is the elastic modulus for undamaged material.

9.2.2 Thermodynamic Approach to CDM

A general framework for damage mechanics was provided by Lemaitre and Desmorat (2005) based on the thermodynamics of irreversible processes. This involved the definition of state variables to represent damage, a state potential and a dissipation potential, to derive the laws of evolution of the state variables. A unified isotropic damage law was proposed in which the damage evolution rate ($\dot{D}$) above a critical value of the plastic strain energy, p_D, was defined by

$$\dot{D} = \left(\frac{Y}{S}\right)^s \dot{p} \quad \text{if} \quad \max \omega_s > \omega_D \quad \text{or} \quad p > p_D \tag{9.4}$$

$\dot{D} = 0$, if not,

$D = D_C \rightarrow$ mesocrack initiation

where Y is the energy density release rate, $\dot{p}$ is the accumulated plastic strain rate, ω_s is the stored energy density, ω_D is the damage threshold stored energy density (a material parameter), S and s are temperature dependent material parameters and D_C is the critical damage parameter. The value of Y for isotropic damage can be derived from the state potential for isotropic damage as

$$Y = \frac{\tilde{\sigma}_{eq}^2 R_v}{2E} \tag{9.5}$$

where $\tilde{\sigma}_{eq}$ is the effective von Mises stress and R_v is a triaxiality function, given by

$$R_v = \frac{2}{3}(1 + v) + 3(1 - 2v)\left[\frac{\sigma_H}{\sigma_{eq}}\right]^2 \tag{9.6}$$

where v is Poisson's ratio and σ_H is the hydrostatic stress. Substitution of (9.2) into (9.5) gives

$$Y = \frac{\sigma_{eq}^2 R_v}{2E(1-D)^2} \tag{9.7}$$

A limiting value of damage, D_c, was also suggested to represent the initiation of meso-cracking. The source of the plastic strain in equation (9.4) enables this approach to be applied to the ductile, creep, fatigue and quasi-brittle damage of materials. In order to apply the CDM approach to fatigue, Lemaitre (1984, 1985) derived the following equation for the rate of change of the damage variable with respect to fatigue cycle, N

$$\frac{\delta D}{\delta N} = \frac{2B_0 \left[\frac{2}{3}(1+\upsilon) + 3(1-2\upsilon)\left(\frac{\sigma_H}{\sigma_{eq}}\right)^2 \right]^{s_o}}{(\beta_o +1)(1-D)^{\beta_o +1}} \left(\sigma_{eq,max}^{\beta_o+1} - \sigma_{eq,min}^{\beta_o+1} \right) \qquad (9.8)$$

where s_o, B_o and β_o are material and temperature dependent coefficients, and $\sigma_{eq,max}$ and $\sigma_{eq,min}$ are the maximum and minimum von Mises equivalent stresses, respectively. Equation (9.8) can be integrated for constant amplitude fatigue loading. Using the boundary conditions ($N = 0 \rightarrow D = 0$) and ($N = N_R$ [number of cycles to rupture] $\rightarrow D = 1$), an expression for the number of cycles to failure is derived as

$$N_R = \frac{(\beta_0 + 1)\left(\sigma_{eq,max}^{\beta_o+1} - \sigma_{eq,min}^{\beta_o+1} \right)^{-1}}{2(\beta_o + 2)B_0 \left[\frac{2}{3}(1+\upsilon) + 3(1-2\upsilon)\left(\frac{\sigma_H}{\sigma_{eq}}\right)^2 \right]^{s_o}} \qquad (9.9)$$

A number of CDM-based phenomenological models have been implemented in commercial finite element codes, including Abaqus® (Dassault Systèmes, Providence, RI, USA) and MSC Marc® (MSC Software Corporation, California, USA). A CDM termed as a ductile damage model is available in Abaqus® that uses a limiting value of strain to predict failure. The damage model can have strain rate and stress triaxiality dependence and the evolution of D can be observed over time.

9.2.3 Micromechanical Approach to CDM

The Gurson, Tvergaard, Needleman (GTN) damage model takes a micromechanical approach to the characterisation of damage and failure. Gurson (1977) proposed a model to describe the growth of microscopic voids in materials by plastic deformation, as illustrated in Fig. 9.2. A set of modified constitutive equations for elastic-plastic materials with a damage variable, f_v, representing the volume fraction of voids was then derived. Needleman and Tvergaard (1984) modified this model to include void initiation and coalescence. In the modified model, damage is represented by an effective void volume fraction, f_v^*, given by

$$f_v^* = \begin{cases} f_v & \text{if} \quad f_v < f_c \\ f_c + \left(\dfrac{f_u^* - f_c}{f_F - f_c}\right)(f_v - f_v) & \text{if} \quad f_v \geq f_c \end{cases} \qquad (9.10)$$

where f_c represents a critical volume fraction for the fast coalescence of voids, f_F is the volume fraction at failure and $f_u^* = 1/q_1$. The parameters q_1 and q_2 (equation 9.11) were introduced to improve the Gurson model at small values of void volume fraction (VVF) and can be considered as material constants. Experimental studies have shown that values of $q_1 = 1.5$ and $q_2 = 1$ are accurate for solids with well-spaced voids. The yield criterion for the GTN model is given by

$$F = \frac{\sigma_{eq}^2}{\sigma_s^2} + 2q_1 f_v^* \cosh\left(\frac{3q_2\sigma_H}{2\sigma_s}\right) - 1 - (q_1 f_v^*)^2 \tag{9.11}$$

where σ_s is the yield stress.

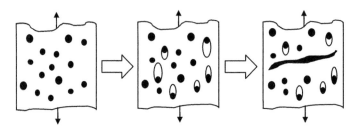

FIG. 9.2 Ductile failure by nucleation, growth and coalescence of voids.

The GTN damage model is available in a number of commercial finite element codes including Abaqus® and MSC Marc®. The current capabilities of the built-in GTN damage model in Abaqus® are demonstrated by modelling a single-lap joint with aluminium adherends and an epoxy adhesive, as shown in Fig. 9.3. Only one quarter of the single-lap joint is modelled. Material definition of the adhesive includes an elastic constitutive law, a plasticity law to capture the hardening behaviour of the material and GTN based yield surface, void nucleation and failure model. The initial VVF of the material can also be defined.

FIG. 9.3 A quarter model of single-lap joint with the adhesive layer modelled using the GTN damage model.

Deformation of the adhesive layer increases with increasing load and growth in VVF can be seen in the adhesive layer in Fig. 9.4. The VVF can be seen to grow from the regions of the adhesive layer where stress concentrations exist. In addition to the VVF, the nucleation of voids (VVFN) and growth of voids (VVFG) can also be tracked during various phases of deformation, as shown in Fig. 9.5.

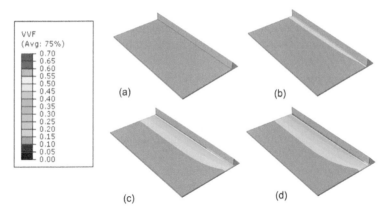

FIG. 9.4 Change in VVF in the adhesive layer of a single-lap joint at various applied displacements: (a) initial value at no load, (b) 0.16 mm, (c) 0.2 mm and (d) 0.22 mm.

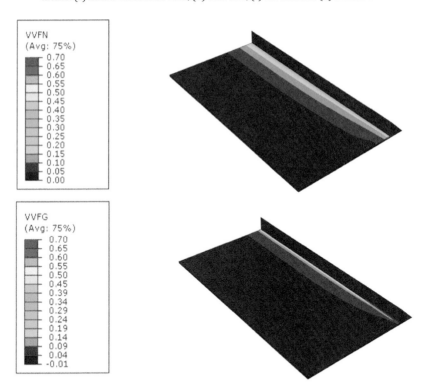

FIG. 9.5 Void nucleation (VVFN) and void growth (VVFG) in the adhesive layer of a single-lap adhesive joint at an applied displacement of 0.22 mm.

The two approaches to CDM discussed above have been applied to the prediction of failure in adhesive joints. Although both have been shown to successfully predict the behaviour of adhesive joints at a macro-scale, their implementation and development in finite element codes is based on very different principles. Also, the number and type of variables that need to be calibrated for a particular model differs significantly. The implementation of these methodologies for adhesive joints under static, fatigue and environmental loading is discussed in the following section.

9.3 APPLICATION OF CONTINUUM DAMAGE MECHANICS TO ADHESIVE JOINTS

One of the main advantages of using CDM models to predict failure in adhesive joints is that prior information about the location of damage initiation or direction of crack propagation is not a requirement. In CZM, the path of the crack needs to be known and a layer of cohesive elements embedded in the finite element mesh at the known location. This information may not always be available and, in fact, another possible use of CDM is in predicting the site (or sites) of damage initiation and subsequent fracture paths. The CDM approach to predicting the initiation and propagation of cracks also allows for the development of multiple cracks that can potentially interact with each other, and the cracks can propagate in any direction in the adhesive domain based on the damage and failure criteria used. This allows for the prediction of interfacial, cohesive and mixed-mode failures in adhesive joints. In addition, CDM-based material models consider the thickness of the adhesive layer and consequently are able to capture the effect of thickness and the properties of the adhesive on failure. Mesh sensitivity is a potential issue in the application of CDM in a finite element framework; however, this may be overcome by using a characteristic length approach. In the following sub-sections the application of various CDM approaches to the prediction of failure in adhesive joints under the conditions of static, fatigue and environmental loading are reviewed.

9.3.1 Static Failure

The simplest approach to implementing CDM is to directly relate damage to the stiffness of the material at a point. This allows for a decrease in material stiffness over time based on the accumulation of damage in the material. A suitable form of stiffness degradation law should be chosen, although a linear stiffness degradation law is mostly used in literature, as illustrated in Fig. 9.6. CDM models are susceptible to mesh sensitivity and, thus, a characteristic length is generally used to overcome this problem. The characteristic length is related to the influence of a Gauss point in an element in a particular direction.

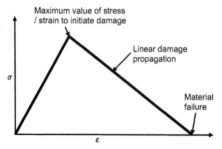

FIG. 9.6 A linear damage propagation law for CDM models.

Based on a CDM approach, Hua et al. (2006) proposed a strain-based damage and failure model. The model was simple to implement and calibrate as it used the maximum equivalent plastic strain to characterize damage initiation and failure. Damage initiation and propagation predicted by Hua et al. (2006) for a single-lap joint is shown in Fig. 9.7. The damage initiates where the corner of the adherend is embedded in the adhesive and then propagates through the adhesive layer close to the adherend interface and also though the adhesive fillet until complete failure at an applied displacement of 0.148 mm when a continuous path of fully damaged adhesive has formed. This model was extended to predict the effects of environmental degradation on joint strength and is, therefore, discussed more fully in Section 9.3.3, which reviews the application of CDM to the prediction of failure in adhesively bonded joints subjected to environmental degradation.

A fracture mode-dependent CDM was developed and implemented by de Moura and Chousal (2006). The model was initially applied to single mode damage and was later developed for mixed-mode damage (de Moura et al. 2008). A linear relationship between the damage and material stiffness was assumed after the initiation of damage. A characteristic length equal to the length of influence of a Gauss point was introduced to convert the displacements into strain and a softening relationship between the stress and strain was defined as follows

$$\sigma = (1 - D)\,C\varepsilon \tag{9.12}$$

where σ is stress, ε is strain and C is the stiffness matrix of undamaged material. Damage initiation was based on a pre-defined value of normal stresses parallel to the crack, and crack propagation was based on a linear fracture energy criterion. The model was applied to a double cantilevered beam bonded joint under mode I loading, where the adhesive layer thickness was varied from 0.1 mm to 0.5 mm. The proposed CDM model was able to predict changes in the fracture process zone owing to changes in the adhesive thickness.

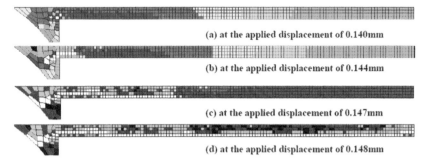

(a) at the applied displacement of 0.140mm

(b) at the applied displacement of 0.144mm

(c) at the applied displacement of 0.147mm

(d) at the applied displacement of 0.148mm

FIG. 9.7 Damage initiation and propagation in a EA9321 adhesive/aluminium adherend single-lap adhesive joint (Hua et al. 2006). The figure shows the adhesive layer with triangular fillet and damage contour, with white indicating fully damaged adhesive.

CDM-based finite element modelling was used to model damage in joggle lap joints of varying overlap lengths by Masmanidis and Philippidis (2015) with an epoxy adhesive and unidirectional glass fibres composite adherends. A nonlinear material model was used for the adherends where the progressive failure was determined by using the Puck criterion along with property degradation. The adhesive was assumed to be isotropic and a parabolic stress surface was used to define the initiation of damage. The damage index was calculated based on a linear degradation relationship between equivalent stress and strain. The failure of the joint was determined based on a maximum equivalent strain value that was dependent on the joint geometry and the adhesive material. The failure strain was determined based on experimental results. The experimental load-displacement response of the joggle lap joints was compared with the predicted response and a good correlation was observed. The models also predicted the experimental observation that joints with overlap length greater than 150 mm showed no further increase in strength.

Chousal and de Moura (2013) used a quadratic stress criterion to determine damage onset whilst crack propagation was then determined using a fracture energy-based criterion. The CDM model accounted for the mode found in adhesive joints by using two mode ratio parameters, which provided the ratio of displacements in mode II and mode I. The damage parameter was determined using the equivalent mixed-mode quantities rather than using the pure-mode quantities. Ban et al. (2008) defined damage in terms of the area of the ultimate strain at the failure load of a double-lap adhesive joint. The steel adherends were modelled as linear elastic and the adhesive was modelled with a piece-wise linear stress-strain curve. The damage zone of the adhesive material was defined based on the ultimate strain value of 7.87% and a reference value for the damage zone was calculated based on the ultimate strain value. This was carried out by applying the experimental failure load to two-dimensional (2D) and three-dimensional (3D) finite element models. The area of the region with value of strain higher than the ultimate strain was calculated and used as the reference zone of the damage at which failure of the joint occurs. This value was used to predict the failure of other double-lap adhesive joints with composite and steel adherends having different adhesive lengths and adherend thicknesses. The same methodology was used for 2D and 3D finite element models; however, the reference damage value was determined separately. Comparison with experimental results did not provide encouraging results where a difference of 62.1% and 54.9% in the experimental and predicted failure loads was reported for 2D and 3D finite element models, respectively. In an attempt to improve the failure load prediction, a damage zone ratio-based criterion was presented where the damage zone was normalised by the adhesive layer length and width. The results of the failure load predictions improved using this criterion, although the predicted failure load was still only within 20.1% of the experimental failure load.

Ren and Li (2014) used the virtual internal bond (VIB) theory to link the molecular theory of adhesive energy with the continuum theory of stress analysis. A constitutive model that could predict cohesive, interfacial and mixed failures of the adhesive in a bonded joint was developed. The 3D constitutive model was based on an empirical adhesive strain energy function. The components of the stress field were calculated using the VIB theory, which provided a relationship between the microscopic and macroscopic levels of a polymer material. At a macro scale, the strain energy at any material point was considered as the sum of all elastic energies of the virtual bonds attached to that point. Adhesive cross-linked chain network was assumed to be a tetrahedron lattice structure, which resembled the face centred cubic (FCC) crystal structure. A hyperelastic constitutive relation from the energy density function was then derived for stress. Based on this linkage of length scales, the modelled was termed a multiscale model.

Another approach that has been seen in recent years is hybrid modelling, where more than one type of failure criterion is considered in a single model. Generally, one criterion is used for modelling the cohesive damage and another is used to model the interfacial damage, such as in Mubashar et al. (2014). The advantage of such an approach is to combine the strengths of various techniques; however, the disadvantage is that it requires calibration or determination of more material model parameters than required by a single criterion. Also, finite element models using more than one damage and failure criteria generally present more convergence problems. Using such an approach, O'Mahoney et al. (2013) applied CZM for interfacial modelling and CDM for bulk adhesive modelling of composite single-lap adhesive joints. The composite adherends were made from symmetric cross-ply laminate consisting of 16 composite plies. A 2D finite element model using plain strain elements was used. The overall response of the bulk adhesive was modelled using an elastic-plastic material model with von Mises plasticity. The CDM model used a linear relationship between the damage variable and the stiffness of the material. The parameters of the finite element model were calibrated using experimental data. Based on the developed finite element model, a Taguchi analysis was performed to determine the sensitivity of joint strength to the parameters of the CZM and CDM models. The implemented model was able to predict the interfacial, cohesive and mixed-mode failure of a single-lap adhesive joint.

The micromechanical approach of continuum damage modelling was applied to adhesive joints by Read et al. (2000), where a modified form of the GTN model for ductile materials was derived for rubber-toughened adhesives. As the yield surface of rubber-toughened adhesives shows hydrostatic stress dependence, parameters were included to account for the effect of pressure on the yield surface. The influence of void interactions on matrix shear banding was also included in the model and the model accounted for the cavitation of rubber particles. This phenomenon is usually visible as stress whitening in the failure of rubber toughened plastics. The modified criterion allows for the changing composition of the polymer matrix during void nucleation. The proposed modified yield function (Φ) is as follows

$$\Phi = \frac{\sigma_e^2}{\sigma_m^2} - (q_1 f)^2 + 2q_1 f \cosh\left(\frac{3\sigma_k}{2\sigma_m}\right) - \left(1 - \frac{\mu\sigma_k}{\sigma_m}\right)^2 = 0 \qquad (9.13)$$

where σ_e is the effective von Mises stress, σ_m is the yield stress, σ_k is the mean normal stress and f is the void volume fraction. μ is a parameter that allows for the dependence of yield stress on the mean normal stress for the unvoided polymer. A detailed framework for the determination of the model material parameters though experimental testing was also presented. Later, Dean and Mera (2004) carried out extensive work on a rubber toughened propylene-ethylene copolymer containing about 17% by weight of filler and used the modified GTN model. The model was shown to predict the stress/strain response of bulk adhesives in tension and compression, and adhesively-bonded tensile-butt joints. A finite element implementation of this model was later used by Jumbo (2007) to predict the crack initiation and growth in an epoxy adhesive used in single and double-lap joints with composite and metal adherends. An example of the development of damage in a single-lap joint using this approach is shown in Fig. 9.8. It can be seen that damage initiates at the embedded corner of the adherend in the adhesive and then propagates in two directions, through the fillet and along the joint, in the adhesive close to the adherend. Although the model is different, the damage evolution is similar to that predicted by Hua et al. (2006).

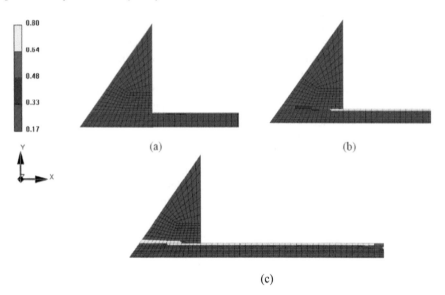

FIG. 9.8 Initiation and propagation of damage in a single-lap joint using a modified GTN material model (Jumbo 2007). The figure shows cross section through adhesive with triangular fillet and damage contours. Load increasing from (a) to (c).

The effect of rubber-toughened adhesive plastic deformation on crack growth was also investigated by Imanaka et al. (2003), in this case with compact tension and single-edge notched bend adhesive joints. The response of the bulk adhesive was modelled using a GTN material model in finite element modelling of the joints. The rubber particles in the adhesive were considered to be initial voids in the adhesive and their extension was considered. The value of the parameter q_1 of the GTN material model was determined by comparing the experimental stress-strain curves obtained using dumbbell and butt joints with those obtained from finite element modelling. Imanaka et al. (2009) also used a GTN model to explain the crack growth resistance behaviour observed in single-edged notched beam and double-cantilever beam joints with rubber-modified epoxy adhesives. Zaïri et al. (2005) used the GTN model coupled with a modified viscoplastic material model to investigate the mechanical response of rubber toughened polymethylmethacrylate (PMMA) at room temperature and compared the results with experimental tests to produce a quantitative agreement with the experimental observations. Some of their experimental observations included stress whitening caused by the cavitation of rubber particles.

Debonding in adhesive joints at cryogenic temperatures was predicted by Lee et al. (2011). The inelastic behaviour of epoxy and polyurethane adhesives was predicted using a modified Bodner-Partom material model that was combined with a GTN material model to predict damage and failure. The elements were not deleted upon failure; rather a very small stiffness was assigned to an element where all the material points in the element exceeded the failure criterion. The proposed finite element model results showed good correlation with the experimental results for lap shear joints.

It can be seen for the referenced work above that the application of CDM models in adhesive joints is still an active research topic. However, the widespread use of micromechanical models in adhesive joints is hindered by the large number of parameters required by these material models such as the GTN material model. The determination of these parameters is not always straightforward and may involve a combination of experimental and numerical methods. This becomes true for CDM models where a non-linear damage propagation law is used, whereas, a linear damage propagation law does not provide a good prediction of material behaviour in all cases. Read et al. (2000) presented a systematic approach to determine the parameters for the modified GTN model given in equation 9.13. The relative simplicity of CZM is also a factor in the widespread adaptation of CZM as the primary technique to model damage and failure in adhesive joints rather than CDM, even though the diffuse damage zones frequently seen in adhesives are, arguably, better represented mechanistically by CDM.

9.3.2   Fatigue Failure

Fatigue is the failure of a structure under a repetitive or cyclic loading regime in which the loads involved are considerably lower than those involved in instantaneous, or quasi-static, failure. Long periods can be spent in an initiation phase of fatigue damage, in which there may be no outward signs of damage. Fatigue damage can be initiated or accelerated by many factors, such as accidental impact, over-loading, corrosion, abrasion etc. and failure can occur rapidly in the final stages. The long time durations, effect of external and internal factors and stochastic nature of fatigue failure makes it difficult to predict accurately. Hence, it is very difficult to design against fatigue failure without resorting to large safety factors, and thus incurring structural inefficiencies. Monitoring fatigue damage can also be difficult, particularly if initiation is in an inaccessible location or if the critical crack size before rapid fracture is very small. However, progressive damage models can help in predicting the sequence of events leading to failure and the associated effects that could be used in structural health monitoring, e.g. changes in stiffness or modal response.

Two of the benefits of adhesive joints are that stress concentrations are more uniformly distributed than in riveted or bolted joints and that the bonding process does not explicitly weaken the adherends, as with bolting, riveting and welding, although there are likely to be stress concentrations in the adherend in the joint area. It might be expected, therefore, that adhesively bonded joints perform well in fatigue, and indeed this is often the case, however, a number of potential problems should also be recognised. It is well known that adhesives and the interfacial region between the adhesive and adherends are sensitive to the environment and this will affect the fatigue resistance of the joint. Adhesives are also generally susceptible to creep under certain conditions and, combined with fatigue, this can lead to accelerated failure (Hart-Smith 1981, Harris and Fay 1992, Ashcroft et al. 2001). It should also be remembered that failure in a bonded joint can occur in the adhesive, in the adherend or in the interfacial region between the two, and that the relative fatigue resistance of the various components is dependent on many factors, such as geometry, environment and loading, and may vary as damage progresses.

The main goals in the modelling of fatigue behaviour are to (i) predict the time (or no. of cycles) for a certain event to occur (such as macro crack formation, critical extent of damage or complete failure) and (ii) to predict the rate of change of a fatigue related parameter such as crack length or 'damage'. The modelling of fatigue can be used to support the design of structures, to ensure that fatigue failure is not likely to occur in service, and to aid in the design of efficient fatigue resistant joints; resulting in safer, cheaper and higher performance structures. However, a progressive damage approach also opens up other opportunities. Another reason for modelling is to aid in understanding the mechanisms involved in fatigue failure. This can be achieved through comparing experimental results with those predicted, and in this a modelling method that can accurately characterise the effects of loading on measurable performance parameters throughout a test samples load history is particularly useful. A third reason for modelling is to support in-service monitoring and re-lifeing of structures. In this case it is necessary to correlate measurable parameters with their consequences, in terms of progressive damage of the structure. For example, in recent works (Zhang and Shang 1995, Crocombe et al. 2002, Graner-Solana et al. 2007, Shenoy et al. 2009a), it has been shown that simple back-face strain measurement can be used to monitor the various stages of fatigue damage in a bonded single-lap joints. Hence, continuous monitoring of such a signal can be used to monitor that the joint is performing as predicted and to initiate timely intervention should the measurements indicate a potentially significant change in the structure. A CDM approach that is able to model the effect of changes in the measurable parameters on the residual strength or remaining life of a structure is valuable in this function.

Fatigue is often divided into initiation and propagation phases. In adhesively-bonded joints, the differentiation between these two phases, and even if there really are two such distinct phases, is questionable. However, in terms of modelling, there is often a distinction between how a propagating crack is analysed and how the number of cycles before a macro crack has formed can be predicted. Another advantage of CDM when applied to fatigue loading is that prediction of both of these phases can be incorporated into a single, integrated methodology, rather than resorting to a hybrid method.

Abdel Wahab et al. (2001) applied the thermodynamic approach proposed by Lemaitre (1984) to the prediction of fatigue in double-lap and lap-strap joints, shown in Fig. 9.9. A simplification to equation (9.8) was derived for the case of small values of stress ratio which, when integrated, assuming an initial condition of $D = 0$, gave the following expression for D

$$D = 1 - [1 - A(\beta + m + 1)\Delta\sigma_{eq}^{\beta+m} R_v^{\beta/2} N)]^{\frac{1}{\beta+m+1}} \qquad (9.14)$$

where $\Delta\sigma_{eq}$ is the von Mises stress range, R_v is the triaxiality function (which is the square of the ratio of the damage equivalent stress to the von Mises equivalent stress), m is the power constant in the Ramberg-Osgood equation and A and β are experimentally determined damage parameters. The number of cycles to failure (N_f) can be determined from equation (9.14) using the condition that, at the fully damaged state, $D = 1$ and $N = N_f$, giving

$$N_f = \frac{\Delta\sigma_{eq}^{-\beta-m} R_v^{-\beta/2}}{A(\beta + m + 1)} \qquad (9.15)$$

As there are two constants in equation 9.15, two data points are required to determine them. Abdel Wahab et al. (2001) used two points from constant amplitude fatigue experiments of CFRP-epoxy double-lap joints to determine these parameters for a particular adhesive at a particular temperature, and showed that equation 9.15 could accurately predict the stress life (S-N) curve. It was also seen that while the material constants in equation (9.15) were temperature-dependent, they could be used with a different substrate if failure remained in the adhesive. Application to a different joint type, the lap-strap joint, was reasonable at low temperatures but poor at elevated temperatures. This was attributed to the high susceptibility of the double-lap joint to creep-fatigue at elevated temperatures compared to the lap-strap joint.

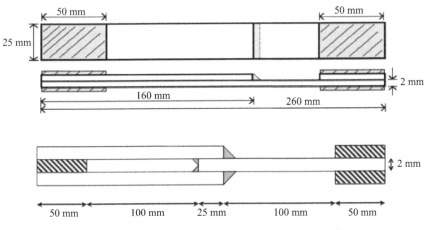

FIG. 9.9 Schematics of lap-strap (top) and double-lap (bottom) joints.

In a later work, Abdel Wahab et al. (2010a) extended the previously described approach to the low cycle fatigue of bulk adhesive where damage evolution curves were derived assuming isotropic damage and a stress triaxiality function equal to one. These were seen to agree well with experimental measurements of damage based on decreasing stress range with increasing cycles under constant displacement amplitude fatigue. Application of this method to single-lap joints (Abdel Wahab et al. 2010b) required determination of the triaxiality function (as given in equation 9.6) to account for the multi-axial stress state in the joint, and it was seen that this value varied along the adhesive layer. The dependency of the triaxiality function on the joint type was further investigated by Abdel Wahab et al. (2011a, 2011b) in a later work.

Although the CDM approach described above enabled the progressive degradation of the adhesive layer to be characterised, it did not allow the initiation and propagation phases of fatigue to be explicitly modelled. Ashcroft et al. (2010) used a simple CDM-based approach to progressively model the initiation and evolution of damage in an adhesive joint, leading to crack formation and growth. In this approach, the damage rate dD/dN was assumed to be a power law function of the localised equivalent plastic strain range, $\Delta\varepsilon_p$, i.e.

$$\frac{dD}{dN} = C_D(\Delta\varepsilon_p)^{m_D} \tag{9.16}$$

where C_D and m_D are experimentally derived constants. The fatigue damage law was implemented in a finite element model. In each element the rate of damage was determined from the finite element analysis using equation (9.16) and the element properties degraded as

$$E = E_0(1 - D)$$
$$\sigma_{yp} = \sigma_{yp0}(1 - D) \tag{9.17}$$
$$\beta = \beta_0(1 - D)$$

where E_0, σ_{yp0} and β_0 are Young's modulus, yield stress and plastic surface modifier constant for the Parabolic Mohr-Coulomb model, respectively, and $D = 1$ represents a fully damaged element, which was used to define the macro-crack length. E, σ_{yp} and β are the values of the Young's modulus, yield stress and plastic surface modifier constant, respectively, after incorporating the material damage. The fatigue life was broken into a number of steps with a specified number of cycles, ΔN, and at each step the increased damage was determined using

$$D_{i+1} = D_i + \frac{dD}{dN}dN \tag{9.18}$$

Shenoy et al. (2010a) showed that this method could be used to predict total-life plots, the fatigue initiation life, fatigue crack growth curves and strength and stiffness wearout plots, and, hence, termed this a unified fatigue methodology (UFM), as illustrated in Fig. 9.10.

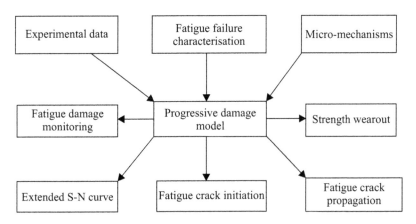

FIG. 9.10 Schematic of unified fatigue methodology (Shenoy et al. 2010a).

Results from applying the method can be seen in Fig. 9.11. Figure 9.11a shows the experimental and predicted reduction in residual strength, i.e. strength wearout, as a function of fatigue cycles for double-lap joints with maximum fatigue loads of 54 and 63% of the static failure load. It can be seen that excellent agreement exists between the experimental and predicted values. Figure 9.11b shows a load-life plot in which it can be seen that the CDM approach predicts the experimental results well. It can also be seen that the damage mechanics approach can also be used to separate the life into initiation and propagation phases, where it is seen that initiation is increasingly dominant at high cycles. The fatigue life predicted using a fracture mechanics approach, which only models the crack propagation phase, has also been added to the figure for comparison. This agrees well with the propagation life predicted from the CDM method but increasingly diverges from the experimental total life results at high cycles. Shenoy et al. (2010b) later showed that this approach could also be applied to variable amplitude fatigue. Previous works (Erpolat et al. 2004a, 2004b) have shown that, in variable amplitude fatigue, the load history affects crack growth with events such as overloads and mean shifts potentially causing accelerated fatigue failure. Although methods of accounting for this in strength wearout (Shenoy et al. 2009b) and fracture mechanics (Ashcroft 2004) approaches to predict fatigue failure have been proposed, the advantage of the CDM method is that it has an inherent capacity to account for load-history without further modification. This is because events such as overloads will increase damage in the process zone ahead of a crack, resulting in a crack acceleration through the weakened material. Shenoy et al. (2010b) demonstrated that the previously described CDM method was able to predict the effects of load interactions in variable amplitude fatigue whereas a standard fracture mechanics method could not.

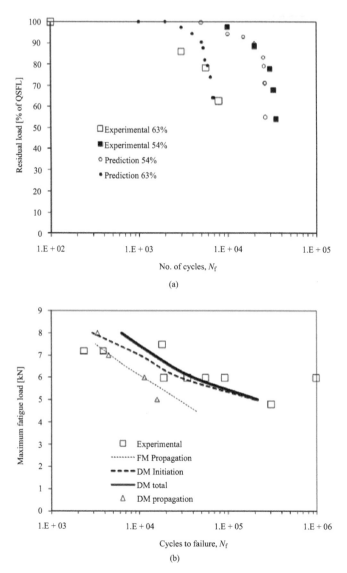

FIG. 9.11 Predicted strength wearout and load-cycle plots using the UFM (Shenoy et al. 2010a).

Walander et al. (2014) experimentally studied Mode-I fatigue crack growth in rubber based and polyurethane based adhesives using a double-cantilever beam specimen. A damage growth law with a constitutive relation for the adhesive material degradation was implemented in a commercial finite element code. The presented damage evolution law was of the form

$$\frac{dD}{dN} = \alpha \left(\frac{\dfrac{\sigma}{1-D} - \sigma_{th}}{\sigma_{th}} \right)^{\beta} \tag{9.19}$$

The law has three material parameters: α, β and σ_{th}, which were determined from the experimental data. Good correlation between the experimental data and the proposed damage law for fatigue was reported.

9.3.3 Environmental Degradation

The sensitivity of adhesive joint failure to environmental effects, such as high temperature and humidity, are well known. However, to date there has been little application of CDM to modelling these effects. This is most likely due to the difficulties of incorporating CDM in a mechanistically accurate model of the effect of the environment on a bonded joint. As mentioned previously, the material parameters in CDM are often difficult to determine and this problem is greatly exasperated if these also need to be functions of environmental parameters such as temperature and moisture content. Therefore, if the effects of the environment on joint strength is all that is required there are simpler methods. However, if an understanding of the process of environmentally enhanced degradation is wanted, the CDM approach can be very useful. However, in order to achieve this, the relevant environmental ageing must also be modelled in a mechanistically accurate manner. In adhesive joints this means taking account of physical processes, such as moisture diffusion, material effects, such as hygro-thermal softening, and transient mechanical effects, such as creep. Therefore, the first part of this section will introduce a general framework for modelling degradation in adhesive joints. After this, the work that has attempted to combine such environmental modelling with CDM to predict the in-service durability of adhesive joints is reviewed.

In order to apply one of the CDM approaches discussed in this chapter to predict environmental degradation in adhesively-bonded joints, it must be integrated into a durability modelling framework. Current state-of-the-art environmental degradation modelling of bonded joints involves multi-physics finite element analysis combined with an appropriate failure criterion and progressive damage modelling approach. This involves three main steps: (i) modelling moisture transport through the joint in order to determine the moisture concentration distribution in the joint as a function of time; (ii) determination of the transient mechanical-hygro-thermal stress-strain state resulting from the combined effects of hygro-thermal effects and applied loads; and (iii) application of a failure criterion to model the progressive failure of the joint and hence enable the residual strength or lifetime to be predicted. A proposed environmental modelling framework incorporating these steps is illustrated in Fig. 9.12. Steps (i) and (ii) can be coupled with various progressive damage modelling methods, such as the fracture mechanics and CZM approaches. As steps (i) and (ii) have been reviewed previously (e.g. Crocombe et al. 2008, Ashcroft and Comyn 2011), this section will concentrate on the application of CDM within a multi-physics finite element-based environmental modelling framework, and the previous work should be referenced for further information on environmental modelling methods for bonded joints.

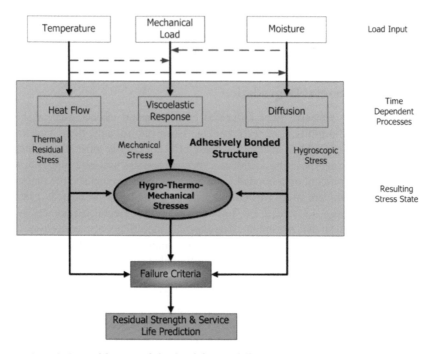

FIG. 9.12 General framework for durability modelling of adhesive joints (Jumbo 2007).

In comparison with CZM, relatively little work has been published on the application of CDM methods to predict failure in adhesively-bonded joints, and even less on the prediction of environmental ageing. As discussed previously, Abdel Wahab et al. (2001) applied Lemaitre's thermodynamic-based damage approach to the fatigue of double-lap and lap-strap joints, including the effect of temperature, by determining temperature dependent damage model parameters. However, no attempt has been made to extend this to include transient thermal or moisture effects. As seen in section 9.3.1, a number of researchers have also applied the GTN damage model to adhesively-bonded joints. However, the only application of the GTN model to durability seen to date is that in Jumbo (2007). In this work a GTN based model was used to predict failure in a variety of single and double-lap joints. In this model, the value of VVF changes due to the growth of existing voids and the nucleation of new voids was characterised by

$$\dot{f}_v = \dot{f}_{\text{growth}} + \dot{f}_{\text{nucleation}} \tag{9.20}$$

The rate of growth of voids is given by

$$\dot{f}_{\text{growth}} = (1 - \dot{f}_v)\dot{p} \tag{9.21}$$

and a statistically-based plastic strain controlled nucleation of voids was used (Chu and Needleman 1980), as given by

$$\dot{f}_{\text{nucleation}} = \frac{f_N}{S_N \sqrt{2\pi}} \exp\left[-\frac{1}{2}\left(\frac{p - \varepsilon_N}{S_N}\right)^2\right] p \tag{9.22}$$

where f_N is the void volume fraction of voids that can nucleate, ε_N is the mean strain at which nucleation occurs, S_N is the standard deviation of the failure strain and p is equivalent plastic strain increment. The direct experimental determination of some of the constants for the GTN model is difficult, hence, in Jumbo (2007) some of the model parameters were determined from experimental data, and others by calibrations from a finite element analysis of a representative joint. The model was applied to both aged and unaged joints, using both 2D and 3D models and accounting for residual stresses from differential hygro-thermal expansions.

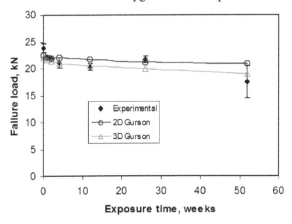

FIG. 9.13 Effect of ageing at 50°C, 95% relative humidity on the failure load of a CFRP-aluminium-epoxy double-lap joint predicted using the Gurson CDM (Jumbo 2007).

Figure 9.13 shows a comparison of the predicted failure load as a function of exposure time for CFRP-aluminium-epoxy double-lap joints conditioned at 50°C and 95% relative humidity. It can be seen that the 3D model results in a good prediction of the experimental trends, whereas the 2D model tends to over-predict the residual strength after ageing.

An obvious drawback to the application of the GTN model to predict the environmental ageing of adhesive joints is the large number of parameters required, many of which are difficult to obtain directly. Thus, these must be deduced using inverse methods or through estimation. A less mechanistically-based, but more readily applicable approach to CDM was suggested by Hua et al. (2008). This was based on a representation of damage based on the stress-plastic displacement response shown in Fig. 9.14, together with the introduction of a damage parameter, D, which varies from 0 at the onset of damage to 1 when the material has lost its load bearing capacity. From points a-c in Fig. 9.14, the material response is dictated by an elastic-plastic constitutive law and the moisture degrades this response. Point c represents the onset of damage, as determined by a damage initiation criterion, after which, along curve c-d, the material response is determined by a damage evolution law. The curve c-d' shows the stress-plastic displacement path without damage. The moisture dependent damage parameter was calibrated from aged mixed-mode flexure (MMF) tests and the model was then shown to be able to predict the effect of ageing on single-lap joints using the same adhesive. The model was also successfully applied to the prediction of durability of joints using a more ductile adhesive, in which case a yield model incorporating hydrostatic stress sensitivity was implemented (Hua et al. 2007). The spatial damage propagation in the 3D CDM for an aged single-lap joint is illustrated using a series of contour plots in Fig. 9.15. The contours represent the damage parameter D. The arrows in Fig. 9.15 indicate the faces exposed to the environment. It can be seen that the damage initiates at the corner of the joint with the saturated edge (A), rather than the slightly less degraded mid-plane section (B), and then propagates from the saturated corner to the middle (B) and the central section (C) of the adhesive layer. The predicted failure load of the joint as a function of environmental ageing time is compared with the experimental results in Fig. 9.16. A good agreement between the predicted and experimental results can be seen and the prediction appears to be largely mesh independent over the range of mesh sizes investigated.

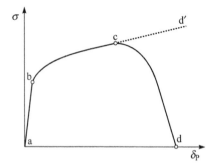

FIG. 9.14 Schematic of stress-plastic strain-based CDM (Hua 2006).

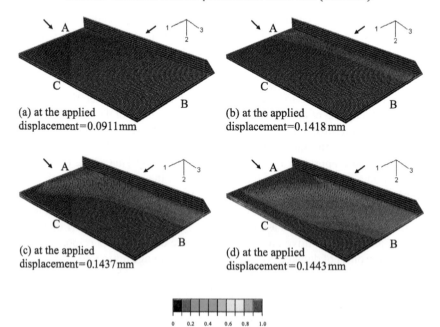

(a) at the applied displacement=0.0911 mm

(b) at the applied displacement=0.1418 mm

(c) at the applied displacement=0.1437 mm

(d) at the applied displacement=0.1443 mm

0   0.2   0.4   0.6   0.8   1.0

FIG. 9.15 3D damage propagation in an epoxy-aluminium single-lap joint aged for 26 weeks using a CDM (Hua 2006).

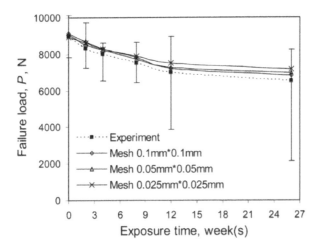

FIG. 9.16 Predicted ultimate failure load of aged epoxy-aluminium single-lap joints using a CDM with different mesh schemes (Hua 2006).

9.4 CONCLUSIONS

Damage mechanics is a tool that is used to model the deterioration of a material under load, ultimately leading to failure. This is achieved through an association of the properties representing the mechanical behaviour of the material with the effects of loads acting upon it via a defined damage variable. The result is the representation of micro-damage as a reduction in the material properties in a damage or process zone in which the effects of the loads are great enough to initiate damage but less than that to cause complete failure. This will affect the response of the structure to further loading and, hence, damage mechanics is often applied in a progressive fashion. In this way, the method is able to predict not just the final failure load, but can also predict the state of damage and response to loading at any point in the load-time history. A number of forms of damage modelling have been applied to adhesively-bonded joints. Damage in the adhesive is generally considered to be isotropic and a scalar damage variable is used, however, in the case of anisotropic damage, a damage tensor may be required to completely define damage in the adhesive joints. As stated previously, the damage variables used in a CDM approach to model failure can be of two main types: the first type can predict the value of damage but does not characterise the damage itself, whereas the second type of damage variable is linked to some physical definition of damage such as porosity or relative area of micro-cavities. The latter type is exemplified by the GTN methods, which have been successfully applied to bonded joints by a number of authors. The advantage of these methods are that direct insights into the role of micro-mechanisms on failure can be developed, such as the role of cavity formation in rubber toughened epoxies. However, the drawback is the number of material constants required, which can be difficult to measure directly. Although inverse methods have been shown to be useful in determining these constants, the loss of a direct connection between the micro-mechanisms and material parameters arguably reduces the method to the first type of damage variable mentioned above. These methods, although not directly related to the micro-mechanisms can still be very powerful in representing the progressive degradation in localised material properties and their effects on structural performance. The damage variables also tend to be easier to calibrate, which probably accounts for their greater popularity and ease with which they have been extended to complex phenomena such as variable amplitude fatigue and environmentally-assisted degradation.

Although, still much less common than the traditional strength of materials and fracture mechanics approaches or CZM, the CDM models have now been demonstrated to be powerful and flexible modelling tools with significant advantages over other methods for certain types of application, materials and type of loading. Although lagging behind the adoption of CZM for adhesive joints, there is no reason why further adoption of this powerful set of modelling methods should not be increasingly used in the future. The XFEM has started to be implemented in commercial finite element codes and, combined with CDM, it can become a tool of significant importance where damage is independent of the finite element mesh and cracks may initiate and propagate based on some micro-mechanical process. For geometries having interfaces, such as adhesive joints, a hybrid approach can be adopted and CDM in combination with CZM can provide a simulation tool that can predict cohesive as well as interfacial cracks. There are still significant research challenges in further developing, validating and applying these methods, which should be of interest to the academic community.

9.5 REFERENCES

Abdel Wahab, M. M., Ashcroft, I. A., Crocombe, A. D., Hughes, D. J. and Shaw, S. J. 2001. Prediction of fatigue threshold in adhesively bonded joints using damage mechanics and fracture mechanics. Journal of Adhesion Science and Technology 15: 763-782.

Abdel Wahab, M. M., Hilmy, I., Ashcroft, I. A. and Crocombe, A. D. 2010a. Evaluation of fatigue damage in adhesive bonding. Part 1: Bulk adhesive. Journal of Adhesion Science and Technology 24: 305-324.

Abdel Wahab, M. M., Hilmy, I., Ashcroft, I. A. and Crocombe, A. D. 2010b. Evaluation of fatigue damage in adhesive bonding. Part 2: Single lap joint. Journal of Adhesion Science and Technology 24: 325-345.

Abdel Wahab, M. M., Hilmy, I., Ashcroft, I. A. and Crocombe, A. D. 2011a. Damage parameters of adhesive joints with general triaxiality. Part 1: Finite element analysis. Journal of Adhesion Science and Technology 25: 903-923.

Abdel Wahab, M. M., Hilmy, I., Ashcroft, I. A. and Crocombe, A. D. 2011b. Damage parameters of adhesive joints with general triaxiality. Part 2: Scarf joint analysis. Journal of Adhesion Science and Technology 25: 925-947.

Ashcroft, I. A., Abdel Wahab, M. M., Crocombe, A. D., Hughes, D. J. and Shaw, S. J. 2001. Effect of temperature on the quasi-static strength and fatigue resistance of bonded composite double lap joints. Journal of Adhesion 75: 61-88.

Ashcroft, I. A. 2004. A simple model to predict crack growth in bonded joints and laminates under variable amplitude fatigue. Journal of Strain Analysis for Engineering Design 39: 707-716.

Ashcroft, I. A., Shenoy, V., Critchlow, G. W. and Crocombe, A. D. 2010. A comparison of the prediction of fatigue damage and crack growth in adhesively bonded joints using fracture mechanics and damage mechanics progressive damage methods. Journal of Adhesion 86: 1203-1230.

Ashcroft, I. A. and Comyn, J. 2011. Effect of water and mechanical stress on durability. pp. 787-822. *In*: Lucas F. M. da Silva, Andreas Öchsner and Robert D. Adams (eds.). Handbook of Adhesion Technology. Springer, Heidelberg.

Ban, Chang-Su, Young-Hwan, Lee, Jin-Ho, Choi and Jin-Hwe, Kweon. 2008. Strength prediction of adhesive joints using the modified damage zone theory. Composite Structures 86: 96-100.

Chousal, J. A. G. and de Moura, M. F. S. F. 2013. Mixed-mode I+II continuum damage model applied to fracture characterization of bonded joints. International Journal of Adhesion and Adhesives 41: 92-97.

Chu, C. C. and Needleman, A. 1980. Void nucleation effects in biaxially stretched sheets. Journal of Engineering Materials and Technology 102: 249-256.

Crocombe, A. D., Ong, A. D., Chan, C. Y., Abdel Wahab, M. M. and Ashcroft, I. A. 2002. Investigating fatigue damage evolution in adhesively bonded structures using backface strain measurement. Journal of Adhesion 78: 745-778.

Crocombe, A. D., Ashcroft, I. A. and Abdel Wahab, M. M. 2008. Environmental degradation. pp. 225-241. *In*: Lucas F. M. da Silva and Andreas Öchsner (eds.). Modelling of Adhesively Bonded Joints. Springer-Verlag, Berlin Heidelberg.

de Moura, M. F. S. F. and Chousal, J. A. G. 2006. Cohesive and continuum damage models applied to fracture characterization of bonded joints. International Journal of Mechanical Sciences 48: 493-503.

de Moura, M. F. S. F., Gonçalves, J. P. M., Chousal, J. A. G. and Campilho, R. D. S. G. 2008. Cohesive and continuum mixed-mode damage models applied to the simulation of the mechanical behaviour of bonded joints. International Journal of Adhesion and Adhesives 28: 419-426.

Dean, G. D. and Mera, R. 2004. Determination of material properties and parameters required for simulation of impact performance of plastics using finite element analysis. NPL Report DEPC-MPR 007, Middlesex.

Erpolat, S., Ashcroft, I. A., Crocombe, A. D. and Abdel-Wahab, M. M. 2004a. Fatigue crack growth acceleration due to intermittent overstressing in adhesively bonded CFRP joints. Compsites Part A: Applied Science and Manufacturing 35: 1175-1183.

Erpolat, S., Ashcroft, I. A., Crocombe, A. D. and Abdel-Wahab, M. M. 2004b. A study of adhesively bonded joints subjected to constant and variable amplitude fatigue. International Journal of Fatigue 26: 1189-1196.

Graner-Solana, A., Crocombe, A. D., Wahab, M. A. and Ashcroft, I. A. 2007. Fatigue initiation in adhesively bonded single lap joints. Journal of Adhesion Science and Technology 21: 1343-1357.

Gurson, A. L. 1977. Continuum theory of ductile rupture by void nucleation and growth: Part 1 Yield criterion and flow rules for porous materials. Journal of Engineering Materials and Technology 99: 2-15.

Harris, J. A. and Fay, P. A. 1992. Fatigue life evaluation of structural adhesives for automotive applications. International Journal of Adhesion and Adhesives 12: 9-18.

Hart-Smith, L. J. 1981. Developments in Adhesives 2. Applied Science Publication, London.

Hua, Y. 2006. Modelling environmental degradation on adhesively bonded joints. PhD thesis, University of Surrey, Guildford, UK.

Hua, Y., Crocombe, A. D., Wahab, M. A. and Ashcroft, I. A. 2006. Modelling environmental degradation in EA9321 bonded joints using a progressive damage failure model. The Journal of Adhesion 82: 135-160.

Hua, Y., Crocombe, A. D., Wahab, M. A. and Ashcroft, I. A. 2007. Continuum damage modelling of environmental degradation in joints bonded with E32 epoxy adhesive. Journal of Adhesion Science and Technology 21: 179-195.

Hua, Y., Crocombe, A. D., Wahab, M. A. and Ashcroft, I. A. 2008. Continuum damage modelling of environmental degradation in joints bonded with EA9321 epoxy adhesive. International Journal of Adhesion and Adhesives 28: 302-313.

Imanaka, M., Nakamura, Y., Nishimura, A. and Iida, T. 2003. Fracture toughness of rubber-modified epoxy adhesives: Effect of plastic deformability of the matrix phase. Composites Science and Technology 63: 41-51.

Imanaka, M., Motohashi, S., Nishi, K., Nakamura, Y. and Kimoto, M. 2009. Crack growth behaviour of epoxy adhesives with rubber and cross-linked rubber particles under mode I loading. International Journal of Adhesion and Adhesives 29: 45-55.

Jumbo, F. S. 2007. Modelling residual stress and environmental degradation in adhesively bonded joints. PhD Thesis, Loughborough University.

Kachanov, L. M. 1958. On rupture time under conditions of creep. Ivestia Akademi Nauk USSR 8: 26-31.

Kachanov, L. M. 1986. Introduction to Continuum Damage Mechanics. Martinus Nijhoff, Dordrecht, Netherlands.

Kattan, P. I. and Voyiadjis, G. X. 2002. Damage Mechanics with Finite Elements. Springer Berlin Heidelberg, New York.

Lee, Chi-Seung, Min-Sung, Chun, Myung-Hyun, Kim and Jae-Myung, Lee. 2011. Numerical evaluation for debonding failure phenomenon of adhesively bonded joints at cryogenic temperatures. Composites Science and Technology 71: 1921-1929.

Lemaitre, J. 1984. How to use damage mechanics. Nuclear Engineering and Design 80: 233-245.

Lemaitre, J. 1985. A continuous damage mechanics model for ductile fracture. Journal of Engineering Materials and Technology 83: 83-89.

Lemaitre, J. and Desmorat, R. 2005. Engineering Damage Mechanics. Springer, Berlin Heidelberg, New York.

Masmanidis, I. T. and Philippidis, Theodore P. 2015. Progressive damage modeling of adhesively bonded lap joints. International Journal of Adhesion and Adhesives 59: 53-61.

Mubashar, A., Ashcroft, I. A. and Crocombe, A. D. 2014. Modelling damage and failure in adhesive joints using a combined XFEM-cohesive element methodology. The Journal of Adhesion 90: 682-697.

Needleman, A. and Tvergaard, V. 1984. An analysis of ductile rupture in notched bars. Journal of Mechanics and Physics of Solids 32: 461-490.

O'Mahoney, D. C., Katnam, K. B., O'Dowd, N. P., McCarthy, C .T. and Young, T. M. 2013. Taguchi analysis of bonded composite single-lap joints using a combined interface–adhesive damage model. International Journal of Adhesion and Adhesives 40: 168-178.

Rabotnov, Y. U. N. 1968. Creep rupture. pp. 342-349. *In*: Proceedings of the XII International Congress on Applied Mechanics. Springer, Stanford.

Read, B. E., Dean, G. D. and Ferriss, D. H. 2000. An elastic-plastic model for the non-linear mechanical behaviour of rubber-toughened adhesives. NPL Report CMMT(A) 289, Middlesex.

Ren, B. and Li, S. 2014. Multiscale modeling and prediction of bonded joint failure by using an adhesive process zone model. Theoretical and Applied Fracture Mechanics 72: 76-88.

Shenoy, V., Ashcroft, I. A., Critchlow, G. W., Crocombe, A. D. and Abdel Wahab, M. M. 2009a. An investigation into the crack initiation and propagation behaviour of bonded single-lap joints using backface strain. International Journal of Adhesion and Adhesives 29: 361-371.

Shenoy, V., Ashcroft, I. A., Critchlow, G. W., Crocombe, A. D. and Abdel Wahab, M. M. 2009b. An evaluation of strength wearout models for the lifetime prediction of adhesive joints subjected to variable amplitude fatigue. International Journal of Adhesion and Adhesives 29: 639-649.

Shenoy, V., Ashcroft, I. A., Critchlow, G. W. and Crocombe, A. D. 2010a. Unified methodology for the prediction of the fatigue behaviour of adhesively bonded joints. International Journal of Fatigue 32: 1278-1288.

Shenoy, V., Ashcroft, I. A., Critchlow, G. W. and Crocombe, A. D. 2010b. Fracture mechanics and damage mechanics based fatigue lifetime prediction of adhesively bonded joints subjected to variable amplitude fatigue. Engineering Fracture Mechanics 77: 1073-1090.

Walander, T., Eklind, A., Carlberger, T. and Stigh, U. 2014. Fatigue damage of adhesive layers – experiments and models. Procedia Materials Science 3: 829-834.

Zaïri, F., Nait-Abdelaziz, M., Woznica, K. and Gloaguen, J. M. 2005. Constitutive equations for the viscoplastic-damage behaviour of a rubber-modified polymer. European Journal of Mechanics - A/Solids 24: 169-182.

Zhang, Z. and Shang, J. K. 1995. A backface strain technique for detecting fatigue crack initiation in adhesive joints. The Journal of Adhesion 49: 23-36.

10

Extended Finite Element Modelling in Static Applications

Raul D.S.G. Campilho* and Tiago A.B. Fernandes

10.1 INTRODUCTION

Adhesive-bonding for the unions in structures is a choice to consider over welding, riveting and fastening, due to issues such as the reduction of stress concentrations, reduced weight penalty and easy manufacturing/automation (Campilho et al. 2005, Wang et al. 2008). As the integrity of bonded structures is usually determined by the strength and durability of their bonds (Messler 1993), it is vital for the design of these structures, the availability of numerical methods and damage models that can be reliably employed to predict their fracture behaviour, to minimize design costs and time to market. Actually, an efficient and quick design relies on the existence of damage models of the adhesives that allow a given bonding solution to be analyzed by numerical methods such as the Finite Element Method (FEM). Different strength prediction techniques are available for bonded structures. Theoretical studies were employed in the early-stages of bonded joint analyses (Volkersen 1938) that allowed the establishment of elastic closed-form solutions for the stress fields in the adhesive layer. With these methods, a quick analysis of the structures behaviour was made possible, although requiring a few simplifications (Panigrahi and Pradhan 2007). The use of FEM codes to simulate bonded structures in the computer became feasible a few decades ago, providing more accurate strength predictions for these structures (Tsai and Morton 1994). Two different trends were available: the strength of materials and fracture mechanics-based methods. The strength of materials approach is based on the evaluation of allowable stresses or strains (Lee and Lee 1992), by theoretical formulations or the FEM, which are compared with the properties of the structure constituents. However, the FEM results are known to be mesh dependent (Feih and Shercliff 2005). Fracture mechanics descriptions

Departamento de Engenharia Mecânica, Instituto Superior de Engenharia do Porto, Instituto Politécnico do Porto, Rua Dr. António Bernardino de Almeida, 431, 4200-072 Porto, Portugal.
* Corresponding author: raulcampilho@gmail.com

provide relatively insensitive results to singular regions in structures, but they depend on the fracture toughness of the structure materials. By these methods, an estimate can be made regarding the tolerable defect sizes in structures, thus avoiding repair of expensive parts. Theoretically, any specimen geometry can be used to measure the critical strain energy release rate (G_c) as long as a data reduction scheme is developed for that particular test geometry. The Compact Tension and Double Cantilever Beam (DCB) tests are the most commonly used test methods to estimate the tensile critical strain energy release rate (G_{Ic}). For shear, the End-Notched Flexure (ENF) test is the most widespread to obtain the shear critical strain energy release rate (G_{IIc}). Cohesive zone models (CZM) have been widely used since the 90's for the prediction of fracture in structures and can be viewed as a fracture mechanics criterion that takes advantage of traction-separation laws between stresses and relative displacements to simulate a continuum media or an interface between two materials along pre-specified paths (Campilho et al. 2007). CZM do not require an initial flaw and are based on cohesive elements connecting plane or three-dimensional solid elements of structures. The mentioned traction-separation laws rely mainly on four parameters: G_{Ic}, G_{IIc}, tensile cohesive traction (t_n^0) and shear cohesive traction (t_s^0). Mainly three techniques are available to define the CZM traction-separation laws for adhesives layers: the property identification method, the direct method and the inverse method. The property identification method consists of building a traction-separation law with a predefined shape based on the known material behaviour, whose values of G_{Ic}, G_{IIc}, t_n^0 and t_s^0 are equalled to the measured quantities. Usually, t_n^0 and t_s^0 are defined by testing to the bulk adhesive in tension and shear, respectively. For t_s^0, Thick Adherend Shear Tests (TAST) are also common. G_{Ic} is commonly characterized by the DCB test and G_{IIc} by the ENF test. The direct method computes the cohesive law of an adhesive layer from the measured data of fracture characterization tests (Campilho et al. 2007). For the tensile CZM law, it consists of the measurement, during a DCB test (mode I), of G_{Ic} and the normal end-opening (w) at the crack tip. Differentiation of the G_{Ic}-w data gives an estimation of the tensile traction (t_n)-w law. An identical procedure can be specified for the shear CZM law. In this case, the ENF test can be used to estimate G_{IIc}, while also measuring the shear end-opening (v) at the crack tip. The shear traction (t_s)-v law is then derived differentiation of the G_{IIc}-v curve. This method was successfully used for the extraction of traction–separation laws for adhesive layers (Sørensen 2002, Zhu et al. 2009) and interfacial properties of fibre-reinforced composites (Sørensen and Jacobsen 2003). By the inverse method, the traction-separation law is built after a few iterative comparisons between experimental data and numerical predictions of the very same data by simulations with the same testing geometry and approximate parameterized cohesive laws. The R-curves (relating the strain energy release rate, G, with the crack length, a (Flinn et al. 1993), the crack opening profile (Mello and Liechti 2004) and most typically the load-displacement (P-δ) curve (Li et al. 2005) can be used for fitting. Högberg et al. (2007) proposed a direct integration scheme to capture the cohesive laws of an adhesive layer by a mixed-mode DCB specimen, which consisted of the differentiation of the J-integral vs. end-opening displacement (J-w) curve. Results showed that substantial differences exist between pure tensile and shear loadings. Andersson and Stigh (2004) used an inverse fitting method to estimate the cohesive parameters of a ductile adhesive layer in a DCB specimen. The results showed that the stress-relative displacement relationship is approximately trapezoidal: linear behaviour followed by deformation at a constant stress and parabolic degradation up to failure. The work of Campilho et al. (2009a) validated with experiments that a trapezoidal CZM applied to tensile loaded scarf repairs on carbon-epoxy laminates. To account for the experimental fractures, the cohesive failure of the adhesive layer and composite interlaminar and intralaminar (in the transverse and fibre directions) failures were described in pure mode I and II laws by an inverse modelling technique applied to the P-δ curves of standardized tests. The predictions were accurate, validating the proposed technique to compute the cohesive laws of different materials/interfaces.

The eXtended Finite Element Method (XFEM) is a recent improvement of the FEM for modelling damage growth in structures. It uses damage laws for the prediction of fracture that are based on the bulk strength of the materials for the initiation of damage and strain for the assessment of failure (defined by G_{Ic}), rather than the cohesive tractions and tensile/shear relative displacements used in CZM. XFEM gains an advantage over CZM modelling as it does not require the crack to follow a predefined path. Actually, cracks are allowed to grow freely within a bulk region of a material without the requirement of the mesh to match the geometry of the discontinuities neither remeshing near the crack (Mohammadi 2008). This method is an extension of the FEM, whose fundamental features were firstly presented in the late 90's by Belytschko and Black (1999). The XFEM relies on the concept of partition of unity, that can be implemented in the traditional FEM by the introduction of local enrichment functions for the nodal displacements near the crack to allow its growth and separation between the crack faces (Moës et al. 1999). Due to crack growth, the crack tip continuously changes its position and orientation depending on the loading conditions and structure geometry, simultaneously to the creation of the necessary enrichment functions for the nodal points of the finite elements around the crack path/tip. It uses damage laws for the prediction of fracture that are based on the bulk strength of the materials for the initiation of damage and strain for the assessment of failure (defined by G_{Ic}), rather than the values of t_n^0/t_s^0 or w/v used in CZM.

Varying applications to this innovative technique were proposed to simulate different engineering problems. Sukumar et al. (2000) updated the method to three-dimensional damage simulation. Modelling of intersecting cracks with multiple branches, multiple holes and cracks emanating from holes was addressed by Daux et al. (2000). The problem of cohesive propagation of cracks in concrete structures was studied by Moës and Belytschko (2002), considering three-point bending and four-point shear scaled specimens. More advanced features such as plasticity, contacting between bodies and geometrical non-linearities, which show a particular relevance for the simulation of fracture in structures, are already available within the scope of XFEM. The employment of plastic enrichments in XFEM modelling is accredited to Elguedj et al. (2006), which used a new enriched basis function to capture the singular fields in elasto-plastic fracture mechanics. Modelling of contact by the XFEM was firstly introduced by Dolbow et al. (2001) and afterwards adapted to frictional contact by Khoei and Nikbakht (2006). Fagerström and Larsson (2006) implemented geometrical nonlinearities within the XFEM. Fatigue applications for XFEM were proposed recently (Xu and Yuan 2009, Sabsabi et al. 2011), but these have not yet been applied to the mixed mode fracture of bonded joints. As a result, only static applications of XFEM applied to bonded joints will be considered in this chapter.

This chapter aims to assess the XFEM feasibility to model crack propagation and to predict the fracture behaviour of adhesively-bonded joints. Initially, the XFEM formulation embedded in Abaqus® is described, with focus on the differences to the conventional FEM formulation and the element division procedure to produce a crack anywhere in the models. Following, most of the few applications of XFEM in bonded joints available in the literature are briefly described, in order to assess the general capabilities of the method, and to discuss advantages and limitations. After this description, two case studies are presented: (1) XFEM modelling of the pure tensile behaviour as a function of temperature (DCB specimen) and (2) XFEM comparison with CZM for single-lap joints and varying ductility adhesives. In the first case study, the main purpose of the XFEM analysis is to validate this numerical tool for strength prediction of adhesive joints under different temperatures and a pure tensile loading. The DCB geometry was considered for tensile fracture characterization. To this end, different data reduction techniques are applied to the experimental data to obtain G_{Ic} at varying temperatures, and the final XFEM parameters under different temperatures are proposed for application to strength predictions at each one of these conditions. The damage laws are then numerically validated to model crack propagation in the adhesive layer at room temperature (RT), 100°C, 150°C and 200°C, by achieving a good correlation to the experimental joint behaviour. In the second study, the main objective is to evaluate CZM and XFEM modelling in predicting the strength of single-lap joints bonded with a brittle (Araldite® AV138), a moderately ductile (Araldite® 2015) and a largely ductile adhesive (Sikaforce® 7888) between aluminium adherends and varying values of overlap length (L_O). The predictions are compared against experimental data. The numerical analysis includes the plot of elastic stress distributions, and strength prediction either based on CZM or XFEM. This procedure enabled assessing in detail the performance of these predictive techniques applied to bonded joints. Moreover, it was possible to evaluate which family of adhesives is more suited for each joint geometry. Following the two case studies, a discussion is included on the suitability of the XFEM for strength prediction of bonded joints, and the general conclusions are presented.

10.2 EXTENDED FINITE ELEMENT METHOD FORMULATION

Although a few static implementations of the XFEM were developed in recent years for scenarios other than bonded joints, the generic Abaqus® embedded formulation will be described in this Section (Abaqus® 2013). The XFEM considers an initial linear elastic behaviour of the materials, which is represented by an elastic constitutive matrix that relates stresses with the normal and shear separations of the cracked elements. Damage and failure are simulated in XFEM by suitable damage initiation criteria and damage laws between the real and phantom nodes of a cracked element (to be detailed further in this Section). The damage initiation criteria can rely on stresses or strains, while the traction-separation laws that simulate material degradation up to failure can be linear or exponential. Abaqus® initiates and propagates damage during the simulation at regions experiencing stresses and/or strains higher than the corresponding limiting values. Six crack initiation criteria are available in Abaqus®. The MAXPS (maximum principal stress) and MAXPE (maximum principal strain) criteria are based on the introduction of the following functions (by the respective order)

$$
f = \left\{ \frac{\langle \sigma_{max} \rangle}{\sigma^0_{max}} \right\} \quad \text{or} \quad f = \left\{ \frac{\langle \varepsilon_{max} \rangle}{\varepsilon^0_{max}} \right\}. \tag{10.1}
$$

σ_{max} and σ^0_{max} represent the current and allowable maximum principal stress. The Macaulay brackets indicate that a purely compressive stress state does not induce damage. ε_{max} and ε^0_{max} represent the current and allowable maximum principal strain. The MAXS (maximum nominal stress) and MAXE (maximum nominal strain) criteria are represented by the following functions, respectively

$$
f = \max \left\{ \frac{\langle t_n \rangle}{t^0_n}, \frac{t_s}{t^0_s} \right\} \quad \text{or} \quad f = \max \left\{ \frac{\langle \varepsilon_n \rangle}{\varepsilon^0_n}, \frac{\varepsilon_s}{\varepsilon^0_s} \right\}. \tag{10.2}
$$

t_n and t_s are the current normal and shear traction components. t^0_n and t^0_s represent the respective limiting values. The strain parameters have identical significance. The quadratic nominal stress (QUADS) and quadratic nominal strain (QUADE) criteria are based on the introduction of the following functions, respectively

$$
f = \left\{ \frac{\langle t_n \rangle}{t^0_n} \right\}^2 + \left\{ \frac{t_s}{t^0_s} \right\}^2 \quad \text{or} \quad f = \left\{ \frac{\langle \varepsilon_n \rangle}{\varepsilon^0_n} \right\}^2 + \left\{ \frac{\varepsilon_s}{\varepsilon^0_s} \right\}^2. \tag{10.3}
$$

All criteria are fulfilled, and damage initiates, when f reaches unity. For damage growth, the fundamental expression of the displacement vector $\mathbf{u}$, including the displacements enrichment, is written as (Abaqus® 2013)

$$
\mathbf{u} = \sum_{i=1}^{N} N_i(x) \left[\mathbf{u}_i + H(x)\mathbf{a}_i + \sum_{\alpha=1}^{4} F_\alpha(x)\mathbf{b}_i^a \right]. \tag{10.4}
$$

$N_i(x)$ and $\mathbf{u}_i$ relate to the conventional FEM technique, corresponding to the nodal shape functions and nodal displacement vector linked to the continuous part of the formulation, respectively. The second term between brackets, $H(x)\mathbf{a}_i$, is only active in the nodes for which any relating shape function is cut by the crack and can be expressed by the product of the nodal enriched degree of freedom vector including the mentioned nodes, $\mathbf{a}_i$, with the associated discontinuous shape function, $H(x)$, across the crack surfaces (Abaqus® 2013)

$$H(x) = \begin{cases} 1 \text{ if } & (x - x^*) \cdot n \geq 0. \\ -1 \text{ otherwise} \end{cases} \qquad (10.5)$$

x is a sample Gauss integration point, x^* is the point of the crack closest to x, and n is the unit vector normal to the crack at x^* (Fig. 10.1). Finally, the third term is only to be considered in nodes whose shape function support is cut by the crack tip and is given by the product of the nodal enriched degree of freedom vector of this set of nodes, b_i^α, and the associated elastic asymptotic crack-tip functions, $F_\alpha(x)$ (Sukumar and Prévost 2003). $F_\alpha(x)$ are only used in Abaqus® for stationary cracks, which is not the current case.

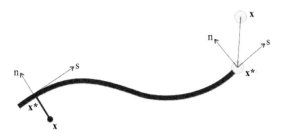

FIG. 10.1 Representation of normal and tangential coordinates for an arbitrary crack (Campilho et al. 2011a).

In the presence of damage propagation, a different approach is undertaken, based on the establishment of phantom nodes that subdivide elements cut by a crack and simulate separation between the newly created sub elements. By this approach, the asymptotic functions are discarded, and only the displacement jump is included in the formulation. Propagation of a crack along an arbitrary path is made possible by the use of phantom nodes that initially have the exactly same coordinates than the real nodes and that are completely constrained to the real nodes up to damage initiation.

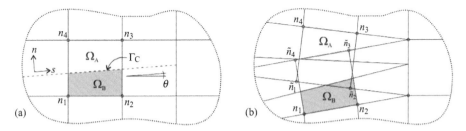

FIG. 10.2 Damage propagation in XFEM using the phantom nodes concept: before (a) and after partitioning (b) of a cracked element into sub elements (Campilho et al. 2011a).

In Fig. 10.2, the highlighted element has nodes n_1 to n_4. After being crossed by a crack at Γ_C, the element is partitioned in two sub-domains, Ω_A and Ω_B. The discontinuity in the displacements is made possible by adding phantom nodes ($\tilde{n}_1$ to $\tilde{n}_4$) superimposed to the original nodes. When an element cracks, each one of the two sub elements will be formed by real nodes (the ones corresponding to the cracked part) and phantom nodes (the ones that no longer belong to the respective part of the original element). These two elements that have fully independent displacement fields replace the original one, constituted by the nodes $\tilde{n}_1$, $\tilde{n}_2$, n_3 and n_4 (Ω_A) and n_1, n_2, $\tilde{n}_3$ and $\tilde{n}_4$ (Ω_B). From this point, each pair of real/phantom node of the cracked element is allowed to separate according to a suitable cohesive law up to failure. At this stage, the real and phantom nodes are free to move unconstrained, simulating crack growth. In terms of damage initiation, Abaqus® allows the user to define initial cracks, but this is not mandatory. Regardless the choice taken, Abaqus® initiates and propagates damage during the simulation at regions experiencing principal stresses and/or strains greater than the corresponding limiting values specified in the traction-separation laws. Crack initiation/propagation will always take place orthogonally to the maximum principal stresses or strains.

10.3 STATIC APPLICATIONS OF THE EXTENDED FINITE ELEMENT METHOD

XFEM applications to bonded joints are scarce and extremely recent in the literature, and this Section actually describes the entire range of applications found while writing this work. Premchand and Sajikumar (2009) studied by the XFEM the variation of the tensile and shear stress intensity factors in adhesively bonded joints with different crack lengths, using the adhesive FM® 300-2 and aluminium adherends. The XFEM model, coupled to the conventional FE method, was implemented in a Matlab® code, for the geometry of Fig. 10.3(a), in which the adhesive holds a flaw with the depicted shape. A series of analyses were carried out under tensile and shear loadings, giving estimations for stress intensity factors in both modes of loading. Figure 10.3b relates to the joint deformed shape after horizontal crack propagation induced by a pure tensile loading applied at the adherend edges.

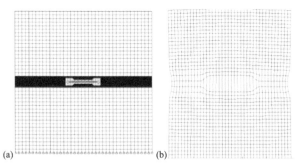

(a)   (b)

FIG. 10.3 Bonded joint geometry for stress intensity factor calculation by XFEM (a) and horizontal crack propagation after a pure tensile load (b) (Premchand and Sajikumar 2009).

The study by Campilho et al. (2011a) evaluated the XFEM feasibility to model crack propagation and to predict the fracture behaviour of a thin bond of two structural epoxy adhesives under tension by the DCB test, under varying restraining conditions: stiff and compliant adherends. The damage laws of the XFEM were built by experimental determination of G_{Ic} and t_n^0, by DCB and bulk tensile tests, respectively. Two DCB configurations were selected to check the suitability of the XFEM in simulating fracture in bonded joints. Configuration A corresponds to testing of a brittle epoxy adhesive (XN1244 by Nagase Chemtex) between stiff steel adherends, while in configuration B a ductile adhesive (Araldite® 2015) with compliant carbon-epoxy adherends was tested. Two methods were employed to evaluate G_{Ic}: the Compliance Calibration Method (CCM) and the Corrected Beam Theory (CBT), including the effects of crack tip rotation and deflection by using a correction factor, since the formulation assumes clamped adherends at the crack tip. The two-dimensional (2D) XFEM analysis was performed in Abaqus® and the damage laws were assumed as triangular, i.e., with linear softening after t_n^0, by using the properties obtained in the DCB and bulk tension tests.

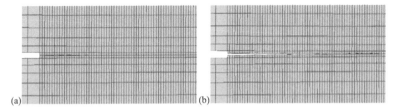

FIG. 10.4 Crack growth by the XFEM algorithm, initiating at the crack tip (a) and growing horizontally along the bondline (b) for configuration A (Campilho et al. 2011a).

The steel and composite adherends were regarded as elastic isotropic and elastic orthotropic, respectively, whilst the adhesive bond was modelled by the XFEM. Validation of the XFEM and the proposed laws was accomplished by comparison of the experimental DCB P-δ curves with the output of the XFEM simulations for the same geometry. Figure 10.4 depicts crack growth by the XFEM algorithm, initiating at the crack tip (a) and growing horizontally along the bondline (b) for configuration A. Figure 10.5 compares the experimental and XFEM P-δ curves of the DCB specimens for configuration A, showing the classical concave shape after the peak load corresponding to crack growth at a constant G_{Ic} value, and an accurate prediction of the specimens behaviour by using the XFEM with the previously characterized parameters.

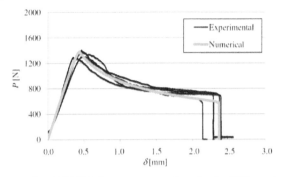

FIG. 10.5 Experimental and XFEM P-δ curves comparison for the DCB specimens, considering configuration A (Campilho et al. 2011a).

For all conditions, the elastic stiffness, peak load, and load during propagation showed a good agreement, which testified the suitability of XFEM to simulate bonded structures for the conditions specified in the analysis, i.e., pure tensile loading.

Campilho et al. (2011b) tested the CZM and XFEM formulations embedded in Abaqus® for the simulation of single and double-lap bonded joints between aluminium adherends and bonded with a brittle adhesive (Araldite® AV138). Values of L_O between 5 and 20 mm were tested. The adhesive was characterized under tension and shear, which allowed the determination of the XFEM damage parameters. The 2D XFEM analysis considered geometrical non-linearities, with plane strain solid elements. Figure 10.6a shows the failure process for a single-lap joint with $L_O = 20$ mm (detail at the overlap edge) using the principal strain criterion for the initiation of damage and estimation of crack growth direction. At this point, the direction of maximum strain led to propagation of damage towards the aluminium adherend.

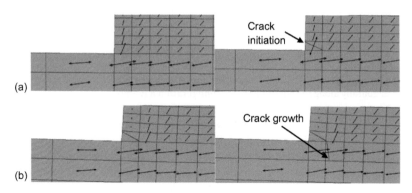

FIG. 10.6 Progressive failure of a single lap joint with $L_O = 20$ mm using XFEM (the arrows represent the directions of maximum principal strain): damage initiation within the adhesive at the overlap edges (a) and damage growth to the aluminium adherend (b) (Campilho et al. 2011b).

When the crack front reached the adherend, damage propagated almost vertically due to the corresponding direction of principal strains at the crack tip (Fig. 10.6b), which clearly does not reflect the real behaviour of single lap joints. The authors concluded that damage propagation along the adhesive bond is thus rendered unfeasible with this technique, as it is currently implemented in Abaqus®, since the XFEM algorithm will always search for maximum stresses/strains at the crack tip, shifting the crack to the adherends, disregarding what happens within the adhesive bond and thus preventing damage propagation along the bondline. As a result of this handicap, a different solution was proposed by the authors, supported by the brittleness of the adhesive used. The joint strength was estimated by the initiation of cohesive cracking of the adhesive bond at the overlap edges, using the maximum principal strain criterion as it showed to be slightly less mesh sensitive than the maximum principal stress equivalent.

Figure 10.7 compares the experimental and XFEM data considering the maximum principal strain criterion, showing that the XFEM is moderately accurate in simulating these structures with brittle adhesives, which undergo a catastrophic failure when the maximum strain of the adhesive is attained anywhere in the structure. However, it was clarified that the chosen methodology was only acceptable due to the brittleness of the adhesive. If a ductile adhesive had been used instead, the predictions would clearly underestimate the experiments.

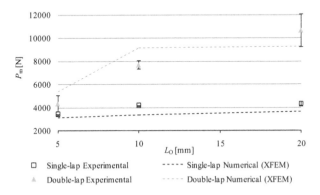

FIG. 10.7 Experimental and XFEM strength comparison as a function of L_0 (Campilho et al. 2011b).

Figure 10.8 shows the results of the presented mesh dependency study, by plotting the values of normalized peak load for element sizes at the overlap edges (equal length and height) between 0.05 and 0.2 mm, showing that, as expected, the predictions are extremely mesh dependent.

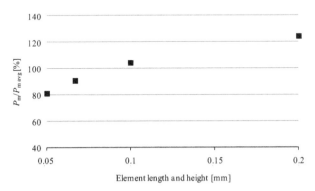

FIG. 10.8 XFEM mesh dependency study for the single-lap joint with $L_0 = 20$ mm (Campilho et al. 2011b).

Globally, the direct comparisons between the experimental data and the output of the simulations revealed fair predictions by the XFEM using the proposed simplification to the original formulation. However, the XFEM did not show to be suited for damage propagation in bonded joints as it is currently implemented in Abaqus®, since the direction of crack growth is ruled by the maximum principal stresses/strains at the crack tip which, in bonded joints, typically leads to damage growth towards and within the adherends.

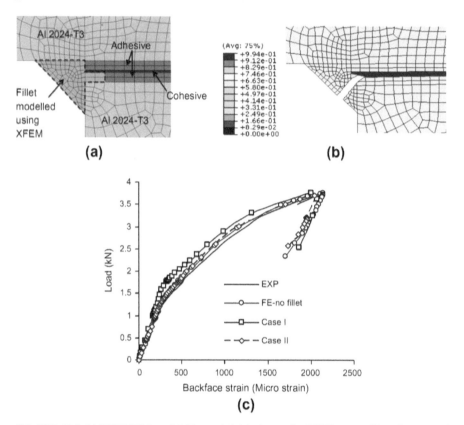

FIG. 10.9 Hybrid XFEM/CZM model (a), crack initiation at the XFEM region (b) and curves of load-backface strain for the experimental and FEM models (Sugiman et al. 2013).

In the work of Sugiman et al. (2013), the CZM technique was used to simulate damage growth in adhesively bonded single lap joints and laminated doublers between aluminium adherends. In an initial phase, the authors used the backface strain technique to follow damage in the adhesive layer, such that the calibration of a unique set of cohesive zone properties for the single lap joint was accomplished. The backface strain technique was also employed to infer the influence of the substrates' plasticity, position of cohesive elements, traction and fracture energy, and adhesive fillet in a single-lap joint. After the attainment of the calibrated CZM parameters, these were employed for strength prediction of the doublers under bending loads. A relevant feature of this work is that, for the simulations of the single-lap joints containing a fillet, the XFEM was used together with CZM modelling to model damage initiation in the fillet, by taking advantage of the non-necessity to define the crack path a priori and automatic mesh partitioning during crack propagation. Thus, an enriched XFEM region was considered for the fillet, while the adhesive bond failure was assessed by CZM (Fig. 10.9a).

As it can be observed in Fig. 10.9a, the XFEM region spans from the fillet until 0.2 mm inside the overlap (region inside the dashed line). The region inside the overlap enabled the continuity of damage between the XFEM and CZM regions. The fillet was considered as a right triangle with side length of 0.5 mm (comprising the full adherends' height). The element size within the fillet was roughly 0.05 × 0.05 mm. The failure path by this hybrid modelling technique is shown in Fig. 10.9b. Regarding the XFEM region, a maximum principal stress criterion with identical elastic stiffness as the adhesive stiffness was considered. Damage evolution was account for by the Benzeggagh-Kenane (BK) criterion. It was experimentally observed that the bonded region between the vertical end of the adherend and the fillet was poorly bonded, since that particular adherend face was not properly treated. Thus, two models were created: one considering a good bond between the fillet and the vertical face (Case I), and another one in which the bonding between these two materials was weak (Case II). In the first scenario, the maximum principal stress and fracture toughness were considered as being equal to the adhesive layer, while in the second, the Young's modulus, maximum principal stress, and normal and shear values of fracture toughness were artificially degraded by the same amount. This quantity was numerically defined as the best match to the experiments and gave 69% of degradation.

Table 10.1 XFEM properties for case I and II (Sugiman et al. 2013).

	Young's modulus	Poisson's ratio	Maximum principal stress [MPa]	G_{Ic} [kJ/m²]	G_{IIc} [kJ/m²]	Damage stabilization [Ns/m]
Case I (good fillet)	2300	0.4	49	2.5	5	0.0001
Case II weak fillet	1314	0.4	15	0.76	1.52	0.0001

Table 10.1 shows the XFEM properties to model the adhesive fillet in both scenarios. Figure 10.9c provides a comparison between the experimental and numerical (CZM/XFEM) load-backface strain curves. Results show that the simulation with case I properties under predicts the experimental strain for a given load. The simulation with case II properties gives a load-backface strain curve similar to that of the experimental results and simulation with no fillet. In the end, the hybrid CZM/XFEM approach was found valid but discarded in the subsequent simulations, since it provides identical results to the simulation without any fillet.

An identical approach was undertaken by Mubashar et al. (2013), considering the hybrid CZM/XFEM approach to model single-lap joints between aluminium adherends, with effective results. The joints were made with two-dimensional models in Abaqus®. The meshes were built with quadrilateral and incompatible mode elements, to improve the mesh response with joint rotations. Only half the joint was modelled, by considering rotational symmetry. The adherends were modelled as purely elastic because of the absence of plastic deformation in the tests. The adhesive layer was modelled by a combination of elasto-plastic continuum elements, CZM elements and XFEM enriched regions. Despite this fact, the hybrid methodology was slightly different from the work of Sugiman et al. (2013). In this work, the adhesive layer including fillets at the overlap ends was modelled with XFEM enriched solid elements, while the adhesive/adherend interfaces were modelled with CZM elements to account for damage growth in the length-wise direction of the joints. It was found that the predicted failure load was within the experimental range of values. The numerical failure began with cracking in the XFEM domain in the fillet at approximately 45° of the horizontal. The crack evolved with this orientation up to attaining the adhesive/adherend interfacial region, when XFEM propagation shifted to the CZM path to continue damage growth within the overlap. Concluding, an accurate representation of the failure process and load were achieved with this hybrid approach.

In the work of Curiel Sosa and Karapurath (2012), the XFEM was applied to simulate delamination damage in Fibre Metal Laminates (FML). The structure was tested in pure tension by means of the DCB specimen, inducing pure mode I crack propagation. The XFEM was evaluated against standard FEM and also experimental tests. The hybrid specimens were composed by outer 4.1 mm thick aluminium layers with a 1.25 mm thick glare laminate in-between (Fig. 10.10).

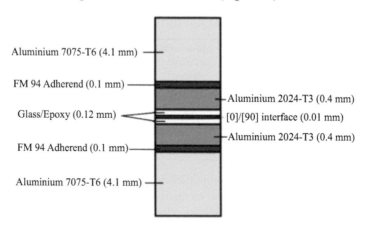

FIG. 10.10 Layout of the hybrid glare DCB specimen (Curiel Sosa and Karapurath 2012).

The initial crack length that is required in DCB testing was introduced in the numerical models by an additional component made of shell elements to create the discontinuity, thus permitting the crack to be created and propagate at any of the existing glare interfaces. Loading was applied by means of a constant velocity of 10 mm/s. The comparative assessment between methods focused on the relationship between crack opening displacement and applied load using the ASTM standardized procedure for DCB testing. Figure 10.11 shows the resulting XFEM P-δ curves, and experimental and CZM simulation data from a previous work (Airoldi et al. 2009). In the CZM simulations, interfacial strengths of 25 and 35 MPa were tested, and it was found that 35 MPa provided the best match to the experiments. For the XFEM, both 25 MPa and 35 MPa were considered and are represented in Fig. 10.11. The XFEM simulations slightly under estimated the area under the curve (i.e., G_{Ic}), unlike the CZM approach that exceeded the experimental values. It was found that the XFEM could produce good results even with a coarse mesh. The authors concluded that the XFEM replicated the experimentally observed delamination disregarding the mesh size and pointed out as important improvements the development of more specific enrichment functions to be applied to composites and FML. Thus, it was concluded that the XFEM is a promising technique in predicting failure of composite structures.

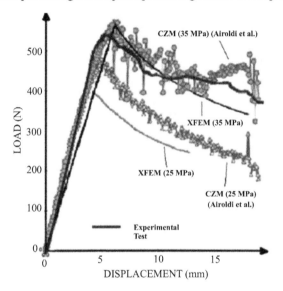

FIG. 10.11 Comparison of XFEM predictions of glare failure with the DCB specimen with experimental and CZM data (Curiel Sosa and Karapurath 2012).

Very few works used the XFEM for bolted connections and, owing to the scarcity in XFEM investigations applied to bonded joints, these works are also briefly presented here. Wang et al. (2012) used the XFEM for failure prediction of single-lap bolted joints between composite adherends. To simplify the numerical models, the laminated composites were modelled as elastic materials. The gradual damage uptake and failure paths of these joints were investigated, and the predicted failure loads by the XFEM of joints with different geometric parameters and lay-ups of $[45°/0°/-45°/90°]_{5S}$ and $[0°]_{40}$ were validated against literature results. The evaluated geometric parameters were the width-to-hole diameter ratio (W/D) and the edge-to-hole diameter ratio (E/D), see Fig. 10.12a.

A high-strength HTA/6376 carbon-fibre composite used in the aeronautical industry was used. The titanium bolt was simulated in the model as elastic isotropic. Both laminates and bolt were modelled by eight-node three-dimensional elements with reduced integration in Abaqus®. A mesh example is presented in Fig. 10.12b. Whilst overall numerical conditions are consistent with modelling of bonded joints, it was necessary to introduce the contact between parts. With this purpose, a penalty function was considered to perform the contact between different components. Figure 10.13 shows the P-δ curve comparison between XFEM modelling and experimental tests of a different work (McCarthy et al. 2005), in which the XFEM error is approximately 13%. It was considered that the error originated from two sources: (1) the washer and bolt are simulated as a single piece, which results in a variation of the friction coefficient between the FEM model with the real joints and (2) the maximum failure stress used in the damage criterion is smaller than the real value. Figure 10.14 shows the progressive failure of a laminate with $E/D = 3$ and $W/D = 6$ under loads and respective von Mises stresses during the process. For this particular selection of geometric parameters, adherend failure is due to net-tension.

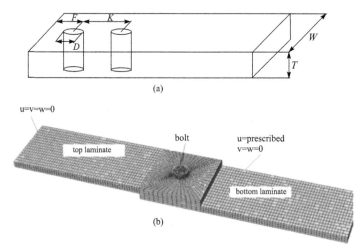

FIG. 10.12 Laminate dimensions (a) and FE model of single-lap bolted joint with one bolt (b) (Wang et al. 2012).

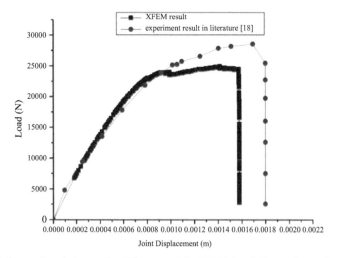

FIG. 10.13 Comparison between the P-δ curves of the XFEM simulation and experimental tests from the literature (Wang et al. 2012).

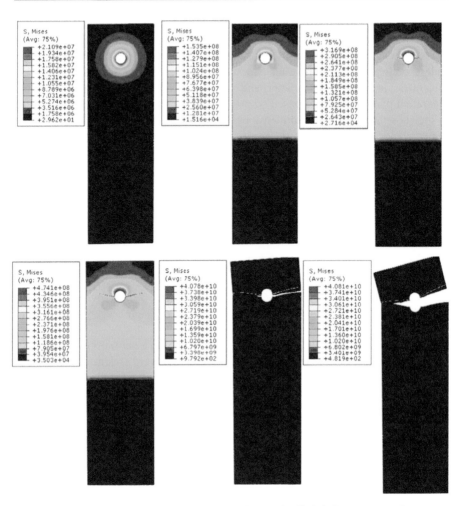

FIG. 10.14 von Mises stress distributions and crack growth of bolt-hole in a joint with geometric parameters $E/D = 3$ and $W/D = 6$ (Wang et al. 2012).

A mesh refinement study was undertaken considering three mesh gradings, and it was found that a smaller mesh increases the accuracy of the failure load predictions, although the failure path is kept identical, even with coarser meshes. In fact, the cracking locations had practically the same locations, showing that the XFEM is accurate in providing the correct failure path disregarding the element size. The predictions were compared against standard FEM with material degradation models. While the XFEM was able to give a clear idea of the failure process, the standard FEM only permitted the visualization of the elements' damage, while crack advance was not possible to track. A numerical parametric study was then carried out, showing that the joints' strength was higher by increasing W/D. Also, joining with two bolts instead of only one does not reflect on the joint strength, although increasing the joint stiffness.

Ahmad et al. (2014) modelled by XFEM damage and fracture by net-tension in woven fabric Carbon-Fibre Reinforced Plastic (CFRP) double-lap bolted joints. Three-dimensional numerical models were built, incorporating the bolt clamp-up and several geometry/fabrication parameters were addressed: joint dimensions, substrates' lay-up, hole diameter and clamp-up torque. The XFEM traction-separation parameters were based on previously measured strength and toughness of the substrates. The three-dimensional (3D) models accounted for smeared-out in-plane and out-of-plane elastic properties and contact interactions between the different components with friction effects, while symmetry conditions were also applied. A higher mesh refinement was applied near the hole because of being the region most likely to suffer failure.

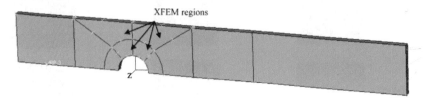

FIG. 10.15 Geometry partitioning and XFEM-assigned regions (Ahmad et al. 2014).

The failure mode encountered in the experiments was a net-tension-like failure. Thus, in the numerical models, XFEM enriched regions were assigned to model this kind of failure (Fig. 10.15), considering a bi-linear traction-separation relationship. The numerical failures were in agreement with the experiments, consisting of a net-tension failure starting from the hole edge. Stress distributions were plotted to describe the failure process and two preparatory studies to the failure analysis were considered: (1) mesh size effect and (2) damage stabilization coefficient. Regarding the 1st study, different mesh sizes were considered at the cracking region, and no evident mesh size effect was encountered, since crack growth is governed by energetic principles. The 2nd study aimed to select the proper damage stabilization coefficient for the analysis. While increasing this parameter enables convergence to be attained easier, at some point the output of the simulations exceed the supposed correct values. A value of 10^{-5} was selected as enough to attain convergence, but without compromising the validity of the results.

Figure 10.16 shows the damage evolution near the hole. Crack initiates at the hole edge at a smaller load that the failure load (a). In (b) the maximum load is attained. Beyond this point, the stiffness undergoes severe degradation accompanied by a load reduction up to complete failure. The comparison between XFEM and the experimental tests gave, depending of the joint configuration, variations up to 20%, but approximately of 10% in average. Given that the XFEM parameters were estimated from independent tests, rather than fitting procedures, the accuracy of the XFEM in modelling is further noted. It was though found that bearing failure presented some difficulties in modelling by the XFEM. In the end, XFEM was considered as a viable manner to design bolted joints.

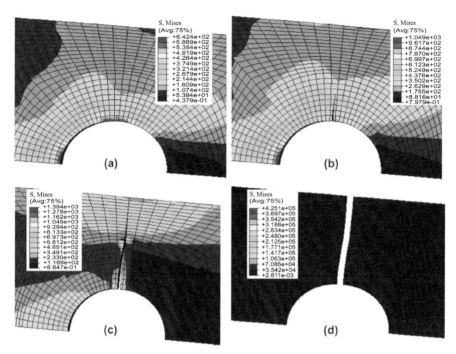

FIG. 10.16 von Mises plots in the adherends and gradual crack growth at the at the hole vicinity for a double-lap woven CFRP joint (Ahmad et al. 2014).

10.4 CASE STUDIES

10.4.1 Pure Tensile Behaviour as a Function of Temperature (DCB Specimen)

The present case study consists of an experimental and XFEM analysis on the tensile behaviour of the high temperature adhesive XN1244 from Nagase ChemteX at 4 different temperatures, up to 200°C, using experimental data from a previously published paper by the authors applied in the context of CZM modelling (Banea et al. 2011). The main purpose of the XFEM analysis is to validate this numerical tool for strength prediction of adhesive joints under different temperatures and a pure tensile loading. The DCB geometry was considered for tensile fracture characterization, which is specified in the D3433-99 (2005) standard. To this end, different data reduction techniques are applied to the experimental data to obtain G_{Ic} at varying temperatures, and the final XFEM parameters under different temperatures are proposed for application to strength predictions at each one of these conditions. The damage laws are then numerically validated to model crack propagation in the adhesive layer at RT, 100°C, 150°C and 200°C, after achieving a good correlation to the experimental joint behaviour.

10.4.1.1 Materials

The one-component epoxy adhesive XN1244, supplied by Nagase Chemtex and suitable for high-temperature applications, was selected for this study. The T_g and bulk tensile properties of this adhesive such as the Young's modulus, t_n^0 and tensile failure strain were estimated in the work of Banea et al. (2012). The Glass Transition Temperature (T_g) was determined to be approximately 155°C by Dynamical Mechanical Thermal Analysis (DMTA). The tensile properties of the adhesive XN1244 were determined using dogbone shape specimens, produced from bulk adhesive plates cured in a steel mould, using a silicone rubber frame, according to the French standard NF 76-142 (1988). Curing of the bulk plates was carried out in a hot plates press (1h at 140°C) at 2 MPa of pressure. The silicone rubber frame allowed the fabrication of 150 mm × 45 mm plates, with a thickness of 2 mm, guaranteeing a good surface finish and demouldability. Tensile testing of the XN1244 specimens was accomplished in a universal testing machine Shimadzu® Autograph (Kyoto, Japan) with a 5 kN load cell, at a constant velocity of 1 mm/min. Strains were measured by a video extensometer Messphysik ME46 (Fürstenfeld, Austria), over a length of 50 mm between hand-painted marks. For the high temperatures, the environmental chamber of the machine was used to apply the thermal load to the specimens. At least three valid results were obtained for each temperature. Table 10.2 summarizes the mechanical properties extracted from these tests, which will be used in this work to define the material parameters for the FEM simulations.

Table 10.2 Properties of the XN1244 adhesive by bulk tensile testing (Banea et al. 2011).

	Young's modulus [GPa]	Tensile strength, [MPa]	Tensile failure strain, [%]
RT	5.87 ± 0.33	68.23 ± 5.06	1.56 ± 0.22
100°C	4.17 ± 0.89	45.16 ± 3.48	1.92 ± 0.42
150°C	0.07 ± 0.01	6.49 ± 0.86	13.71 ± 1.46
200°C	0.04 ± 0.02	1.44 ± 0.17	3.33 ± 0.2

Figure 10.17 shows, as an example, representative stress-strain (σ-ε) curves for the tensile tests as a function of temperature of testing. These results report a decrease of strength with the increase of testing temperature, together with an increase in the ductility of the adhesive. For the DCB tests used in this work to characterize G_{Ic}, hard tool steel (DIN 40CrMnMo7) adherends were employed, to guarantee a fully elastic behaviour of the adherends during the tests. The most relevant mechanical and physical properties of the steel DIN 40CrMnMo7 are as follows (as reported by the manufacturer): tensile failure strength of 1000-1068 MPa, yield Stress of 861-930 MPa and elongation of 14-17% at room temperature. Young's modulus variation with temperature (in GPa): 205 at 20°C, 200 at 200°C and 185 at 400°C.

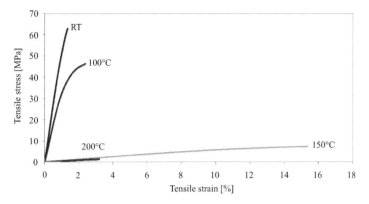

FIG. 10.17 σ-ε curves of representative tensile bulk tests to the adhesive XN1244 adhesive as a function of temperature (Banea et al. 2011).

10.4.1.2 Experimental Work

The DCB specimens' dimensions are presented in Fig. 10.18. The following values were selected for this work: adherends length L_T = 192.7 mm, adherends thickness t_P = 12.7 mm, adhesive thickness t_A = 0.2 mm and initial crack $a_0 \approx 65$ mm.

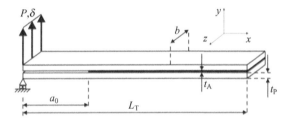

FIG. 10.18 Geometry and dimensions of the DCB specimens.

Fabrication of the specimens followed the standardized procedures (de Moura et al. 2008). The DCB specimens were then tested at RT, 100°C, 150°C and 200°C in a hydraulic testing machine Instron® model 8801 (Norwood, Massachusetts), under a constant crosshead rate of 0.5 mm/min. For the tests at high temperatures, an environmental chamber was used. The P-δ curves were registered during the tests, and high resolution photographs were also taken during the specimens testing with 5 s intervals using a 10 MPixel digital camera, for a posterior correlation with the P-δ data, allowing the determination of G_{Ic}. This was carried by the knowledge of the elapsed time from the beginning of the test, which can be easily and accurately related to the P-δ data, as well as the current value of a, visually inspected in the photographs. Before the beginning of the tests, a thermocouple was attached to one of the steel adherends of the specimen to assure the consistency between the specimen temperature and the air temperature inside the chamber (used for the machine readings). For each temperature, four specimens were considered.

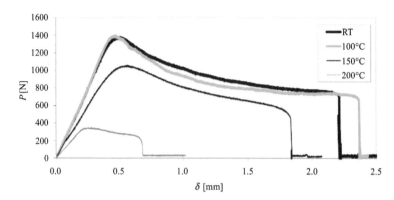

FIG. 10.19 Experimental P-δ curves of the DCB tests for the different temperatures, representative of the global results (Banea et al. 2011).

Figure 10.19 provides a comparison between representative P-δ curves at each one of the testing temperatures, showing the classical concave shape corresponding to crack growth at a constant G_{Ic} value (Banea et al. 2010). Results at RT and 100°C were quite close regarding the initial slope and maximum load, although a slight increase of failure displacement was found at 100°C. At 150°C, a moderate reduction of the maximum load took place, accompanied with a reduction of the failure displacement, whilst at 200°C the degradation of the adhesive properties owing to the surpassing of T_g was evident. All tested specimens showed a cohesive failure, disregarding the temperature of testing. All fracture surfaces showed a substantial amount of adhesive in each of the specimens' sides, giving indication of a cohesive failure, although some shifts in the crack path towards one of the adhesive/adherend interfaces were detected in some specimens.

10.4.1.3 Data Reduction Methods for G_{Ic}

The most common data reduction schemes to derive G_{Ic} rely on Linear-Elastic Fracture Mechanics (LEFM) principles. The CCM uses the Irwin-Kies equation (Trantina 1972), in which G_{Ic} at a given testing time depends on the current value of P, specimen width, b, and current value of dC/da ($C = \delta/P$ is the specimen compliance). Equally available as LEFM techniques are the Direct Beam Theory (DBT), based on elementary beam theory (Kanninen and Popelar 1985), and the CBT, including the effects of crack tip rotation and deflection (Bonhomme et al. 2009). As it becomes clear in Fig. 10.17, the adhesive XN1244 is quite brittle at RT, but it becomes clearly more ductile with increasing temperatures, reaching nearly 13% of failure strain at 150°C. This brings some issues regarding the applicability of LEFM-based methods, since the fracture process zone can become quite large, and this can turn the predicted values of G_{Ic} not accurate because of the large-scale plasticity around the crack tip (Wang 1983). Despite this fact, for the failure strains observed in Fig. 10.17, previous works showed that LEFM methods still behave well (de Moura et al. 2008). Two different methods were employed in this work to evaluate G_{Ic}: the CCM and the CBT.

Figure 10.20 pictures the influence of the testing temperature on G_{Ic}. This parameter attains its maximum at 100°C, although it is only approximately 10% higher than at RT. Rationale for this apparently odd behaviour is given by the slight increase on the adhesive ductility (Fig. 10.17) that increases plastic straining at the crack tip prior to propagation, although the peak strength decreases. At a temperature of 150°C, the reduction of G_{Ic} suggests the vicinity of T_g, which is afterwards confirmed by the results obtained at 200°C that show a complete degradation of the adhesive properties. The R-curves obtained by the two methods at RT, 100°C, 150°C and 200°C showed crack growth at a nearly constant value of G_{Ic}. The results were quite close between the CCM and CBT.

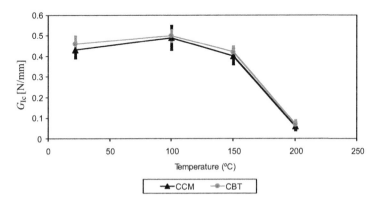

FIG. 10.20 G_{Ic} as a function of temperature by the CCM and CBT (Banea et al. 2011).

10.4.1.4 Numerical Work

The FEM software Abaqus® (Providence, Rhode Island), which includes a XFEM module, was considered for the analysis, with the main objective of estimating a damage law for the thin layer of adhesive XN1244 at varying temperatures. Each one of these laws can then be used to simulate the adhesive layer in bonded structures, for an effective strength prediction. Triangular damage laws were considered, i.e., with linear decrease of the transmitted loads after damage initiation, built from the experimental results of the DCB and bulk tension tests (the input parameters for the simulations are the Young's modulus, Poisson's ratio, t_n^0 and G_{Ic}). The steel adherends were modelled with elastic isotropic conditions, using the generic values of Young's modulus previously specified and making linear extrapolations for each testing temperature as a simplification, owing to the small variations of Young's modulus. These values are further presented in this description. The simplification of the steel behaviour to purely elastic is also only feasible because of the absence of the adherends plastic flow. The adhesive layer was considered as an elastic isotropic material up to the attainment of the damage initiation condition. At that point, damage occurred according to the triangular damage law up to complete failure. The numerical analysis considered non-linear geometrical effects to account for 2nd order effects and 2D conditions, which are usually carried out in test geometries that show a constant width-wise shape and that provide an accurate representation of the joint mechanics involved, including damage growth (de Moura et al. 2008). Figure 10.21 shows the mesh for the analysis, including details at regions of mesh size grading, as well as the boundary and loading conditions. Restraining and loading conditions consisted of clamping the edge node of the lower adherend, while the edge node of the upper adherend was horizontally restrained and pulled vertically. An initial crack was not considered in the numerical models for their simplification, as cracking initiates at the locus of higher magnitude of stresses/strains, as previously mentioned in Section 10.2. Plane strain models were built for the simulations with general purpose solid elements (CPE4 from Abaqus®). The XFEM formulation that was adopted for this work is implemented in Abaqus® CAE (Abaqus® 2013) and was briefly discussed in Section 10.2 (Campilho et al. 2011a).

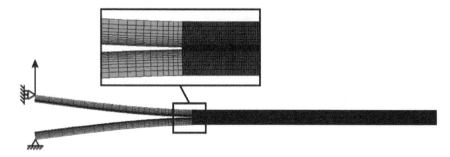

FIG. 10.21 Deformed shape of the DCB specimen at the beginning of crack propagation, with boundary and loading conditions (enhancement of the deformations by 20×) (Campilho et al. 2011a).

The input parameters of the adhesive layer and steel damage laws introduced in Abaqus® were as follows. For the steel adherends, a Poisson's ratio of 0.3 was considered for all testing temperatures, while values of Young's moduli [GPa] of 205, 202.8, 201.4 and 200°C were defined for RT, 100°C, 150°C and 200°C, respectively. The adhesive's properties were specified based on the data of Table 10.2 (average values of Young's modulus and tensile strength – used for t_n^0) and equating for G_{Ic} [N/mm] values of 0.47, 0.50, 0.42 and 0.07 for increasing temperatures as previously specified for the steel properties (Banea et al. 2011).

10.4.1.5 Results

The objective of this study was to validate the damage laws and respective parameters, defined for the adhesive at the different testing temperatures by bulk tensile or DCB tests. The validation of the XFEM laws for the adhesive is accomplished if a good correlation is found between the P-δ curves from the DCB tests and the XFEM results. Figure 10.22 shows the XFEM crack propagation at RT, with crack onset and growth taking place cohesively along the adhesive layer. As previously mentioned, when no initial crack is present, it initiates orthogonally to the maximum principal stress/strain direction, when the maximum principal stress/strain surpasses the material property defined for the analysis. In this example, since this direction is vertical, owing to the pure tensile loading, the crack grows horizontally along the adhesive layer up to failure of the specimen. It should also be mentioned that, if failure for any of the conditions had occurred adhesively at one of the interfaces, this would not be captured by the numerical models, because crack onset and growth is always ruled by stresses and the maximum principal stress direction, providing a cohesive propagation along the middle of the adhesive layer.

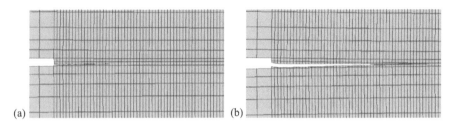

(a)   (b)

FIG. 10.22 Crack growth at RT by the XFEM algorithm, initiating at the crack tip (a) and growing horizontally along the bondline (b) (Campilho et al. 2011a).

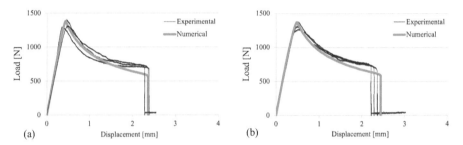

(a)   (b)

FIG. 10.23 Experimental/numerical P-δ curves comparison for the DCB specimens at RT (a) and 100°C (b).

Figure 10.23 plots the experimental/numerical P-δ curve comparisons for the DCB specimens at RT (a) and 100°C (b). The curves for 150°C (a) and 200°C (b) reveal a similar agreement. By comparing the maximum loads of the XFEM predictions and experiments (average values), the following deviations were found: 3.5% (RT), 4.0% (100°C), 2.6% (150°C) and 6.8% (200°C). On the other hand, the deviations in the failure displacement were as follows: 1.7% (RT), 6.5% (100°C), 3.2% (150°C) and 11.2% (200°C). In summary, the elastic stiffness, maximum load, load during propagation and failure displacement for all conditions using the previously characterized parameters showed a good agreement, which testifies the suitability of this technique to simulate bonded structures. As a final remark regarding the suitability of the XFEM in reproducing the behaviour of adhesives with different characteristics (i.e. values of strength and G_{Ic}) it is expected that, provided that the adhesives characterization is carried out under identical conditions to the structure to be simulated, the predictions are accurate. Actually, on one hand, t_n^0 is used in the elastic FEM analyses to detect damage onset region/s (when the maximum principal stresses attain its magnitude anywhere in the model), giving an accurate prediction if the models are properly defined. On the other hand, crack propagation is mainly ruled by G_{Ic} of the adhesive, which defines the extent of the damage process zone, and whose correct estimation gives a faithful representation of the fracture process taking place near the growing crack, either for small or large values of G_{Ic}.

10.4.1.6 Conclusions of the Study

In this study, damage laws were derived to model crack propagation of a thin layer of a structural epoxy adhesive in bonded structures at different temperatures (RT, 100°C, 150°C and 200°C) using the XFEM, after determination of the model parameters, G_{Ic} and t_n^0, by DCB and bulk tensile tests, respectively. The DCB specimens used to define G_{Ic} were considered for numerical validation of the XFEM procedure to simulate bonded structures, by using the damage laws defined in the characterization tests to reproduce the experimentally obtained P-δ curves of the DCB tests. The simulation response for the various tested temperatures matched with high accuracy the experimental results, regarding the most relevant features of the joints' fracture, as the elastic stiffness, maximum load sustained, transmitted loads during crack growth and displacement at failure. As a result of these findings, the numerical procedure was validated for pure tensile behaviour.

10.4.2 XFEM Comparison with CZM for Single-Lap Joints and Varying Ductility Adhesives

In this 2nd case study, the performance of a brittle (Araldite® AV138), a moderately ductile (Araldite® 2015) and a largely ductile adhesive (Sikaforce® 7888) was tested in single-lap joints between aluminium adherends with varying values of L_O. The experimental work carried out is accompanied by a detailed numerical analysis by FEM, starting with the plot of elastic stress distributions, and strength prediction either based on CZM or XFEM. This procedure enabled assessing in detail the performance of these predictive techniques applied to bonded joints. Moreover, it was possible to evaluate which family of adhesives is more suited for each joint geometry.

10.4.2.1 Materials

The adherends were cut from a high strength aluminium alloy sheet (AA6082 T651) by precision disc cutting. This material was characterized in bulk tension in previous works by the authors (Campilho et al. 2011b, Pinto et al. 2011) using dogbone specimens and the following mechanical properties were obtained: Young's Modulus of 70.07 ± 0.83 GPa, tensile yield stress (σ_y) of 261.67 ± 7.65 MPa, tensile failure strength (σ_f) of 324 ± 0.16 MPa and tensile failure strain (ε_f) of 21.70 ± 4.24%. The experimental curves and the numerical approximation, to be used further in work, are presented in Fig. 10.24. Three structural adhesives, ranging from brittle to highly ductile, were considered: the brittle epoxy Araldite® AV138, the ductile epoxy Araldite® 2015 and the high strength and ductile polyurethane Sikaforce® 7888. The mechanical and toughness properties of these adhesives were obtained in previous works by the authors by experimental testing (Neto et al. 2012, Campilho et al. 2013a, 2013b). Bulk specimens were tested in a servo-hydraulic machine to obtain the Young's Modulus, σ_f and ε_f. The DCB test was selected to obtain G_{Ic} and the ENF test was used for G_{IIc}. The collected data of the adhesives is summarized in Table 10.3.

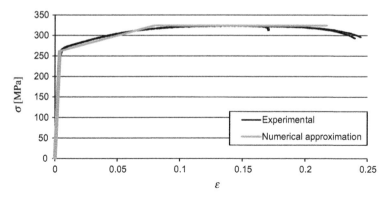

FIG. 10.24 Experimental and numerical σ-ε curves of the aluminium (Campilho et al. 2011b).

Table 10.3 Properties of the adhesives Araldite® AV138, Araldite® 2015 and SikaForce® 7888 (Neto et al. 2012, Campilho et al. 2013a, 2013b).

Property	AV138	2015	7888
Young's modulus, E [GPa]	4.89 ± 0.81	1.85 ± 0.21	1.89 ± 0.81
Poisson's ratio, v	0.35 *	0.33 *	0.33 *
Tensile yield strength, σ_y [MPa]	36.49 ± 2.47	12.63 ± 0.61	13.20 ± 4.83
Tensile failure strength, σ_f [MPa]	39.45 ± 3.18	21.63 ± 1.61	28.60 ± 2.0
Tensile failure strain, ε_f [%]	1.21 ± 0.10	4.77 ± 0.15	43.0 ± 0.6
Shear modulus, G [GPa]	1.56 ± 0.01	0.56 ± 0.21	0.71 b
Shear yield strength, τ_y [MPa]	25.1 ± 0.33	14.6 ± 1.3	–
Shear failure strength, τ_f [MPa]	30.2 ± 0.40	17.9 ± 1.8	20 *
Shear failure strain, γ_f [%]	7.8 ± 0.7	43.9 ± 3.4	100 *
Toughness in tension, G_{Ic} [N/mm]	0.20 *	0.43 ± 0.02	1.18 ± 0.22
Toughness in shear, G_{IIc} [N/mm]	0.38 *	4.70 ± 0.34	8.72 ± 1.22

* manufacturer's data
a estimated in reference (Campilho et al. 2011b), b estimated from Hooke's law

10.4.2.2 Experimental Work

The geometry and dimensions of the single-lap joints are described in Fig. 10.25. The joint dimensions are as follows: t_P = 3 mm, t_A = 0.2 mm, L_O = 12.5, 25, 37.5 and 50 mm, and joint total length between grips L_T = 180 mm. To fabricate the specimens, the adherends were initially cut from a bulk plate in an automatic disc cutter and then machined to the final dimensions. The bonding process consisted of grit blasting with corundum sand, debris cleaning with acetone and assembly in a steel mould for the correct alignment between the adherends. Calibrated wires with a diameter of 0.2 mm were inserted between the adherends at the overlap edges to assure the correct value of t_A (Campilho and Rocha 2011). The correct positioning of the adherends to produce the different values of L_O was performed with a digital caliper. Curing of the specimens was carried out according to the manufacturer's specifications for complete curing, i.e. for at least 48 hours at room temperature. Aluminium tabs were glued at the specimens' edges for a correct alignment in the testing machine. The specimens were tested in a Shimadzu AG-X 100 testing machine with a 100 kN load cell, at RT and under displacement control (1 mm/min). Four valid results were always provided for each condition.

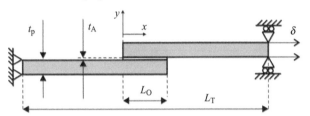

FIG. 10.25 Geometry and dimensions of the single-lap joints.

The FEM software Abaqus® was selected to perform the numerical analysis, which has CZM and XFEM embedded modules (Abaqus® 2013). An initial stress analysis was performed to better understand the observed behaviour. The adherends were modelled as elasto-plastic solids with an approximated curve to the real σ-ε curve of the aluminium (Fig. 10.24). The adhesive was modelled with CZM elements, or solid elements with enriched formulation for the XFEM. Geometrical non-linearities were considered. The joints were modelled as 2D, with plane-strain solid elements (CPE4 from Abaqus®). Different mesh refinements were considered for the stress and failure analyses (Fig. 10.26 shows a representative mesh for the failure analyses by CZM and XFEM).

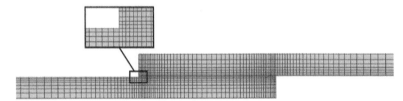

FIG. 10.26 Representative mesh for the failure analysis by CZM and XFEM.

In both scenarios, the meshes were constructed with bias effects from the adhesive centre regions towards the overlap edges, and towards the adhesive layer in the adherends' thickness direction, to reduce the computational effort but without compromising the precision of the results. The meshes for the stress analysis models are highly refined, with 0.02 mm × 0.02 mm elements in the adhesive layer, to accurately capture the peak stresses at the overlap ends, which theoretically are singular regions (Campilho et al. 2005). For the CZM and XFEM failure analyses, only one element was considered in the adhesive thickness direction. Thus, the element size in the adhesive was 0.2 mm × 0.2 mm. Restraining and loading conditions consisted of clamping the joint at one edge and applying a vertical restraint and tensile displacement at the opposite edge. In the CZM analysis, the adhesive was modelled by the continuum approach, with a single row of cohesive elements and a traction-separation law including the adhesive layer stiffness. For the XFEM models, the adhesive layer was modelled by the same elements used for the adherends, considering one layer of solid elements. The common triangular CZM approach with a quadratic stress criterion to infer damage initiation and a linear energetic criterion for failure was used. For full details of this formulation, the reader can refer to the work of Campilho et al. (2011b). The XFEM formulation followed in this study is the one described in Section 10.2. For both modelling techniques, the required data for the adhesives was taken from Table 10.3.

10.4.2.3 Results

10.4.2.3.1 Failure Modes

After performing the experimental tests, all failures were cohesive in the adhesive layer. However, plastic deformation was found in the adherends for some of the test conditions, namely for the conditions with bigger failure loads. All $P\text{-}\delta$ curves were predominantly linear up to failure, except for the $L_O = 50$ mm joints bonded with the Sikaforce® 7888. This behaviour was consistent with the numerical results.

10.4.2.3.2 Stress Analysis

A peel (σ_y) and shear (τ_{xy}) stress analysis is carried out at the adhesive mid-thickness that enables further discussions on the obtained strength results. The plots of σ_y and τ_{xy} stress distributions in the adhesive layer as a function of percentile L_O are presented in Fig. 10.27 and Fig. 10.28, respectively. A normalization procedure was carried out, dividing σ_y and τ_{xy} stresses by τ_{avg}, the average shear stress along the overlap for the respective value of L_O. An identical normalization procedure was carried out for L_O by plotting stresses vs. x/L_O ($0 \leq x \leq L_O$, see Fig. 10.25). The results are relative to the joints bonded with the Araldite® 2015 and they are representative of all joints.

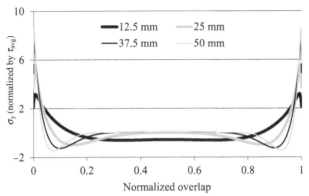

FIG. 10.27 σ_y stress distributions at the adhesive mid-thickness as a function of the normalized overlap (x/L_O).

σ_y stresses are typically much lower in magnitude than τ_{xy} stresses, except at the bond edges, in which σ_y stress singularities build up owing to the square-edge geometry (Radice and Vinson 2006, Taib et al. 2006). At the inner overlap region, these stresses are compressive. The shape of σ_y stresses is mainly due to the loading asymmetry, giving separation at the overlap edges and compression in-between. The peak peel stresses at the bond edges increase with L_O, which is a negative aspect for the joint strength. These peak stresses are responsible for a significant strength reduction of bonded joints (Campilho et al. 2005, 2009b).

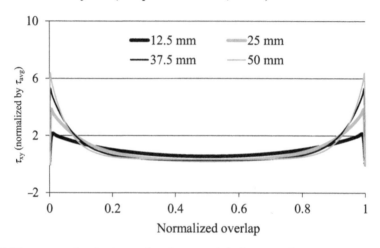

FIG. 10.28 τ_{xy} stress distributions at the adhesive mid-thickness as a function of the normalized overlap (x/L_O).

The obtained τ_{xy} distributions are also consistent with the available literature results, with a smaller load bearing potential at the overlap inner region and peaking towards the overlap edges (Vable and Reddy Maddi 2006, Luo and Tong 2007). This behaviour is due to the differential deformation of each one of the adherends along the overlap, from their free overlap edge towards the other overlap edge (Campilho et al. 2005, Volkersen 1938, Adams et al. 1997). At the central region of the overlap these effects are cancelled, with τ_{xy} stresses developing solely by the tensile pulling of the specimen. τ_{xy} stress gradients increase with L_O because of the increasing gradient of longitudinal strains in the adherends caused by the bigger bonding areas and loads. This markedly affects the strength improvement of bonded joints with L_O for brittle adhesives, which do not allow plasticization at the overlap edges (Grant et al. 2009). Oppositely, ductile adhesives enable the redistribution of stresses at these regions while the inner region of the overlap is gradually put under loads, giving a bigger increase of the joint strength with L_O (Campilho et al. 2005, Davis and Bond 1999).

10.4.2.3.3 Joint Strength Predictions by CZM

In the numerical analysis by CZM the adhesive layer was modelled by a single row of cohesive elements, i.e. by the continuum approach, and the adherends were considered as elasto-plastic using the properties defined in Section 10.4.2.1. In all simulations, damage initiated at the overlap edges with propagation to the inner regions of the bond (Campilho et al. 2012). Figure 10.29 gives an example of the overall correlation obtained between the experimental and numerical (CZM) P-δ curves, by showing as an example L_O = 25 mm Araldite® AV138 (a) and L_O = 50 mm Araldite® 2015 joints (b). The overall correlation is good, especially for the joints bonded with the Araldite® 2015, regarding the failure load (in this case study addressed as P_m), stiffness or failure displacement. For the Sikaforce® 7888, the numerical results under predict the experiments, but this will be specifically addressed further in this chapter. Figure 10.30 gives the summarized results of P_m-L_O for the three adhesives, considering the experimental values of average and respective error bars, and also the numerical predictions by CZM (dots with deviation and lines, respectively).

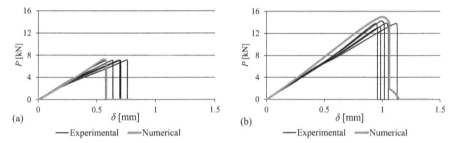

(a) ——Experimental ——Numerical   (b) ——Experimental ——Numerical

FIG. 10.29 Experimental and numerical (CZM) P-δ curves for the L_O = 25 mm Araldite® AV138 (a) and L_O = 50 mm Araldite® 2015 (b).

The experimental results show a distinct behaviour between the three bonded systems, which is highly dependent on the adhesive properties (stiffness, strength and ductility). The adhesive stiffness has a large effect on stress distributions: a low stiffness adhesive provides a more uniform stress distribution compared to a stiff adhesive (Adams et al. 1997), which puts the Araldite® AV138 in disadvantage in view of the values of Young's modulus presented in Table 10.3. The adhesive strength has more preponderance for short values of L_O, in which σ_y and τ_{xy} stresses are more flat along the overlap (Fig. 10.27 and Fig. 10.28). A ductile adhesive is able to redistribute the load and make use of the less stressed parts of the overlap, whereas a brittle adhesive concentrates the load at the ends of the overlap without giving the possibility of plasticization, giving a low average shear stress at failure (Adams and Peppiatt 1974). In view of this, the Araldite® AV138, which is a very strong but brittle adhesive, performs slightly better than the Araldite® 2015 (less strong but ductile) for short overlaps (experimental difference of 4.9% for L_O = 12.5 mm). This occurs because, for short values of L_O, the strength of the adhesive rules P_m. However, by increasing L_O, the Araldite® 2015 quickly surpasses the Araldite® AV138 since it is moderately ductile and, thus, it is able to sustain some plasticity before failure, unlike the Araldite® AV138. Figure 10.30 shows a marked change between L_O = 12.5 and 25 mm, because the plasticity of the Araldite® 2015 enables the plasticization of the adhesive layer and corresponding increase of P_m, which also reflects on the higher values of L_O. For L_O = 50 mm, the Araldite® 2015 surpasses the Araldite® AV138 by 49.1%. Compared with these two adhesives, the Sikaforce® 7888 is simultaneously strong and highly ductile, in such a way that failure with this adhesive approaches the conditions of generalized failure up to substantially long overlaps (Davis and Bond 1999). Thus, compared to the Araldite® 2015 and the Araldite® AV138, it excels in P_m values from small to large values of L_O. The percentile improvement over the Araldite® 2015 is 46.7% for $L_O = 12.5$ mm and 58.7% for $L_O = 50$ mm. Regarding the Araldite® AV138, the improvements are 39.8% and 136.5%, by the same order. In view of these results, recommendation goes to using less strong but ductile adhesives (if it is not possible to combine both features), except for very small values of L_O, in which strong adhesives also perform well. Adhesives that combine these two characteristics are undoubtedly the best choice. For non-uniform loadings like peel, cleavage or thermal internal stresses, a joint with a ductile adhesive will also behave better (Adams et al. 1997).

Comparing the experiments with the CZM strength predictions (Fig. 10.30), the results were very close for the Araldite® 2015 and Araldite® AV138, while for the Sikaforce®7888 the numerical strengths clearly under predicted P_m. The largest error was 9.8% for the Araldite® AV138, 7.7% for the Araldite® 2015 and 19.7% for the Sikaforce®7888. The obtained results for the Araldite® AV138 and Araldite® 2015 show consistent predictions, in line with previously obtained results by this method for brittle and moderately ductile adhesives (Campilho et al. 2011b).

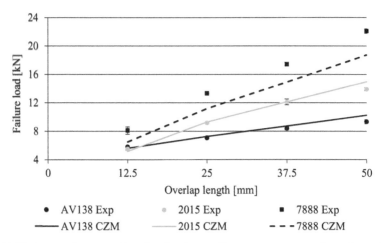

FIG. 10.30 P_m as a function of L_O for the three adhesives: experimental results and CZM predictions.

The under prediction of the P_m values obtained for the Sikaforce®7888 are related to its large plasticity, which is not accurately modelled by the triangular laws considered in this analysis. In fact, a trapezoidal law would fit better the experimental data (Campilho et al. 2013a). Nonetheless, the proposed solution enables a rough prediction of the joints' behaviour. A mesh dependency study was also performed to estimate the influence of the mesh refinement for the cohesive elements that model the adhesive layer in the strength predictions. This study considered the joints with

L_O = 12.5 mm bonded with the Araldite® AV138 and Sikaforce® 7888, as the limiting adhesives in which concerns the ductility. In this analysis, a constant value of t_A was used (0.2 mm), while the cohesive elements' length varied between 0.05 and 0.4 mm. Results showed a maximum deviation to the average P_m of all mesh sizes of 0.06% (Sikaforce®7888) and 0.33% (Araldite® AV138). Actually, mesh independency is a characteristic of CZM modelling (Campilho et al. 2010), since damage growth is ruled by energetic criteria, based on the values of G_{Ic} and G_{IIc} of the material. Since the energy required for propagation is averaged over the damaged area, opposed to the use of a discrete value of maximum stress/strain as it happens for the strength of materials criteria, results are mesh independent provided that a minimum refinement is used (Kafkalidis and Thouless 2002).

10.4.2.3.4 Joint Strength Predictions by XFEM

The XFEM was also evaluated to predict the joints' strength, although the mixed-mode propagation in single-lap joints gives rise to some intrinsic limitations of the method. Regarding the initiation criteria, these are stress or strain-based, and this has the handicap of mesh dependency (Campilho et al. 2011a). Regarding the MAXPE and MAXPS criteria, these are based on one parameter (the maximum principal strain or stress, respectively), and this is not consistent with the fracture process of thin adhesive layers, due to the constraining effects imposed by the surrounding stiff adherends (Campilho et al. 2013a). When analysing the fracture process of bonded joints, the tensile and shear behaviour should be considered simultaneously to infer damage onset and growth. Moreover, during damage propagation, the crack follows the orthogonal direction of the principal stress/strain direction at the crack tip (Moës and Belytschko 2002). It was discussed in a previous work (Campilho et al. 2011b) that these criteria are not the most suited for bonded joints under mixed-mode crack propagation, because the mixed-mode conditions lead to crack growth up to the adhesive/adherend interface and further into the adherend. This clearly does not reflect the real behaviour of single-lap joints. In these conditions, XFEM is only suitable for the identification of the locus of damage initiation in adhesive bonds and predict the respective load, which may or not be a rough estimator of the joint strength, depending of the adhesive characteristics. However, it does not take into account the prospect of damage growth under mixed- mode or along interfaces between different materials. Opposed to this behaviour, the other initiation criteria described in Section 10.2 enable the user selection of the growth direction, either normal or parallel to the local direction 1 and 2 of the enriched finite elements that comprise the adhesive. Given the layout of the finite elements depicted in Fig. 10.26, this allows the choice between horizontal (i.e. length-wise in the adhesive layer) and vertical crack propagation. In the present scenario, horizontal propagation was selected for all 4 criteria described in Section 10.2 other than MAXPE or MAXPS in an attempt to model crack growth along the adhesive layer. Despite the possibility to induce lengthwise crack propagation, due to the preponderant shear damage propagation, large transverse deflection and associated rotations of the adhesive layer, damage propagation was not possible to accomplish. This impossibility was related to large convergence difficulties, even after applying all available convergence-assisting tools (da Silva and Campilho 2011).

In view of this, for all the six initiation criteria depicted in Section 10.2, the strength prediction of the XFEM was limited to associating P_m to crack onset. In a previous work (Campilho et al. 2011b), dealing with single and double-lap joints bonded with a brittle adhesive, a mesh sensitivity study was performed and it was found that, with this approach, an element size in the adhesive layer with the value of t_A provided a rough estimator of P_m when using strain-based criteria. Thus, in this work, a mesh size of 0.2 mm in the adhesive layer was selected (Fig. 10.26). Results showed a distinct behaviour between stress and strain-based criteria, with the former clearly underestimating P_m. Percentile variations were very large: maximum of 74.8% for the MAXPS, 71.4% for the MAXS and 67.3% for the QUADS criterion. These variations are justified by the singularity that makes the limiting stresses of the adhesives to be reached very swiftly during the loading process. These results are consistent with previous works, were the strain criteria showed to be less sensitive than stress criteria (Campilho et al. 2011b). Thus, only the strain-based criteria are used here to predict the joints strength. In the scope of these criteria, the results for the Sikaforce® 7888 over predicted the experiments by a large amount, showing that for adhesives that are highly ductile the proposed technique does not work well (depending on L_O and the criterion, over predictions of up to 5 times the experimental value were obtained).

Figure 10.31 compares the experimental and XFEM data considering the MAXPE, MAXE and QUADE criteria and the adhesives Araldite® AV138 and Araldite® 2015. The experimental data includes the average values and respective error bars. For these adhesives, especially for the former, the crack initiation loads are rough indicators of P_m. Between the three criteria, the maximum differences to the experiments were of 16.6% for the MAXPE criterion, 6.33% for the MAXE criterion and 6.67% for the QUADE criterion (Araldite® AV138) and 25.4% for the MAXPE criterion, 62.1% for the MAXE criterion and 53.5% for the QUADE criterion (Araldite® 2015). To be mentioned that for the Araldite® 2015, by excluding the results for L_O = 12.5 mm, the maximum differences substantially reduce.

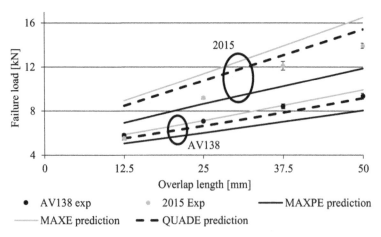

FIG. 10.31 P_m as a function of L_O for the Araldite® AV138 and Araldite® 2015: experimental results and XFEM predictions.

The obtained results show that the adopted procedure is accurate in simulating bonded joints with brittle adhesives like the Araldite® AV138, roughly accurate in modelling adhesives with a moderate ductility (e.g. Araldite® 2015) but not recommended for largely ductile adhesives (e.g. Sikaforce® 7888). To have a clear idea of the size effects, a mesh dependency study was performed (Fig. 10.32) for the adhesives Araldite® AV138 and Sikaforce® 7888 using mesh sizes in the adhesive layer between 0.02 and 0.2 mm (square elements) and $L_O = 12.5$ mm. In both cases, the predicted P_m values increase with the mesh size. The accuracy for the Araldite® AV138 is best for the element size equal to t_A (0.2 mm), and this is a feature for brittle adhesives (Campilho et al. 2011b). Joints bonded with largely ductile adhesives are not well predicted by this approximation.

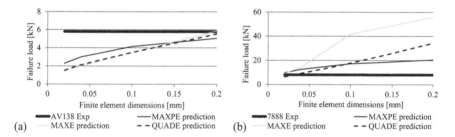

FIG. 10.32 XFEM mesh dependency study for the Araldite® AV138 (a) and Sikaforce® 7888 (b).

10.4.2.3.5 Conclusions of the Study

In this study, FEM-based CZM and XFEM techniques were considered for strength prediction. The experimental analysis revealed a distinct behaviour between the adhesives: (1) the brittle adhesive showed a small improvement of P_m with L_O because of the lack of plasticization at the damage initiation sites, (2) the moderately ductile adhesive showed a smaller value of P_m for the smallest L_O, but was able to endure plasticization at the overlap edges, and thus to increase strength at a higher rate than the former brittle adhesive, and finally (3) the high strength and high ductility polyurethane showed the highest strength and improvement rate of P_m, on account of failure under conditions approaching generalised failure of the adhesive. The FEM stress analysis enabled the detailed justification of this behaviour. CZM modelling with triangular shape damage laws revealed to be highly accurate for the brittle and moderately ductile adhesives, while underestimating by a non-negligible amount P_m for the polyurethane adhesive (maximum of 19.7%). A mesh dependency study confirmed that this numerical tool is highly stable to the mesh size. The XFEM analysis revealed in first hand that it is not possible to model damage propagation on account of the shear-predominant failure mode along the adhesive line. Failure was estimated by crack initiation using different stress and strain-based criteria, based on previous evidence that a mesh size in the adhesive layer equal to t_A and a strain-based initiation criterion gives a rough estimator of P_m for brittle adhesives. The stress-based criteria were found to under estimate P_m by a significant amount, unlike the strain-based criteria, which revealed to be a rough indicator of P_m, except for the polyurethane adhesive, representative of largely ductile adhesives. The mesh size study confirmed the P_m variations with the size of the adhesive representative elements, which is a clear limitation of the method. As a result of this work, it was possible to evaluate in detail these two numerical tools for strength prediction and provide an indication of the behaviour of different types of adhesives applied to single-lap joints.

10.5 SUITABILITY OF THE XFEM FOR BONDED JOINTS

The limitation of CZM is the requirement of the cohesive elements placement along predefined paths where damage can occur, which can be hard to identify. The XFEM allows overcoming this limitation, since it propagates cracks arbitrarily within solid FEM elements by using enrichment functions for the displacements. However, some of the initiation criteria in the current XFEM implementation in Abaqus® are restricted to only one value of maximum strength/strain leading to the initiation of damage (by the maximum principal stress or strain criterion, respectively), which can be a limitation as damage in thin adhesive bonds is not consistent with that of bulk materials, due to the constraining effects imposed by the adherends (Xie and Waas 2006). This does not allow the separation of the adhesive behaviour into tensile and shear, which can be in some cases mandatory for the accuracy of the results if large constraining effects are present in the bond (Campilho et al. 2009a). This clearly does not reflect the behaviour of bonded joints and can be attributed to an algorithm for propagation not still suited to multi material structures as it does not search for failure points outside the crack tip nor following the interfaces between different materials. Thus, the separation of the adhesive behaviour into tensile and shear behaviour that is performed for CZM is not possible to accomplish, which can be mandatory if large constraining effects exist (Campilho et al. 2009a). Apart from this feature, the current implementation of the method itself involves an even more important handicap. It is known that, if no initial cracks are introduced in the models, the XFEM algorithm will automatically search for the maximum principal stresses/strains in each one of the structure materials (in the present scenario, in both the adhesive and adherends), to initiate damage propagation in the first locus in which these stresses/strains surpass the respective material properties. During damage propagation, the XFEM algorithm continuously searches for the principal stress/strain direction at the crack tip, to specify the direction of subsequent crack growth (Moës et al. 1999, Moës and Belytschko 2002). For the specific case of single or double-lap joints, cracking initiates in the adhesive bond orthogonally to the direction of principal stresses/strains, growing up to the adhesive/adherend interface. Restriction of damage propagation only for the adhesive bond is also rendered unfeasible to surpass this limitation as crack propagation halts when the crack attains the adherends (Campilho et al. 2011b).

As a result, under these conditions, slight inconsistencies between the predictions and experiments can occur. As a means to prevent this limitation of the current XFEM implementation in Abaqus®, a modification that could account for the distinction between tension and shear would increase the accuracy of the simulations (Campilho et al. 2011a). Concerning the mesh dependency, the XFEM behaves similarly to CZM, i.e., it is almost mesh independent for the simulation of fracture propagation, due to the fact that G_{Ic} is averaged over a finite region. Thus, for a large range of mesh sizes, provided that a minimum refinement is used, all the relevant features of the failure process are accurately captured (Campilho et al. 2011b). However, the predictions of damage initiation load are mesh dependent, since the state of stress at a region with large stress concentrations largely depends of the mesh size. Apart from this, owing to the crack propagation method, i.e. orthogonally to the maximum stress/strain, mixed mode propagation is not allowed, although it is frequent in bonded joints. This will invariably lead to a wrong interpretation of the structures behaviour. Under these circumstances, assessment of the structures strength can be carried out by the initiation stress/strain criterion. However, this simplification is only expected to be sufficiently accurate for brittle adhesives that fail catastrophically when the peak stress/strain is achieved. For ductile adhesives, it will underestimate the structures strength, as the adhesive will endure plasticization and redistribution of stresses before collapsing. Nonetheless, the failure load predictions, using the described simplified technique, are expected to be mesh size dependent because of the stress/strain gradients at the sharp corners of bonded structures (Panigrahi and Pradhan 2007).

From this discussion it becomes clear that the XFEM, as it is currently implemented, is only suitable for the identification of the locus of damage initiation in adhesive bonds, by comparing the maximum principal stress/strain in each of the constituent materials to the respective maximum values. However, it does not show to be suited for the simulation of damage growth, as the principle for defining the crack direction (orthogonal to the maximum principal stress/strain) does not model accurately the propagation of damage in multi-material structures. Moreover, XFEM does not it take into account the prospect of damage growth along interfaces between different materials. For the specific case of bonded joints, a modification of the XFEM algorithm that would consider these possibilities would bring a significant breakthrough for the simulation of these structures, with the accuracy of CZM but eliminating the major handicap of this method to follow the damage paths specified by the placement of the cohesive elements. Nonetheless, the XFEM is still a promising method that can be used in a large number of applications for the strength prediction of structures.

10.6 CONCLUSIONS

XFEM applied to bonded joints is still a very recent subject. This chapter described the main available works regarding this advanced modelling technique to model crack propagation and to predict the fracture behaviour of adhesively-bonded joints. Although more than one XFEM formulation exists, the XFEM formulation embedded in Abaqus® was initially described, since it is the formulation that is readily available to designers or researchers and, thus, with broader application. The described formulation as it was presented explains the main differences to the conventional FEM formulation and the element division procedure to produce a crack anywhere in the models. After this description, the most pertinent applications to the authors' knowledge of XFEM for bonded joints were briefly described, along with few bolted joint investigations, due to the scarcity of works on this matter. From these works, it is possible to realize that the XFEM is not yet a completely suitable tool for the analysis of bonded joints, mainly because of the mode-mixity in crack growth of bonded joints, which is not properly addressed by the current implementation of the method. This is because crack should grow perpendicularly to principal stresses or strains, which usually leads to cracking towards the adherends. However, as depicted by some studies, tensile pure-mode crack growth generally works very well. Adherend net failure in bolted joints is also well modelled by XFEM. The two case studies corroborated these assumptions. The first study addressed strength prediction of DCB (pure tensile mode) joints as a function of temperature and revealed accurate predictions of the entire P-δ curves for a wide range of temperatures, showing that, under pure mode, the XFEM works properly. In the second study, the case study aimed at evaluating CZM and XFEM modelling in predicting the strength of single-lap joints bonded with a brittle (Araldite® AV138), a moderately ductile (Araldite® 2015) and a largely ductile adhesive (Sikaforce® 7888) between aluminium adherends and varying values of L_O. CZM modelling with triangular shape damage laws revealed to be highly accurate for the brittle and moderately ductile adhesives, while underestimating by a non-negligible amount P_m for the polyurethane adhesive (maximum of 19.7%). A mesh dependency study confirmed that this numerical tool is highly stable to the mesh size. The XFEM analysis revealed in first hand that it is not possible to model damage propagation on account of the shear-predominant failure mode along the adhesive line, which results in crack propagation towards the adherends at the overlap region. Failure was then estimated by crack initiation using different stress and strain-based criteria, based on previous evidence that a mesh size in the adhesive layer equal to t_A gives a rough estimator of P_m for brittle adhesives. The stress-based criteria were found to under estimate by a significant amount P_m, unlike the strain-based criteria, which revealed to be a rough indicator of P_m, except for the polyurethane adhesive, representative of largely ductile adhesives. The mesh size study confirmed the P_m variations with the size of the adhesive layer elements, which is a clear limitation of the method. As a result of the findings reported in this chapter, it can be concluded that the XFEM is a valuable tool for general purpose crack propagation analysis and strength prediction, but some issues still have to be addressed so that it can easily applied for the analysis of bonded joints.

10.7 REFERENCES

76-142, N. T. 1988. Méthode de preparation de plaques d'adhésifs structuraux pour la réalisation d'éprouvettes d'essai de caractérisation.

Abaqus® 2013. Documentation. D. Systèmes. Vélizy-Villacoublay.

Adams, R. D. and Peppiatt, N. A. 1974. Stress analysis of adhesive-bonded lap joints. The Journal of Strain Analysis for Engineering Design 9: 185-196.

Adams, R. D., Comyn, J. and Wake, W. C. 1997. Structural Adhesive Joints in Engineering. Chapman & Hall, London.

Ahmad, H., Crocombe, A. D. and Smith, P. A. 2014. Strength prediction in CFRP woven laminate bolted double-lap joints under quasi-static loading using XFEM. Composites Part A: Applied Science and Manufacturing 56: 192-202.

Airoldi, A., Vesco, M., van der Zwaag, S., Baldi, A. and Sala, G. 2009. Damage in GLARE laminates under indentation loads: Experimental and numerical results. Proceedings of the 17th international conference on composite materials, Edinburgh, UK.

Andersson, T. and Stigh, U. 2004. The stress–elongation relation for an adhesive layer loaded in peel using equilibrium of energetic forces. International Journal of Solids and Structures 41: 413-434.

Banea, M. D., da Silva, L. F. M. and Campilho, R. D. S. G. 2010. Temperature dependence of the fracture toughness of adhesively bonded joints. Journal of Adhesion Science and Technology 24: 2011-2026.

Banea, M. D., da Silva, L. F. M. and Campilho, R. D. S. G. 2011. Mode I fracture toughness of adhesively bonded joints as a function of temperature: Experimental and numerical study. International Journal of Adhesion and Adhesives 31: 273-279.

Banea, M. D., da Silva, L. F. M. and Campilho, R. D. S. G. 2012. Effect of temperature on tensile strength and mode I fracture toughness of a high temperature epoxy adhesive. Journal of Adhesion Science and Technology 26: 939-953.

Belytschko, T. and Black, T. 1999. Elastic crack growth in finite elements with minimal remeshing. International Journal for Numerical Methods in Engineering 45: 601-620.

Bonhomme, J., Argüelles, A., Viña, J. and Viña, I. 2009. Numerical and experimental validation of computational models for mode I composite fracture failure. Computational Materials Science 45: 993-998.

Campilho, R. D. S. G., de Moura, M. F. S. F. and Domingues, J. J. M. S. 2005. Modelling single and double-lap repairs on composite materials. Composites Science and Technology 65: 1948-1958.

Campilho, R. D. S. G., de Moura, M. F. S. F. and Domingues, J. J. M. S. 2007. Stress and failure analyses of scarf repaired CFRP laminates using a cohesive damage model. Journal of Adhesion Science and Technology 21: 855-870.

Campilho, R. D. S. G., de Moura, M. F. S. F., Pinto, A. M. G., Morais, J. J. L. and Domingues, J. J. M. S. 2009a. Modelling the tensile fracture behaviour of CFRP scarf repairs. Composites Part B: Engineering 40: 149-157.

Campilho, R. D. S. G., de Moura, M. F. S. F. and Domingues, J. J. M. S. 2009b. Numerical prediction on the tensile residual strength of repaired CFRP under different geometric changes. International Journal of Adhesion and Adhesives 29: 195-205.

Campilho, R. D. S. G., de Moura, M. F. S. F., Ramantani, D. A., Morais, J. J. L. and Domingues, J. J. M. S. 2010. Buckling strength of adhesively-bonded single and double-strap repairs on carbon-epoxy structures. Composites Science and Technology 70: 371-379.

Campilho, R. D. S. G. and Rocha, J. M. M. 2011. Strength prediction of adhesively bonded repairs on carbon-epoxy laminates by the single and double-strap techniques. Polymer Composites 32: 1254-1264.

Campilho, R. D. S. G., Banea, M. D., Chaves, F. J. P. and Silva, L. F. M. d. 2011a. eXtended Finite Element Method for fracture characterization of adhesive joints in pure mode I. Computational Materials Science 50: 1543-1549.

Campilho, R. D. S. G., Banea, M. D., Pinto, A. M. G., da Silva, L. F. M. and de Jesus, A. M. P. 2011b. Strength prediction of single- and double-lap joints by standard and extended finite element modelling. International Journal of Adhesion and Adhesives 31: 363-372.

Campilho, R. D. S. G., Banea, M. D., Neto, J. A. B. P. and da Silva, L. F. M. 2012. Modelling of single-lap joints using cohesive zone models: Effect of the cohesive parameters on the output of the simulations. The Journal of Adhesion 88: 513-533.

Campilho, R. D. S. G., Banea, M. D., Neto, J. A. B. P. and da Silva, L. F. M. 2013a. Modelling adhesive joints with cohesive zone models: Effect of the cohesive law shape of the adhesive layer. International Journal of Adhesion and Adhesives 44: 48-56.

Campilho, R. D. S. G., Moura, D. C., Gonçalves, D. J. S., da Silva, J. F. M. G., Banea, M. D. and da Silva, L. F. M. 2013b. Fracture toughness determination of adhesive and co-cured joints in natural fibre composites. Composites Part B: Engineering 50: 120-126.

Curiel Sosa, J. L. and Karapurath, N. 2012. Delamination modelling of GLARE using the extended finite element method. Composites Science and Technology 72: 788-791.

D3433-99 ASTM Standard 2005. Standard test method for fracture strength in cleavage of adhesives in bonded metal joints.

da Silva, L. F. M. and Campilho, R. D. S. G. 2011. Advances in Numerical Modelling of Adhesive Joints. Springer, Heidelberg.

Daux, C., Moës, N., Dolbow, J., Sukumar, N. and Belytschko, T. 2000. Arbitrary branched and intersecting cracks with the extended finite element method. International Journal for Numerical Methods in Engineering 48: 1741-1760.

Davis, M. and Bond, D. 1999. Principles and practices of adhesive bonded structural joints and repairs. International Journal of Adhesion and Adhesives 19: 91-105.

de Moura, M. F. S. F., Campilho, R. D. S. G. and Gonçalves, J. P. M. 2008. Crack equivalent concept applied to the fracture characterization of bonded joints under pure mode I loading. Composites Science and Technology 68: 2224-2230.

Dolbow, J., Moës, N. and Belytschko, T. 2001. An extended finite element method for modeling crack growth with frictional contact. Computer Methods in Applied Mechanics and Engineering 190: 6825-6846.

Elguedj, T., Gravouil, A. and Combescure, A. 2006. Appropriate extended functions for X-FEM simulation of plastic fracture mechanics. Computer Methods in Applied Mechanics and Engineering 195: 501-515.

Fagerström, M. and Larsson, R. 2006. Theory and numerics for finite deformation fracture modelling using strong discontinuities. International Journal for Numerical Methods in Engineering 66: 911-948.

Feih, S. and Shercliff, H. R. 2005. Adhesive and composite failure prediction of single-L joint structures under tensile loading. International Journal of Adhesion and Adhesives 25: 47-59.

Flinn, B. D., Lo, C. S., Zok, F. W. and Evans, A. G. 1993. Fracture resistance characteristics of a metal-toughened ceramic. Journal of the American Ceramic Society 76: 369-375.

Grant, L. D. R., Adams, R. D. and da Silva, L. F. M. 2009. Experimental and numerical analysis of single-lap joints for the automotive industry. International Journal of Adhesion and Adhesives 29: 405-413.

Högberg, J. L., Sørensen, B. F. and Stigh, U. 2007. Constitutive behaviour of mixed mode loaded adhesive layer. International Journal of Solids and Structures 44: 8335-8354.

Kafkalidis, M. S. and Thouless, M. D. 2002. The effects of geometry and material properties on the fracture of single lap-shear joints. International Journal of Solids and Structures 39: 4367-4383.

Kanninen, M. F. and Popelar, C. H. 1985. Advanced Fracture Mechanics. Oxford University Press, Oxford, UK.

Khoei, A. R. and Nikbakht, M. 2006. Contact friction modeling with the extended finite element method (X-FEM). Journal of Materials Processing Technology 177: 58-62.

Lee, S. J. and Lee, D. G. 1992. Development of a failure model for the adhesively bonded tubular single lap joint. The Journal of Adhesion 40: 1-14.

Li, S., Thouless, M. D., Waas, A. M., Schroeder, J. A. and Zavattieri, P. D. 2005. Use of a cohesive-zone model to analyze the fracture of a fiber-reinforced polymer–matrix composite. Composites Science and Technology 65: 537-549.

Luo, Q. and Tong, L. 2007. Fully-coupled nonlinear analysis of single lap adhesive joints. International Journal of Solids and Structures 44: 2349-2370.

McCarthy, C. T., McCarthy, M. A., Stanley, W. F. and Lawlor, V. P. 2005. Experiences with modeling friction in composite bolted joints. Journal of Composite Materials 39: 1881-1908.

Mello, A. W. and Liechti, K. M. 2004. The effect of self-assembled monolayers on interfacial fracture. Journal of Applied Mechanics 73: 860-870.

Messler, R. W. 1993. Joining of Advanced Materials. Butterworths/Heinemann, Stoneham, USA.

Moës, N., Dolbow, J. and Belytschko, T. 1999. A finite element method for crack growth without remeshing. International Journal for Numerical Methods in Engineering 46: 131-150.

Moës, N. and Belytschko, T. 2002. Extended finite element method for cohesive crack growth. Engineering Fracture Mechanics 69: 813-833.

Mohammadi, S. 2008. Extended Finite Element Method for Fracture Analysis of Structures. Blackwell Publishing, New Jersey, USA.

Mubashar, A., Ashcroft, I. A. and Crocombe, A. D. 2013. Modelling damage and failure in adhesive joints using a combined XFEM-cohesive element methodology. The Journal of Adhesion 90: 682-697.

Neto, J. A. B. P., Campilho, R. D. S. G. and da Silva, L. F. M. 2012. Parametric study of adhesive joints with composites. International Journal of Adhesion and Adhesives 37: 96-101.

NF 76-142, French Standard. Méthode de preparation de plaques d'adhésifs structuraux pour la réalisation d'éprouvettes d'essai de caractérisation 1988.

Panigrahi, S. K. and Pradhan, B. 2007. Three dimensional failure analysis and damage propagation behavior of adhesively bonded single lap joints in laminated FRP composites. Journal of Reinforced Plastics and Composites 26: 183-201.

Pinto, A. M. G., Campilho, R. D. S. G., Mendes, I. R., Aires, S. M. and Baptista, A. P. M. 2011. Effect of hole drilling at the overlap on the strength of single-lap joints. International Journal of Adhesion and Adhesives 31: 380-387.

Premchand, V. P. and Sajikumar, K. S. 2009. Fracture analysis in adhesive bonded joints with centre crack. NCTT09 10th National Conference on Technological Trends, Trivandrum, India.

Radice, J. and Vinson, J. 2006. On the use of quasi-dynamic modeling for composite material structures: Analysis of adhesively bonded joints with midplane asymmetry and transverse shear deformation. Composites Science and Technology 66: 2528-2547.

Sabsabi, M., Giner, E. and Fuenmayor, F. J. 2011. Experimental fatigue testing of a fretting complete contact and numerical life correlation using X-FEM. International Journal of Fatigue 33: 811-822.

Sørensen, B. F. 2002. Cohesive law and notch sensitivity of adhesive joints. Acta Materialia 50: 1053-1061.

Sørensen, B. F. and Jacobsen, T. K. 2003. Determination of cohesive laws by the J integral approach. Engineering Fracture Mechanics 70: 1841-1858.

Sugiman, S., Crocombe, A. D. and Aschroft, I. A. 2013. Modelling the static response of unaged adhesively bonded structures. Engineering Fracture Mechanics 98: 296-314.

Sukumar, N., Moës, N., Moran, B. and Belytschko, T. 2000. Extended finite element method for three-dimensional crack modelling. International Journal for Numerical Methods in Engineering 48: 1549-1570.

Sukumar, N. and Prévost, J. H. 2003. Modeling quasi-static crack growth with the extended finite element method Part I: Computer implementation. International Journal of Solids and Structures 40: 7513-7537.

Taib, A. A., Boukhili, R., Achiou, S. and Boukehili, H. 2006. Bonded joints with composite adherends. Part II. Finite element analysis of joggle lap joints. International Journal of Adhesion and Adhesives 26: 237-248.

Trantina, G. G. 1972. Fracture mechanics approach to adhesive joints. Journal of Composite Materials 6: 192-207.

Tsai, M. Y. and Morton, J. 1994. An evaluation of analytical and numerical solutions to the single-lap joint. International Journal of Solids and Structures 31: 2537-2563.

Vable, M. and Reddy Maddi, J. 2006. Boundary element analysis of adhesively bonded joints. International Journal of Adhesion and Adhesives 26: 133-144.

Volkersen, O. 1938. Die nietkraftoerteilung in zubeanspruchten nietverbindungen konstanten loschonquerschnitten. Luftfahrtforschung 15: 41-47.

Wang, J., Kang, Y. L., Qin, Q. H., Fu, D. H. and Li, X. Q. 2008. Identification of time-dependent interfacial mechanical properties of adhesive by hybrid/inverse method. Computational Materials Science 43: 1160-1164.

Wang, S. S. 1983. Fracture mechanics for delamination problems in composite materials. Journal of Composite Materials 17: 210-223.

Wang, Z., Zhou, S., Zhang, J., Wu, X. and Zhou, L. 2012. Progressive failure analysis of bolted single-lap composite joint based on extended finite element method. Materials & Design 37: 582-588.

Xie, D. and Waas, A. M. 2006. Discrete cohesive zone model for mixed-mode fracture using finite element analysis. Engineering Fracture Mechanics 73: 1783-1796.

Xu, Y. and Yuan, H. 2009. Computational analysis of mixed-mode fatigue crack growth in quasi-brittle materials using extended finite element methods. Engineering Fracture Mechanics 76: 165-181.

Zhu, Y., Liechti, K. M. and Ravi-Chandar, K. 2009. Direct extraction of rate-dependent traction–separation laws for polyurea/steel interfaces. International Journal of Solids and Structures 46: 31-51.

11

C H A P T E R

Numerical Modelling of Bonded Joints and Repairs

Raul D.S.G. Campilho* and Raul D.F. Moreira

11.1 INTRODUCTION

Adhesive bonded joints have become more efficient in the last few decades due to the developments in adhesive technology, which has resulted in higher peel and shear strengths and also in allowable ductility up to failure. As a result of the reported improvement in the mechanical characteristics of adhesives, adhesive bonding has progressively replaced traditional joining methods such as bolting or riveting (Campilho et al. 2008). Bonded joints are frequently expected to sustain static or cyclic loads for considerable periods of time without any adverse effect on the load-bearing capacity of the structure. Adhesive bonding has found applications in various areas from high technology industries such as aeronautics, aerospace, electronics and automotive to traditional industries such as construction, sports and packaging. Adhesive bonding is a valid option due to its numerous advantages over the conventional bolting or riveting methods (Kimiaeifar et al. 2013), e.g., more uniform stress distributions, reduced weight penalty, minimal aerodynamic disturbance and fluid sealing characteristics. It is common knowledge that stress concentrations still subsist in bonded joints along the bond length owing to the gradual transfer of load between the two adherends in the overlap region (also known as differential straining along the overlap), especially in single-lap joints (Campilho et al. 2008, Pearson and Mottram 2012). As a result, shear stresses (τ) concentrate at the overlap edges, with only a very small amount of load being carried in the central region. Peel stresses (σ) also develop in the same regions owing to the joint rotation and curvature of the adherends (Pinto et al. 2010). Both of these can be harmful to the structure, especially when using relatively brittle adhesives, which do not allow redistribution of stresses at the loci of higher concentrations, i.e., the overlap edges,

Departamento de Engenharia Mecânica, Instituto Superior de Engenharia do Porto, Instituto Politécnico do Porto, Rua Dr. António Bernardino de Almeida, 431, 4200-072 Porto, Portugal.
* Corresponding author: raulcampilho@gmail.com

leading to premature failures (Pinto et al. 2010). Bonding of patches with adhesives at the damaged region, which provides durable and resistant unions (Assih et al. 2001), is being increasingly used in structural repairs. Typically, the initial strength and stiffness of the damaged components cannot be restored using this technique without a significant weight penalty. Thus, a substantial amount of research has been carried out in the last decades on the development of efficient adhesively-bonded repairs and on adhesives' technology (Kelly 1983, Soutis et al. 1999). The larger bond areas and the reduction of stress concentrations at the bond edges due to the adherend tapering effect justify the higher efficiency of the scarf repairs, compared to the easy-execution strap repairs (Campilho et al. 2011a). The outcome of the optimization of stresses is a higher strength for the same bond area than strap repairs (Bikerman 1968), which renders scarf repairs more suited to critical applications.

A lack of suitable material models and failure criteria has resulted in a tendency to 'overdesign' adhesive joints. Safety considerations often require that adhesively-bonded structures, particularly those employed in primary load-bearing applications, include mechanical fasteners (e.g. bolts) as an additional safety precaution. These practices result in heavier and more costly components. The development of reliable design and predictive methodologies can be expected to result in a more efficient use of adhesives. In order to design structural joints in engineering structures, it is necessary to be able to analyse them. Initially, theoretical methods (mainly closed-form) were proposed for stress distributions in the adhesive for simple geometries such as the single or double-lap joint and failure estimation was carried out by comparison of the maximum stresses with the material strengths. The theoretical analysis of adhesively-bonded joints started almost 80 years ago, with the simple closed-form model of Volkersen (1938) that considers the adhesive and adherends as elastic and that the adhesive deforms only in shear. The equilibrium equation of a single-lap joint led to a simple governing differential equation with a simple algebraic equation. However, the analysis of adhesive joints can be highly complex if composite adherends are used, the adhesive deforms plastically or if there is an adhesive fillet. In those cases, several differential equations of high complexity might be obtained (non-linear and non-homogeneous). Some decades later, the Finite Element (FE) method initiated its incursion in the analysis of adhesively-bonded joints (Carver and Wooley 1971), by consideration of stress/strain (i.e., continuum mechanics) or fracture mechanics criteria for failure prediction (Tsai and Morton 1994). To predict the joint strength, one must have the stress distribution and a suitable failure criterion. The stress distribution can be obtained by a FE analysis or a closed-form model. For complex geometries and elaborated material models, the FE method is preferable. One of the simplest failure models is that based on a stress or strain limit state, i.e. based on a continuum mechanics approach. Fracture mechanics principles can also be used within a FE analysis. This can be based on either the stress intensity factor or energy approaches. Adams and co-workers are among the first to have used the FE method for analyzing adhesive joint stresses (Adams and Peppiatt 1974, Crocombe and Adams 1981). One of the first reasons for the use of the FE method was to assess the influence of the spew fillet. The joint rotation and the adherends and adhesive plasticity are other aspects that are easier to treat with a FE analysis. The study of Harris and Adams (1984) is one of the first FE analyses taking into account these three aspects. Even though these analyses were promising, they had few limitations: stress/strain predictions depend on the mesh size at the critical regions, while fracture criteria such as the Virtual Crack Closure Technique (VCCT) are restricted to Linear Elastic Fracture Mechanics (LEFM) and need an initial crack. Cohesive Zone Models (CZM) have been used in the last decade for the strength prediction of adhesive joints, as an add-on to FE analyses that allows simulation of damage growth within bulk regions of continuous materials or interfaces between different materials (da Silva and Campilho 2011, Feraren and Jensen 2004). Compared to conventional FE, a much more accurate prediction is achieved, since different shapes can be developed for the cohesive laws, depending on the nature of the material or interface to be simulated. The triangular and trapezoidal CZM shapes are most commonly used for strength prediction of typical structural materials. For the application of this technique, traction-separation laws with a pre-defined shape are established at the failure paths and the values of energy release rate in tension and shear (G_I and G_{II}, respectively) along the fracture paths and respective critical values or toughness (G_{Ic} and G_{IIc}) are required. The cohesive strengths in tension and shear (t_n^0 and t_s^0, respectively) are equally needed and they relate to damage initiation, i.e. end of the elastic behaviour and beginning of damage. Different techniques are nowadays available for the definition of cohesive parameters (G_{Ic}, G_{IIc}, t_n^0 and t_s^0), such as the property identification technique, the direct method and the inverse method. These methods usually rely on Double-Cantilever Beam (DCB), End-Notched Flexure (ENF) or single-lap specimens, generally with good results. In damage mechanics techniques, a damage parameter is considered to modify the response of the structure's materials typically by reducing the stiffness or strength, e.g. for adhesive layers (Khoramishad et al. 2010), or composite delaminations or matrix failure (Daudeville and Ladevèze 1993), to model damage during loading. Moreover, it is possible to model different damage mechanisms at the same time, each of them by an independent damage variable. Damage growth can be defined as a function of the load for static loads (Raghavan and Ghosh 2005) or the number of cycles for fatigue modelling (Wahab et al. 2001, Imanaka et al. 2003). For bonded joints, little work is published in this field. Chen et al. (2011), which used the average plastic strain energy to predict crack initiation and propagation in single-lap joints, and also the static failure load by damage mechanics, is one of the few exceptions. Damage mechanics is particularly recommended if the failure path is not known (Shenoy et al. 2010). Another method for modelling cracks in materials is the eXtended Finite Element Method (XFEM), which uses enriched shape functions to represent a discontinuous displacement field (Moës et al. 1999). The main advantage of XFEM is that the crack may initiate at any point in the material and propagate based on loading conditions. No remeshing is required as the crack can grow within an element, and it does not need to follow element boundaries (Mohammadi 2008).

This chapter was prepared for engineers and scientists that have already some background in adhesive joints and numerical modelling. A description with selected examples is given for the most relevant strength prediction techniques. Amongst these techniques are continuum mechanics, fracture mechanics, CZM and damage mechanics. XFEM is not addressed in this chapter. CZM and damage mechanics are divided into static and fatigue applications. Between these techniques, CZM are addressed in more detail, since this is the most powerful and accurate method. A case study is then presented regarding static CZM modelling of adhesively-bonded scarf repairs with external reinforcements. In this case study, experimental data is used to validate the CZM methodology by direct comparison of the failure loads for several repair configurations. All the obtained results are also justified with a detailed stress analysis to the scarf and external reinforcement adhesive layers.

11.2 STRENGTH PREDICTION OF BONDED JOINTS AND REPAIRS

11.2.1 Continuum Mechanics Approach

In the continuum mechanics approach, the maximum values of stress, strain or strain energy, predicted by the FE analyses, are usually used in the failure criterion and are compared with the corresponding material allowable values. Initially, the maximum principal stresses were proposed for very brittle materials whose failure mode is at right angles to the direction of maximum principal stress. This criterion ignores all the other principal stresses, even though they are not nil. Establishing the failure modes in lap joints bonded with brittle adhesives, Adams et al. (1997) have extensively used this criterion to predict the joint strength with success. However, because of the singularity of stresses at re-entrant corners of joints, stresses depend on the mesh size used and how close to the singular points the stresses are taken. Values of stresses calculated at Gauss points near the singularity or extrapolation of Gauss point values to the singularity were, in fact, used. Therefore, care must be taken when using this criterion. Although the criterion is sensitive to the mesh size used, the physical insight into the failure process is very clear, as the maximum principal stress is the most responsible for the failure of joints bonded with brittle adhesives. However, it should be noted that the adherend corners are usually not sharp in practice. There is, in general, a small amount of rounding at the adherend corner due to the production process. This may affect the stress distributions in the region of the adherend corner and, therefore, the joint strength, because stresses in this area are very sensitive to the change in the geometry. One consequence of the adherend rounding is the non-existence of the singularity, which facilitates the application of a stress or strain limit criterion. Adams and Harris (1987) theoretically and experimentally demonstrated that the strength of single-lap joints with rounded adherends with a toughened adhesive increased substantially compared with joints with sharp adherend corners. More recently, Zhao et al. (2011a) and Zhao et al. (2011b) have also studied the effect of adherend rounding by continuum mechanics. von Mises proposed a yield criterion, which states that a material yields under multi-axial stresses when its distortion energy reaches a critical value, that is

$$\sigma_{VM}^2 = (\sigma_1 - \sigma_2)^2 + (\sigma_2 - \sigma_3)^2 + (\sigma_3 - \sigma_1)^2 = \text{constant}, \qquad (11.1)$$

where $\sigma_i (i = 1, 2, 3)$ are the principal stresses. Such a criterion has been used by Ikegami et al. (1990) to study the strength of scarf joints between glass fibre composites and metals. It should be noted that this criterion is more applicable to material yielding than strength.

τ stresses have been extensively used to predict lap joint strength, especially in closed-form analyses, considering a limiting maximum shear stress equal to the bulk adhesive shear strength. Greenwood (1969) used the maximum τ stress calculated by Goland and Reissner (1944) to predict the joint strength. da Silva et al. (2009a) and da Silva et al. (2009b) showed for single-lap joints that this criterion is only valid for brittle adhesives and short overlaps. This approach ignores the normal stresses existing in lap joints and therefore it over-estimates the joint strength. When ductile adhesives are used, criteria based on stresses are not appropriate because joints can still endure large loads after adhesive yielding. For ductile adhesives, Harris and Adams (1984) used the maximum principal strain as failure criterion for predicting the joint strength. This criterion can also predict the failure mode. However, it is equally sensitive to the mesh size, as previously discussed for the maximum principal stress approach. Hart-Smith (1973a) proposed that the maximum shear strain might be used as a failure criterion when plastic deformation was apparent. Other analyses go beyond that of Hart-Smith, which allows both shear and peel contributions to plasticity, such as that by Adams and Mallick (1992). Crocombe (1989) studied the failure of cracked and uncracked specimens under various modes of loading and used a critical σ stress at a distance from the singularity with some success. An alternative method was also proposed to use an effective stress, matched to the uniaxial bulk strength, at a distance. However, it was found for the latter criterion that the critical distance at which it should be applied varied with different modes of loading because of the change in the plastic zone size. No general criterion for a given adhesive was presented. Strain energy is the area under the axial stress-strain (σ_a-ε) curve. It is equally possible to use criteria based on strain energy, which take into account all the stress and strain components. As a result, they are more suitable as a failure criterion than either stresses or strains alone. Plastic energy density has also been used as a failure criterion (Adams and Harris 1987), being similar to the total strain energy criterion but it only takes the plastic part of the deformation into account. In the analysis of lap joints, Crocombe (1989) found that, for largely ductile adhesives, the whole overlap yielded before failure. A new failure criterion was then proposed based on the yielding of adhesive in the whole overlap. Such a criterion is useful for very ductile adhesives in which the adhesive bond cannot support any larger load once it yields globally. However, it should be noted that the adhesives need to be extremely ductile (more than 20% of failure strain in shear) for the whole adhesive bond to yield before final failure. It should be realized that all the mentioned criteria are applicable to continuous structures only. They run into difficulty when defects occur or more than one material is present, since stresses or strains are not well defined at the singular points. As a result, new criteria or modified versions of the above criteria need to be developed.

11.2.2 Fracture Mechanics Approach

Continuum mechanics assumes that the structure and its material are continuous. Defects or two materials with re-entrant corners obviously are not consistent with such an assumption. Consequently, continuum mechanics gives no solution at these singular points resulting in stress or strain singularities. Cracks are the most common defects in structures, for which fracture mechanics has been developed. In fracture mechanics, it is well accepted that stresses calculated by using continuum mechanics are singular (infinite) at the crack tip. With current theories on mechanics, such a singularity always exists when the crack angle is less than 180°. This result was found by Williams (1959) for stress singularities in a wedged notch. This argument is also applicable to the stress singularity in two materials bonded together with a re-entrant corner. Actually, the stress discontinuity still exists, although the free surfaces do not exist.

Fracture mechanics has been successfully applied to many engineering problems in recent years. The damage tolerance design concept, originally adopted in the aircraft industry, was based mainly on the well-established concept of LEFM, and it has gradually gained ground in other engineering fields. Many studies dealing with adhesive joints use the strain energy release rate, G, and respective critical value or fracture toughness, G_c (Fernlund and Spelt 1991a, b), instead of stress intensity factors because these are not easily determinable when the crack grows at or near to an interface. However, the fracture of adhesive joints inherently takes place under mixed mode because of the varying properties between different materials and the complex stress system. Failure criteria for mixed mode fracture can be developed in a way analogous to the classical failure criteria, although the fracture surface (or envelope) concept must be introduced. Various mathematical surface functions have been proposed to fit the experimental results, such as the 3D criterion (Dillard et al. 2009)

$$\left(\frac{G_I}{G_{Ic}}\right)^\alpha + \left(\frac{G_{II}}{G_{IIc}}\right)^\beta + \left(\frac{G_{III}}{G_{IIIc}}\right)^\gamma = 1, \tag{11.2}$$

where G_I, G_{II} and G_{III} are the values of G under pure tension, shear and tearing modes, respectively, and G_{Ic}, G_{IIc} and G_{IIIc} the respective values of G_c. The linear ($\alpha = \beta = \gamma = 1$) and quadratic energetic criteria ($\alpha = \beta = \gamma = 2$) are the most used. The law parameters may be chosen to best fit the experimental data, or they may be prescribed based on some assumed relationship.

The use of a generalized stress intensity factor, analogous to the stress intensity factor in classical fracture mechanics, to predict fracture initiation for bonded joints at the interface corners has also been investigated. Groth (1988) assumed that initiation of fracture occurs when the generalized stress intensity factor reaches its critical value, initially tuned experimentally. Gleich et al. (2001) carried out a similar study by calculating the singularity strength and intensity for a range of adhesive thickness (t_A) values. These approaches work well for the joints that were used to determine the critical stress intensity factor but their application is questionable for extrapolation to other types of geometries. Fracture mechanics can thus be used to predict joint strength or residual strength if there is a crack tip or a known and calibrated singularity. When materials deform plastically, the LEFM concepts have to be extended into elasto-plastic fracture mechanics. The J-integral proposed by Rice (1968) is suited to deal with these problems. This approach has been used by a variety of researchers to predict the joint strength of cracked adhesive joints with good results (Banea et al. 2010, Choupani 2008).

11.2.3 Cohesive Zone Modelling

The concept of cohesive zone was initially proposed by Dugdale (1960) to describe damage under static loads at the cohesive process zone ahead of the apparent crack tip. CZM were largely refined and tested since then to simulate crack initiation and propagation in cohesive and interfacial failure problems or composite delaminations. CZM are based on spring or more typically cohesive elements (Feraren and Jensen 2004), connecting two-dimensional (2D) or three-dimensional (3D) solid elements of structures. An important feature of CZM is that they can be easily incorporated in conventional FE softwares to model the fracture behaviour in various materials, including adhesively-bonded joints. CZM are based on the assumption that one or multiple fracture interfaces/regions can be artificially introduced in structures, in which damage growth is allowed by the introduction of a possible discontinuity in the displacement field. The technique consists of the establishment of traction-separation laws (addressed as CZM laws) to model interfaces or finite regions.

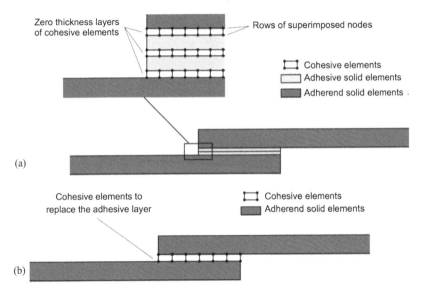

FIG. 11.1 Cohesive elements to simulate zero thickness failure paths – local approach (a) and to model a thin adhesive bond between the adherends – continuum approach (b) in an adhesive bond.

The CZM laws are established between paired nodes of cohesive elements, and they can be used to connect superimposed nodes of elements representing different materials or different plies in composites, to simulate a zero thickness interface (local approach; Fig. 11.1a; (Campilho et al. 2005)), or they can be applied directly between two non-contacting materials to simulate a thin strip of finite thickness between them, e.g. to simulate an adhesive bond (continuum approach; Fig. 11.1b; (Campilho et al. 2013)). CZM provide a macroscopic reproduction of damage along a given path, disregarding the microscopic phenomena on the origin of failure, by the specification of a traction-relative displacement (t-δ) response between paired nodes at both crack faces along the pre-defined crack path, by specification of large scale parameters ruling the crack growth process such as G_{Ic}, G_{IIc} or G_{IIIc} (Zhao et al. 2011b). The traction-separation laws are typically represented by linear relations at each one of the loading stages (Yang and Thouless 2001). Figure 11.2 represents the 2D triangular CZM model actually implemented in Abaqus® for static damage growth, which is detailed here. The 3D version is similar, but it also includes the tearing component. More details on the 3D CZM implemented in Abaqus® can be found in Abaqus® (2013).

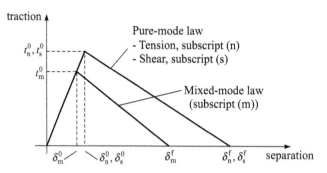

FIG. 11.2 Triangular CZM law.

The subscripts n and s relate to pure normal (tension) and shear behaviours, respectively. t_n and t_s are the corresponding current stresses, and δ_n and δ_s the current values of δ. CZM require the knowledge of G_I and G_{II} along the fracture paths and respective values of G_{Ic} and G_{IIc}. Additionally, t_n^0 and t_s^0 must be defined, relating to the onset of damage, i.e., cancelling of the elastic behaviour and initiation of stress softening. δ_n^0 and δ_s^0 are the peak strength displacements, and δ_n^f and δ_s^f the failure displacements. The values of δ_n^f and δ_s^f are defined by G_{Ic} or G_{IIc}, respectively, as these represent the area under the CZM laws. As for the mixed mode CZM law (Fig. 11.2), t_m^0 is the mixed mode cohesive strength, δ_m^0 the corresponding displacement, and δ_m^f the mixed mode failure displacement.

Under pure mode loading, the respective t-δ response attains its peak at the cohesive strength (t_n^0 or t_s^0), corresponding to damage initiation by the induced reduction of stiffness of the cohesive element. Softening follows and, when the values of t are completely cancelled, the crack propagates up to the adjacent set of paired nodes in the failure path, permitting the gradual debonding between crack faces. Under mixed loading (i.e., when two or three modes of loading are simultaneously present), stress and/or energetic criteria are often used to combine the pure mode laws, thus simulating the typical mixed mode behaviour inherent to bonded assemblies. By the mentioned principles, the complete failure response of structures can be simulated. For the estimation of the cohesive law parameters, a few data reduction techniques are available (e.g. the property identification technique, the direct method and the inverse method) that enclose varying degrees of complexity and expected accuracy of the results (Liljedahl et al. 2006).

CZM have been largely used in recent years to simulate the behaviour of structures up to failure, as they allow to include in the numerical models multiple failure possibilities, within different bulk regions of materials or between materials interfaces, e.g. at the adhesive bonding interfaces. The knowledge of the loci of damage occurring in a structure is not required as an input parameter, as the CZM globally searches damage initiation sites at specified failure paths satisfying the established criteria. However, cohesive elements must exist at the planes where damage is prone to occur, which, in several applications, can be difficult to know in advance. Several CZM law shapes have been presented in the literature, depending on the nature of the material or interface to be simulated. The triangular, exponential and trapezoidal shapes are the most commonly used for strength prediction of typical materials.

11.2.3.1 Static Applications

The majority of CZM studies relate to static applications. Kafkalidis and Thouless (2002) performed a FE analysis of symmetric and asymmetric single-lap joints using a continuum CZM approach that included the adhesive plasticity by means of trapezoidal CZM laws. Using CZM parameters determined for the particular combination of materials used, the FE predictions for different bonded shapes (varying the overlap length and t_p) showed an excellent agreement with the experimental observations. Figure 11.3 compares the measured and FE peak load values for a series of symmetric single-lap joints with aluminium adherends having an adherend thickness (t_p) of 2.0 mm and an adhesive layer of Ciba® XD4600 with $t_A = 0.25$ mm.

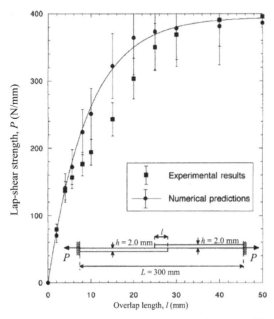

FIG. 11.3 Experimental and FE single-lap joint strength as a function of the overlap length for symmetric specimens (Kafkalidis and Thouless 2002).

The slight discrepancies on the predicted strengths of Fig. 11.3 were justified by the sensitivity of the joint behaviour to the boundary conditions. In fact, the experimental configuration results in a slight misalignment of the grips, which is not accounted for in the simulations. Figure 11.4 shows the FE predictions for the sequence of deformation and crack growth in an asymmetric joint with a t_p ratio of two, which was consistent with the experimental behaviour. Overall, the FE models predicted accurately the failure loads, displacements and deformations of the joints.

FIG. 11.4 FE deformation sequence and crack growth in an asymmetric single-lap joint (Kafkalidis and Thouless 2002).

The analysis of Karac et al. (2011) extends the typical applications of CZM to rate dependent simulations (local approach), by using an implicit finite volume method coupled to the CZM analysis. In the work of Crocombe et al. (2006), a local CZM with a triangular shape was developed to predict the strength of environmentally degraded adhesively-bonded single-lap joints. Apart from the CZM parameters, the model included two moisture dependent fracture parameters, all of these experimentally calibrated by mixed-mode flexure (MMF) tests. The mechanical diffusion events were also implemented within Abaqus®, resulting on overall accurate predictions of the degraded strength of the joints.

Campilho et al. (2009a) experimentally tested carbon-epoxy scarf repairs in tension, using scarf angles (α) between 2 and 45°, considering CZM for strength prediction. To account for the ductile behaviour of the adhesive (Huntsman Araldite® 2015), a trapezoidal CZM including the adhesive plasticity was used. The cohesive laws of the adhesive layer – continuum approach – and composite interlaminar and composite intralaminar (in the transverse and fibre direction) – local approach – necessary to replicate numerically the experimental failure paths, were previously characterized with DCB and ENF tests for tension and shear, respectively, using an inverse methodology. The layout of cohesive elements in the models is depicted in Fig. 11.5.

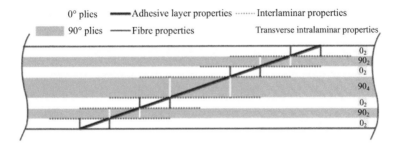

FIG. 11.5 Layout of cohesive elements with different CZM laws in the scarf repair FE models (Campilho et al. 2009a).

The adhesive representative elements were inserted along the scarf profile replacing the adhesive bond, the interlaminar elements were positioned between different oriented plies, the transverse intralaminar elements were used vertically in the 90° plies to simulate the intralaminar matrix cracking, and the fibre elements were placed vertically in the 0° plies to simulate fibre cracking. Two distinct failure modes were experimentally observed: type A and type B failures. Type A failure was observed for the repairs with $\alpha = 15°$, 25° and 45°, consisting of a cohesive failure of the adhesive. Type B failure occurred for $\alpha = 2°$, 3°, 6° and 9°, representing a mixed cohesive and interlaminar/intralaminar failure (Fig. 11.6). A detailed discussion for this failure mode modification was provided, based on a stress analysis of the repairs. The CZM simulations accurately reproduced the experimental fracture modes, and the failure mode modification between types A and B, showing the effectiveness of CZM modelling for static applications.

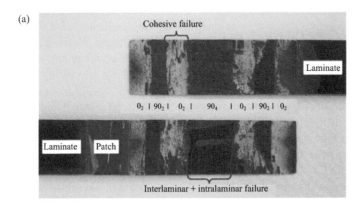

FIG. 11.6 Type B failure for a $\alpha = 3°$ repair: experimental fracture surfaces (a) and FE prediction (distorted image for a clear visualization) (b) (Campilho et al. 2009a).

Figure 11.7 reports the strength results, showing an exponential increase with the reduction of α, related to the increase of available shear area for load transfer between the adherends and the patch, accompanied by a reduction of σ stresses along the scarf length. On account of the described results, the authors concluded that CZM can be successfully applied to predict the mechanical behaviour of bonded structures.

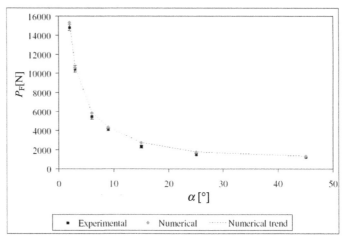

FIG. 11.7 Experimental and FE strength of the scarf repairs as a function of α (Campilho et al. 2009a).

11.2.3.2 Fatigue Applications

The fatigue strength prediction with CZM is a recent possibility (Turon et al. 2007a). Important issues possible to account for with fatigue CZM are the accurate prediction of cyclic damage initiation, and the inclusion of fatigue life-time prediction, because under cyclic loads either one of these phases can be dominant depending on the specimens' geometry, materials, load range and testing/loading conditions. However, the main advantage of CZM for modelling fatigue crack growth is that damage initiates within un-cracked structures when the cyclic cohesive stresses exceed a pre-determined critical value (Moroni and Pirondi 2011). With this purpose, the static formulation of CZM is adjusted to account for fatigue damage. In fact, for cyclic loadings, damage is not only dependent on the relative separations, δ_n and δ_s, but also on the load cycle count. An associated limitation to some fatigue CZM formulations is the geometry dependency, which restricts the applicability of the models to specific test specimens (Khoramishad et al. 2010). Regarding the CZM approach, continuum modelling is often selected for fatigue. A few lines of analysis for fatigue applications are currently applied: (1) a cycle-by-cycle analysis for the damage progression, (2) cyclic extrapolation techniques and (3) analysis based on the maximum fatigue load. The 2nd and 3rd approaches are less demanding in which regards to computational resources, whilst the former allows a more precise damage evolution to be implemented, predicting more accurately features such as crack growth retardation after overload (Roe and Siegmund 2003), although not being viable for high cycle fatigue (Khoramishad et al. 2010).

The cycle-by-cycle fatigue CZM (i.e., 1st approach) proposed by Maiti and Geubelle (2005) for crack growth in polymeric materials was upgraded from a static CZM, considering an evolution law relating the cohesive stiffness, the rate of crack opening displacement and the cyclic count from the onset of failure. The fatigue component relied on two parameters that were easily defined from the Paris curve, between the crack growth per cycle and the range of applied stress intensity factor. The proposed fatigue CZM of Turon et al. (2007b), following the 2nd described line of analysis for fatigue crack modelling of composites, does not require calibration of the fatigue parameters, solely requiring the static CZM parameters and the Paris-like law coefficients from fatigue crack growth tests. However, the proposed formulation is restricted to test geometries in which G can be computed analytically and it is independent of the crack length during the test, such as in the DCB test. Khoramishad et al. (2010) developed a fatigue CZM for adhesively-bonded joints with 2024-T3 aluminium adherends and the adhesive FM® 73 M OST (Cytec). The model was based on the maximum fatigue loading conditions (3rd line of analysis described) that is independent of the joint geometry, relying only on the adhesive system. Triangular damage laws were considered, and a strain-based fatigue damage model was developed and coupled to a static CZM for simulation of the fatigue influence on the structures' behaviour. The development and calibration of the proposed fatigue model were carried out for single-lap joint tests, which was subsequently applied without any modification to doublers in bending. Static tests were initially performed on both geometries for the definition of the static CZM parameters. The static CZM was tuned by an iterative technique to match the experimental responses. Fatigue tests were subsequently carried out at 5 Hz and with a fatigue load ratio (R) of 0.1. For the assessment of damage evolution in the adhesive bond during fatigue loading, the back-face strain technique and in situ-microscopy were used. The back-face strain data was afterwards used for validation of the CZM by comparison with the predicted-back-face strains. Fatigue damage simulation was implemented by degradation of the CZM response. A fatigue damage parameter was established, which evolved during the cyclic loading. Fatigue crack growth was divided into two steps: (1) a static CZM analysis was carried out considering the maximum fatigue load, using the static damage variable and providing the stress/strain state at the beginning of the fatigue test, and the maximum principal strain in the cohesive element (ε_{max}) was calculated for all cohesive elements; and (2) a cyclic damage variable was established at each integration point of the cohesive elements, which was numerically integrated in each increment, corresponding to a specified value of cycle increment (ΔN). The values of t_n^0, t_s^0, G_{Ic} and G_{IIc} were then linearly reduced proportionally to the damage variable. The increment-by-increment implementation of this process, with the individual calculation of ε_{max} and damage variable until the structure no longer sustained the maximum fatigue load, gave the predicted fatigue life.

11.2.4 Damage Mechanics

In these methodologies, a damage parameter is established to modify the constitutive response of materials by the depreciation of stiffness or strength, e.g. for thin adhesive bonds (Khoramishad et al. 2010), or composite delaminations or matrix failure, to represent the severity of material damage during loading. The damage variables can be categorized under two main groups: (1) variables that predict the amount of damage by the redefinition of the material constitutive properties but that do not directly relate to the damage mechanism, and (2) variables linked to the physical definition of a specific kind of damage, such as porosities or relative area of micro-cavities (Voyiadjis and Kattan 2005). The 2nd group of damage variables is based on macroscopic material properties whose evolution is governed by a state equation. By damage mechanics, the growth of damage is defined as a function of the load for static modelling or the cyclic count for fatigue analyses. Generic models for bulk materials are available in the literature (Lemaitre and Desmorat 2005), and respective modifications for particular kinds of damage modelling, e.g. damage nucleation from voids and the formation of micro cracks (Kattan and Voyiadjis 2005). For bonded joints, little work is published in this field. Compared to fatigue CZM, damage mechanics techniques do not provide a clear distinction between fatigue initiation and propagation phases, although they can give a basis for the predictive analysis (Khoramishad et al. 2010).

11.2.4.1 Static Applications of Damage Mechanics

A few works are currently available using phenomenological damage mechanics techniques for static failure of structures, by the inclusion of one or more state variables affecting the macroscopic constitutive response of materials up to failure. The work of Sampaio et al. (2004) focused on the static strength prediction of butt adhesive joints of aluminium adherends bonded with a thin epoxy adhesive bond (Dow® DH331) by an analytical damage mechanics model accounting for the value of t_A. The model combined a relatively straightforward mathematical background with the possibility to describe complex non-linear mechanical behaviour of bonded joints, including strain softening and size effects. The model was built on the assumption of two elastic bars (aluminium adherends) and a damageable bar to model the adhesive bond, considering an approximation of the 3D fracture behaviour to a one-dimensional and equivalent model with a one-material homogeneous bar. An energy based formulation was established as a function of the axial strain, ε, and an auxiliary damage variable, D, was associated with the dissipative mechanism of failure to account for the reduction of elastic energy. To account for t_A effects in bonded joints, a quadratic energy function was defined based on previous experimental evidence. For validation of the model, the predicted values of failure stress were plotted against experimental data for different values of t_A, giving accurate representations of the tests and a decreasing strength of the butt joints with t_A up to a bulk representative constant value. Chen et al. (2011) used the average plastic strain energy to predict crack initiation and propagation in single-lap joints, and also the failure load by a damage mechanics technique. Two adhesives were used, designated by MY750 and CTBN. The physical fracture process was simulated by a FE analysis as follows. When a converged solution was obtained after each applied load increment, a check was made to see whether the failure condition had been attained anywhere in the joint. Under these conditions, the values of Young's modulus (E) and Poisson's ratio (v) of the material within the failed region were reset to zero (or near zero), such that it could deform almost freely without transferring any load. The stresses within this region were also reset to zero and then the whole system of equations (stiffness) was reassembled and solved. This FE model thus accounted for damage by voiding the cracked elements. All the checking and property depreciation were implemented in the FE program and done automatically. A new load increment was not applied until the damage process had stopped.

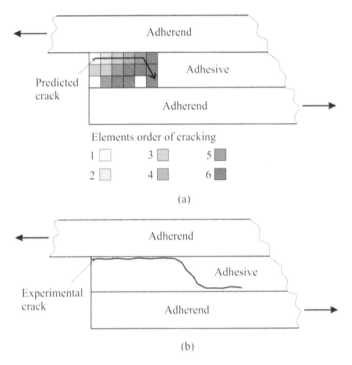

FIG. 11.8 Failure progress for MY750 adhesive/steel single-lap joints; (a) schematic failure step by step obtained numerically; (b) crack patch observed in the experiments (Chen et al. 2011).

The comparison between experimental and FE predicted failure loads showed that the specific energy criterion used was fairly successful for fracture prediction, as shown in the example of Fig. 11.8, related to the MY750. The experimental failure loads compared very well with the experimental results with or without a fillet of adhesive.

11.2.4.2 Fatigue Applications of Damage Mechanics

Fatigue damage mechanics applications are not as widespread as static implementations, although a few formulations were proposed. Wahab et al. (2001)

compared fracture mechanics and damage mechanics approaches for fatigue life-time prediction of bonded double-lap joints with carbon-epoxy adherends, showing that the proposed continuum model was more suited for constant amplitude fatigue than fracture mechanics. Variable amplitude fatigue was not addressed in this work, but there is ample evidence of damage or crack growth accelerations or decelerations occurring under variable amplitude due to load interaction effects, depending on the type of material (Shenoy et al. 2010). Currently, scarce works are published on the effect of variable amplitude fatigue on bonded joints by damage mechanics.

Shenoy et al. (2010) investigated the variable amplitude fatigue behaviour of single-lap joints by damage mechanics, made of 7075 T6 aluminium adherends bonded with the epoxy adhesive film FM® 73M (Cytec), considering various types of variable amplitude conditions. Significant load interaction effects were found by the application of the Palmgren–Miner law to the experimental data, showing that a smaller amount of fatigue cycles at higher fatigue loads gives a significant reduction in the fatigue life by accelerating damage growth. Two different lines of analysis were conducted for life prediction: fracture mechanics and damage mechanics. Fracture mechanics results consistently under predicted the joints fatigue life for the entire range of fatigue spectra. This was because of the perceived inability of fracture mechanics models to account for the crack initiation phase that showed to be more dominant than the load interaction effects. The damage mechanics approach used a relation between localised damage and plastic strain in an empirical continuum damage law, allowing modelling the initiation and propagation phases of the joints fatigue life, and also accounting for load history effects by the accumulation of incremental damage. Damage mechanics predicted the fatigue life within the experimental scatter, oppositely to fracture mechanics.

11.3 CASE STUDY: CZM MODELLING OF SCARF REPAIRS ON ALUMINIUM STRUCTURES

The substantial or full strength recovery achieved by scarf repairs, provided that the repair is correctly designed, usually makes them permanent (Das et al. 2008, Wang and Gunnion 2008). Thus, in high responsibility structures requiring a full or significant strength recovery, or when a flush surface is imposed by aerodynamic or stealth reasons, an adhesively-bonded scarf repair is often used. However, it should be noted that small values of α, necessary to obtain higher efficiencies, may not be applicable since they require a larger repair area (Campilho et al. 2011a). The scarf repair technique has become particularly important in the last decades, due to the increasing use of sandwich panels in aircraft structures (Das et al. 2008).

The majority of the works on the strength of scarf joints or repairs focus on their tensile behaviour (Campilho et al. 2009a). However, compression/buckling behaviour (Campilho et al. 2009b) and bending (Campilho et al. 2011a) are also addressed. The most common design approaches for scarf repairs are analytical methods (Mortensen and Thomsen 1997, Gleich et al. 2000), experimental strain-measurement based methods (Found and Friend 1995, Fredrickson et al. 2008) and FE analyses (Harman and Wang 2006, Tzetzis and Hogg 2008, Kimiaeifar et al. 2012). The analytical methods developed in this area typically focus on displacement and stress analyses, which do not capture the stress gradients along the adhesive thickness. Few works address composite repairs, but these do not account for different oriented plies of the adherends. The work of Ahn and Springer (1998) is an exception, adapting the failure models of Hart-Smith (1981) to take into account each individual ply of the composite. The strain-measurement methods are often used to monitor the condition of in-service-bonded structures and also for strength prediction (Campilho et al. 2009b). The FE was also extensively used to obtain the stress fields and predict failure of these repairs, using appropriate failure criteria. In tension, experimental and FE studies showed an exponentially increasing strength with the reduction of α, due to the corresponding increase in the bond area (Campilho et al. 2009a, Hart-Smith 1973b). Odi and Friend (2004) performed a 2D FE stress and failure analysis of tensile loaded scarf repairs, using α values ranging from 1.1 to 9.2°. A quasi-isotropic $[0_2/\pm45/90/\pm45/0_2]_s$ composite plate and patch lay-up was considered. The numerical model captured the τ stress variations along the bond length arising from different ply orientations. The numerical failure loads were obtained using different failure criteria for the laminates and the Average Shear Stress Criterion for the adhesive. The most accurate laminate failure predictions were obtained with the Maximum Stress Criterion, agreeing with the experimental results. Even though these numerical methodologies have a significant usefulness in predicting the failure loads and identifying damage onset locations, they cannot account for the progressive damage evolution and identify failure paths, as CZM are able to (Yang et al. 1999, Teng et al. 2013). In addition, in these repairs, a significant difference exists between damage initiation and ultimate loads, due to σ and τ peak stresses developing at bond edges (Campilho et al. 2007). Thus, CZM implemented within FE models are a more accurate tool to predict the failure path and strength of adhesively-bonded repairs. Campilho et al. (2007) numerically studied the tensile strength and failure modes of carbon-fibre reinforced plastics (CFRP) scarf repairs using a triangular CZM, for values of α from 2 to 45°. The authors concluded that the model successfully predicted the strength and failure modes in these repairs, and also that the adhesive fracture properties present a smaller influence on the repairs strength than the mechanical ones. In another work (Campilho et al. 2009a), a developed trapezoidal CZM was applied to tensile loaded 2D scarf repairs on CFRP laminates, for values of α between 2 and 45°. Validation was carried out by comparison with experiments. To account for the experimental fractures, the cohesive failure of the adhesive layer and composite interlaminar and intralaminar (in the transverse and fibre directions) failures were considered at different regions. The corresponding tensile and shear CZM laws were estimated with an inverse modelling technique. The accurate predictions of the fracture loads and failure mechanisms validated the proposed technique.

Another feasible technique to increase the strength of scarf repairs, opposed to that of using very small values of α, consists of the application of external doublers adhesively-bonded at the scarf repaired region to protect the patch tips and to provide a larger cross-sectional area at the repaired region (Whittingham et al. 2009). These doublers are generally very thin and should follow the repaired structure contour as closely as possible (Savage and Oxley 2010). The most efficient solution is to bond these reinforcements on both the structure faces (Charalambides et al. 1998a, 1998b, Zimmerman and Liu 1995). However, a more practical solution consists on their application only on the outer face of the repair (Tzetzis and Hogg 2008, Hamoush et al. 2005). This repair configuration can also be imposed by accessibility difficulties to the inner face of the structure, or be rendered unfeasible for sandwich structures (Baker et al. 1999). Gunnion and Herszberg (2006) investigated the effect of a composite reinforcement on both faces of scarf joints with 16 and 32 ply CFRP laminates, applied to the full length and width of the joints. This technique resulted in a significant drop of σ and τ peak stresses within the bond, which otherwise develop near the scarf edges. Different lay-ups and increasing the number of reinforcement plies from the initial analysis (2 plies) provided no significant differences in reducing σ and τ peak stresses. The most comprehensive study regarding this issue consisted on a CZM-based numerical analysis on composite repairs with values of $\alpha = 5°$, $10°$, $15°$, $25°$ and $45°$ (Pinto et al. 2010). This technique showed a significant reduction of σ and τ peak stresses at the scarfed region. The analysis evaluated the use of a single reinforcement to the scarf repair (overlaminating at the outer side of the repair) and double reinforcement (overlaminating at the outer and inner sides of the repair). Overlaminating lengths of 2.5 and 5 mm were considered. For all conditions the strength improved exponentially with the reduction of α. The study showed that increasing the overlaminating length highly increases the repair strength. On the other hand, double reinforcement is also recommended over single reinforcement, because of the improved stress reduction effect at the scarfed region on account of the larger reinforcement areas and suppression of the structure transverse deflection.

This case study reports on an experimental and numerical study of the tensile behaviour of 2D scarf repairs of aluminium structures bonded with the ductile epoxy adhesive Araldite® 2015 (from Huntsmann, Basel, Switzerland). The numerical analysis (by FE) was performed in Abaqus® and used CZM for the simulation of damage onset and growth in the adhesive layer, thus enabling the strength prediction of the repairs. A parametric study was performed on the value of α and different configurations of external reinforcement (applied on one or two sides of the repair, and also different reinforcement lengths, c).

11.3.1 Experimental Work

The material used for the adherends is the high strength aluminium alloy AW6082 T651. This alloy was chosen due to its wide utilization in structural applications under various extruded forms (Dang Hoang et al. 2013). For the repaired structure, a sheet with 3 mm thickness was used and, for the reinforcements, a sheet with 1 mm thickness was considered. The σ_a-ε curves for the aluminium alloy were obtained according to the ASTM-E8M-04 standard. The resulting mechanical properties are showed in Table 11.1 (Campilho et al. 2011b).

Table 11.1 Properties for the aluminium alloy AW6082 T651 (Campilho et al. 2011b).

Properties	Alluminium alloy AW6082 T651
Young's modulus, E [GPa]	70.07 ± 0.83
Poisson's ratio, v	0.3
Tensile yield strength, σ_y [MPa]	261.67 ± 7.65
Tensile failure strength, σ_f [MPa]	324.00 ± 0.16
Tensile failure strain, ε_f [%]	21.70 ± 4.24
Vickers hardness, [HV]	100

The adhesive Araldite® 2015, used in this work, is a two-part structural adhesive allowing large plastic flow prior to failure, which is an important feature for bonded structures as it allows redistribution of stresses at stress concentration regions, occurring because of the sharp edges at the overlap ends and also asymmetry/ distinct deformation of the adherends along the bonded region. Details regarding the property characterization of this adhesive are given further in this work.

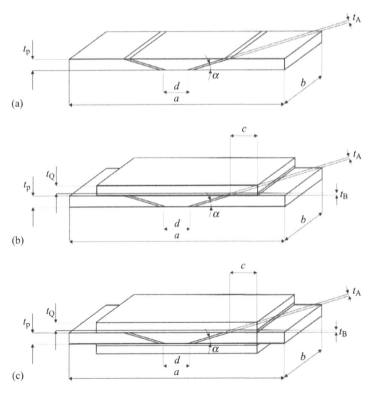

FIG. 11.9 Geometry and characteristic dimensions of the scarf repairs with configuration C1 (a), C2 (b) and C3 (c).

Figure 11.9 shows the geometry of the scarf repairs without reinforcement (configuration C1), with an outer reinforcement (configuration C2), and with outer and inner reinforcements (configuration C3). The relevant dimensions are the length between testing grips, $a = 200$ mm, $c = 5$ and 10 mm, plates separation, $d = 10$ mm, adhesive thickness in the scarf, $t_A = 0.2$ mm, adhesive thickness in the reinforcement, $t_B = 0.2$ mm, structure thickness, $t_P = 3$ mm, reinforcement thickness, $t_Q = 1$ mm, structure width, $b = 25$ mm and $\alpha = 10°$, 15°, 20°, 30° and 45°. The experimental work only considered $c = 5$ mm, while in the FE analysis values of $c = 5$ and 10 mm were tested. The scarfed specimens for the structure and patches were obtained from aluminum bars with $b = 25$ mm and $t_P = 3$ mm cut by an automated disc saw. The scarf profiles were produced by milling with High-Speed Steel (HSS) mills. The reinforcements were cut in a guillotine with the required dimensions. Before bonding, the surfaces were prepared by sand blasting and cleaned with a proper degreaser. The specimens were bonded with the ductile adhesive Araldite® 2015, using a methodology that guaranteed the specimens' alignment in both the thickness and width directions. The value of $t_A = t_B = 0.2$ mm was achieved by using calibrated wire between the repair components. The patches were initially bonded in a sole step, and afterwards the reinforcements were bonded (separately in both structure faces for configuration C3). After each bonding step the exceeding adhesive was removed with a stone grinding wheel, and the following step was only undertaken after a new surface treatment by the aforementioned procedure. The tensile tests were carried out up to complete failure of the specimens in a Shimadzu-Autograph AG-X machine equipped a 100 kN load cell. The testing setup guaranteed that the grips were perfectly aligned during the test and that no rotation of the grips occurred. The specimens were tested at room temperature under displacement control (1 mm/min). The load-displacement (P-δ) data was taken for further analysis and 5 specimens were tested for each geometry. At least four valid results were always obtained.

11.3.2 Numerical Work

11.3.2.1 Numerical Conditions

The FE simulations were carried out in Abaqus®/standard. The aluminium adherends were modelled by continuum elements as an isotropic plastic material with the elastic-perfectly plastic approximation defined in a previous work (Campilho et al. 2009a). The adhesive layers were modelled with a triangular CZM defined in the following Section, considering a single row of cohesive elements. This corresponds to the continuum approach, i.e., using the CZM elements to model the entire volume representative of the adhesive layer. The considered CZM incorporated mixed-mode coupling criteria for damage initiation and growth. The proposed modelling technique is currently implemented within Abaqus® CAE. The FE analyses were carried out in 2D and plane-strain conditions, which are the most suited to the proposed geometries.

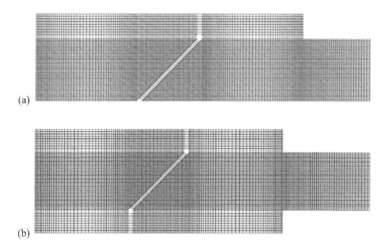

FIG. 11.10   FE mesh details for configuration C2, $\alpha = 45°$ and $c = 5$ mm (a) and configuration C3, $\alpha = 45°$ and $c = 5$ mm (b).

Two FE mesh examples are given in Fig. 11.10, regarding configuration C2, $\alpha = 45°$ and $c = 5$ mm (a) and configuration C3, $\alpha = 45°$ and $c = 5$ mm (b). The FE mesh was constructed through the partition of the scarf adhesive layer on its length with spacing between elements of 0.1 mm. Under this assumption, the number of CZM elements modelling the adhesive layer varies with α. Based on this initial model partition, the adherends and patch mesh was created in a structured manner, i.e., with horizontal and vertical edges, which enabled the construction of a distortion-free mesh (Bruggi et al. 2014). On the remaining areas of the repair the mesh was generated automatically, with bigger mesh refinement at the reinforcement edges because of the expected stress concentrations (Matthews et al. 2014). To produce a non-distorted mesh at the adhesive layer (Fig. 11.10), a small voided region was necessary at one (configuration C2) or two (configuration C3) scarf edges. Despite this simplification and consequent effect on stress distributions, an analysis was made to ensure that the variations are not too significant. The FE models were built with 4 node plane strain solid elements with reduced integration (CPE4R from Abaqus®), with the exception of the scarf boundary elements of the plates and patches, for which 3 node plane strain solid elements were considered (CPE3 from Abaqus®). The adhesive layers were modelled with 4 node cohesive elements (COH2D4 from Abaqus®), compatible with the aforementioned solid elements. The boundary conditions were defined to simulate as faithfully as possible the actual testing conditions in the testing machine. Thus, the joints were clamped at one of the edges, while the other edge was subjected to a tensile displacement concurrently with transverse restraining (Reis et al. 2011). Vertical symmetry was also applied at the middle of the repairs length (Nazir and Dhanasekar 2014). Figure 11.11 shows both boundary and symmetry conditions. The nomenclature described in Fig. 11.11 will be referred to further in this case study.

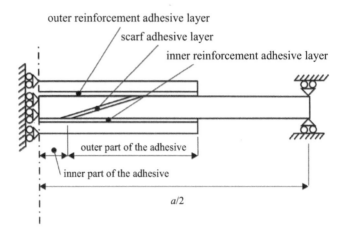

FIG. 11.11 Boundary and symmetry conditions used in the numerical models (example for configuration C3).

11.3.2.2 CZM Description

The CZM description that follows provides additional information to that presented in Section 11.2.3 regarding the triangular CZM formulation used in this case study, including the description of the mixed-mode criteria. CZM are based on a relationship between stresses and relative displacements (in tension or shear) connecting paired nodes of cohesive elements (Fig. 11.2), to simulate the elastic behaviour up to t_n^0 (tension) or t_s^0 (shear) and subsequent softening, to model the degradation of material properties up to failure. The areas under the pure-mode traction-separation laws in tension or shear are equalled to G_{Ic} or G_{IIc}, by the respective order. Under pure loading, damage grows at a specific integration point when stresses are released in the respective damage law. Under a combined loading, stress and energetic criteria are often used to combine tension and shear (Campilho et al. 2007). The triangular law (Fig. 11.2) assumes an initial linear elastic behaviour followed by linear degradation. Elasticity is defined by a constitutive matrix (**K**), containing the stiffness parameters and relating stresses (t) and strains (ε) across the interface (Abaqus® 2013)

$$t = \begin{Bmatrix} t_n \\ t_s \end{Bmatrix} = \begin{bmatrix} K_{nn} & K_{ns} \\ K_{ns} & K_{ss} \end{bmatrix} \cdot \begin{Bmatrix} \varepsilon_n \\ \varepsilon_s \end{Bmatrix} = \mathbf{K}\varepsilon. \tag{11.3}$$

ε_n and ε_s are the current tensile and shear strains, respectively. A suitable approximation for thin adhesive layers is provided with $K_{nn} = E$, $K_{ss} = G$ and $K_{ns} = 0$, G being the shear modulus (Campilho et al. 2008). Damage initiation can be specified by different criteria. In this case study, the quadratic nominal stress criterion was considered for the initiation of damage, already shown to give accurate results (Campilho et al. 2008) and expressed as (Abaqus® 2013)

$$\left\{\frac{\langle t_n\rangle}{t_n^0}\right\}^2 + \left\{\frac{t_s}{t_s^0}\right\}^2 = 1. \tag{11.4}$$

$\langle\ \rangle$ are the Macaulay brackets, emphasizing that a purely compressive stress state does not initiate damage. After t_m^0 is attained (Fig. 11.2) by the fulfilment of equation (11.4), the material stiffness is degraded. Complete separation is predicted by a linear power law form of the required energies for failure in the pure modes (Abaqus® 2013)

$$\frac{G_I}{G_{Ic}} + \frac{G_{II}}{G_{IIc}} = 1. \tag{11.5}$$

11.3.2.3 Cohesive Parameters

In order to fully characterize the cohesive laws in tension and shear it is necessary to know beforehand the values of t_n^0, t_s^0, G_{Ic} and G_{IIc}. In the past some studies were published that, by approximation, assumed that these parameters were identical to the ones obtained in bulk tests (Pocius 2002). However, the deformation restraints in the adhesive layer caused by the adherends and the mixed-mode crack propagation give rise to differences in the adhesive behaviour between bulk or as a thin layer (Andersson and Stigh 2004). Thus, proper estimation methods are required for the CZM parameters in the real loading conditions that they are going to be employed. For the adhesive Araldite® 2015, the cohesive laws of the adhesive layer were obtained in a previous work (Campilho et al. 2011c) by an inverse method to determine the parameters t_n^0 and t_s^0. This methodology is explained in detail in another work (de Moura et al. 2008). Basically, it consists on the determination of G_{Ic} and G_{IIc} through DCB and ENF tests, respectively, using a suited method or theory. Then the values of G_{Ic} and G_{IIc} are used to construct CZM laws in both pure modes (tension and shear) that initially have a typical value of t_n^0 and t_s^0 estimated in function of the adhesive characteristics. The obtained CZM law is used on the corresponding numeric model (DCB for tension and ENF for shear), with the same dimensions of the model that will be simulated. Following, the value of t_n^0 and t_s^0 is defined with an adjustment procedure between the numerical and experimental P–δ curves of the respective fracture characterization test, so that the estimated cohesive law can reproduce faithfully the observed behaviour of the adhesive layer. Table 11.2 shows the obtained parameters by this technique used to simulate under mixed mode conditions the adhesive layer of the adhesive Araldite® 2015 (Campilho et al. 2009a). The values of E and G were obtained experimentally with bulk (Marques and da Silva 2008) and Thick Adherend Shear Tests (TAST) (Campilho et al. 2009c), respectively.

Table 11.2 Cohesive parameters for the adhesive Araldite® 2015 (Campilho et al. 2009a).

E [MPa]	1850	G [MPa]	650
t_n^0 [MPa]	23.0	t_s^0 [MPa]	22.8
G_{Ic} [N/mm]	0.43	G_{IIc} [N/mm]	4.7

11.3.3 Results

11.3.3.1 Stress Analysis

The stress analysis was carried out with solid FE elements to model the adhesive layers instead of CZM elements. Twelve elements were considered in the adhesive layers through-thickness, and stresses were taken from the mid-thickness of the adhesive layers. To study the influence of reinforcement with configuration C2 or C3, and the values of α and c in the stress fields along the adhesive layer, the distribution of σ and τ stresses is presented for the repairs with $\alpha = 45°$ and $10°$ regarding the scarf adhesive layer and the outer and inner reinforcement adhesive layers (Fig. 11.11). All stress plots are normalized by the average τ stress (τ_{avg}) at the scarf adhesive layer for the respective repair configuration. Identically, x-coordinates are normalized to the scarf adhesive layer length (L_S) for each α value. Figure 11.12 shows the σ stress distributions at the scarf adhesive layer considering $\alpha = 45°$ (a) and $10°$ (b).

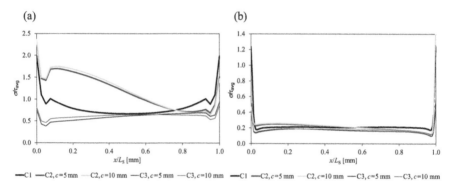

FIG. 11.12 σ stresses at the mid-thickness of the scarf adhesive layer for $\alpha = 45°$ (a) and $\alpha = 10°$ (b).

For configuration C1, σ stresses are practically uniform along the most of the scarf length, apart from its edges where σ stresses peak (Campilho et al. 2007). Considering configuration C2, the unreinforced scarf edge shows an increase of σ peak stresses, while the opposite occurs at the other scarf edge. This happens because this configuration introduces the repair transverse deflection induced by the modification of the load path. Using configuration C3 enables a reduction of σ peak stresses throughout the scarf. These effects are much more significant for bigger values of α, for which σ stresses have comparable magnitude to τ stresses. For the smaller values of α, σ stresses are minimal compared to τ stresses, except at the scarf edges. The reduction of σ peak stresses with configuration C3 is highly advantageous because it takes place at the potential crack initiation sites. Figure 11.13 presents the distribution of τ stresses at the scarf adhesive layer for the repairs with $\alpha = 45°$ (a) and $10°$ (b).

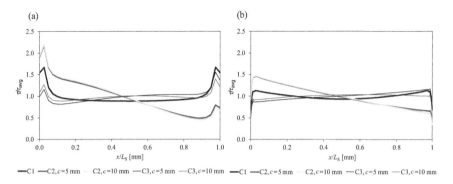

FIG. 11.13 τ stresses at the mid-thickness of the scarf adhesive layer for $\alpha = 45°$ (a) and $\alpha = 10°$ (b).

Figure 11.13 shows only minor stress concentrations at the ends of the scarf for configuration C1, which constitutes a major evolution from the strap repair configurations and is due to the adherend tapering effect at the adherends' edges (Campilho et al. 2007, Objois et al. 2005). Comparing configurations C1 and C2, the latter shows a slight increase of τ stresses at the inner (e.g., unreinforced) scarf edge and a decrease at the opposite edge. Configuration C3 shows a decrease of τ peak stresses at the scarf ends. τ stresses become more uniform along the scarf with the reduction of α. In view of these results, configuration C2 is prone to potentiate premature failure at the unreinforced scarf edge. Configuration C3, on account of the reduction of σ and τ peak stresses at both scarf ends, should provide a significant strength improvement of the repair.

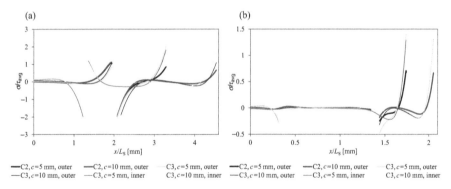

FIG. 11.14 σ stresses at the mid-thickness of the outer and inner reinforcement layers for $\alpha = 45°$ (a) and $\alpha = 10°$ (b).

Figure 11.14 shows σ stresses at the mid-thickness of the outer and inner reinforcement layers (configurations C2 and C3, where applicable) for $\alpha = 45°$ (a) and $\alpha = 10°$ (b). All curves are divided into two segments: from the symmetry line to the scarf adhesive layer (inner part) and from the scarf adhesive layer to the reinforcement edge (outer part). Figure 11.11 exemplifies these segments. This also applies to the upcoming τ stresses. σ stresses at the inner part are generally not significant, except at the outer reinforcement adhesive layer for the bigger values of α and near the scarf adhesive layer. For configuration C3, at the inner reinforcement adhesive layer compressive σ stresses develop, but these do not potentiate failure. At the outer part of the reinforcement adhesive layers, major σ peak stresses occur at the reinforcement edge, showing that this is the probable site of damage initiation. Configuration C2 also shows smaller normalized stresses than configuration C3, occurring on account of less transverse restraining of the repair. Between different c values, σ peak stresses have identical magnitude, despite $c = 10$ mm having a larger bonded area, which will likely increase the repairs' strength. Figure 11.15 represents τ stresses at the mid-thickness of the outer and inner reinforcement layers for $\alpha = 45°$ (a) and $\alpha = 10°$ (b). The inner parts of the adhesive layers once again shows reduced τ stresses, showing smaller load bearing under load than the outer parts of the adhesive layers. Though, noteworthy τ peak stresses occur for the bigger values of α at the edge near the scarf. At the outer parts of the adhesive layers, stresses tend to be highest at the edge of the reinforcement, supporting the idea that damage is prone to initiate at these regions. Notwithstanding, the edge near the scarf also exhibits τ peak stresses, while the inner region is stressed but with lower magnitude. Configuration C3 reveals higher normalized τ stresses than configuration C2, showing that a larger amount of load is transferred to the reinforcements, which is consistent with the previously reported reduction of σ and τ stresses at the scarf adhesive layer for configuration C3 (Fig. 11.12 and Fig. 11.13, respectively). By increasing c from 5 to 10 mm, normalized τ values tend to be slightly reduced which, added to the larger bonding area, should improve the load bearing capability of the repairs accordingly.

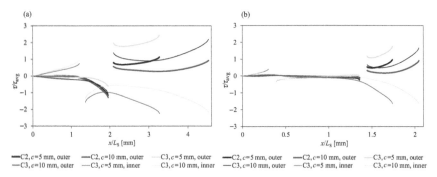

FIG. 11.15 τ stresses at the mid-thickness of the outer and inner reinforcement layers for $\alpha = 45°$ (a) and $\alpha = 10°$ (b).

11.3.3.2 Failure Path

Table 11.3 describes the numerical failure modes for all repair configurations, which revealed to be consistent with the experimental fractures. Some figures with the failure process are also presented to ease the interpretation of Table 11.3. The presented nomenclature regards to: A – failure at the scarf adhesive layer, B – failure at the outer reinforcement adhesive layer and C – failure at the inner reinforcement adhesive layer. Failure modes in the same horizontal line refer to simultaneous failure at all respective adhesive layers, while failures in different lines occur sequentially from top to bottom. C1 configuration repairs initiate failure at the scarf edges, mainly on account of the high σ peak stresses at those regions (Fig. 11.12), while τ stresses also have less significant concentrations, especially for the bigger α values (Fig. 11.13). For the C2 configuration repairs, failure initiates at the lower edge (i.e., unreinforced) of the scarf adhesive layer and only after grows to the reinforcement edge of the outer adhesive layer (for all α and c values). This behaviour is accredited to the significant σ and τ peak stress concentrations at the unreinforced edge of the scarf adhesive layer for all α values (Fig. 11.12 and Fig. 11.13, respectively), which makes this the damage initiation site. Afterwards, damage shifts to the outer reinforcement adhesive layer, beginning at the reinforcement edge because of the highest values of σ peak stresses (Fig. 11.14), while τ peak stresses are slightly higher than at the inner edge (Fig. 11.15).

Table 11.3 Failure modes of the repairs with configurations C1, C2 and C3.

Configuration	C1	C2 $c = 5$ mm	C2 $c = 10$ mm	C3 $c = 5$ mm	C3 $c = 10$ mm
10	A	A B	A B	A and B C	A and B C
15	A	A B	A B	A and B C	A and B C
20	A	A B	A B	A and B C	A and B C
30	A	A B	A B	A B C	A B C
45	A	A B	A B	A B C	A B C

By considering configuration C3 repairs, for $\alpha = 10$, 15 and 20° and $c = 5$ and 10 mm, failure occurs simultaneously at the scarf and outer reinforcement adhesive layers, subsequently propagating to the inner reinforcement adhesive layer. On the other hand, for the repairs with $\alpha = 30$ and 45°, failure initiates at the scarf adhesive layer, grows to the outer reinforcement adhesive layer and finally extends to the inner reinforcement adhesive layer. Damage initiation at the scarf adhesive layer, concurrently or not with the outer reinforcement adhesive layer, relates to the at least equal normalized magnitude of σ and τ stresses at that region comparing to the reinforcement adhesive layers (Fig. 11.12 and Fig. 11.13 compared against Fig. 11.14 and Fig. 11.15). For the smaller values of α, this occurs simultaneously to failure at the outer reinforcement adhesive layer, which occurs because of smaller values of normalized σ stresses (Fig. 11.12b) that turn the scarf adhesive layer more resistant up to damage onset at the outer reinforcement adhesive layer. The inner reinforcement adhesive layer is always the last one to fail, mainly because of the reduction of τ peak stresses at the outer part of the adhesive layer (Fig. 11.15) and the larger bonded length between the scarf edge and reinforcement edge.

For visualization of the adhesive layers' failure modes the SDEG parameter from Abaqus® is used. This parameter gives a quick perception of the cohesive elements degradation and crack propagation. SDEG = 0 specifies that a cohesive element is in the elastic region (before t_m^0 in Fig. 11.2) and SDEG = 1 indicates that a cohesive element is 100% degraded (δ_m^f in Fig. 11.2). Two examples with the limiting values of α are presented. Figure 11.16 represents the failure process for configuration C3 with $\alpha = 10°$ and $c = 5$ mm, which shows that failure started simultaneously at the scarf adhesive layer and at the edge of the outer reinforcement. Figure 11.17 shows the failure process for configuration C2 with $\alpha = 45°$ and $c = 10$ mm. In this case, failure initiates at the free end of the scarf adhesive layer and grows to the outer reinforcement.

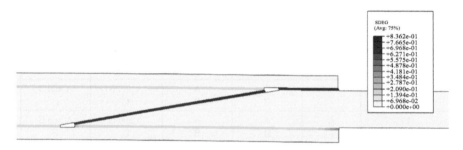

FIG. 11.16 Failure process for configuration C3 with $\alpha = 10°$ and $c = 5$ mm.

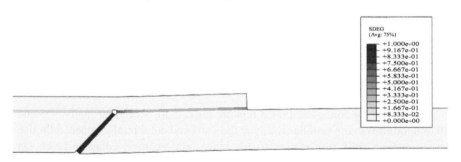

FIG. 11.17 Failure process for configuration C2 with $\alpha = 45°$ and $c = 10$ mm.

11.3.3.3 Strength Analysis

In this Section, a comparison is made between the experimental strengths and the CZM predictions. This will enable validating the numerical methodology and the qualitative results regarding the effect of the different repair configurations. In Fig. 11.18(a) a comparison is made between the experimental and numerical P-δ curves for the configuration C2 repair with $\alpha = 10°$ and $c = 5$ mm. Fig. 11.18(b) relates to the configuration C3 repair with $\alpha = 10°$ and $c = 5$ mm. Figures 11.19(a) and (b) are identical plots but considering $\alpha = 45°$ instead of 10°.

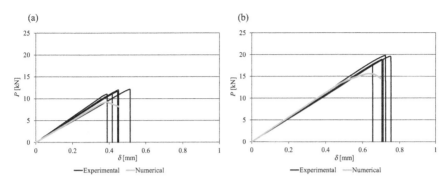

FIG. 11.18 Experimental and numerical P-δ curves for configuration C2 (a) and C3 repair (b) with $\alpha = 10°$ and $c = 5$ mm.

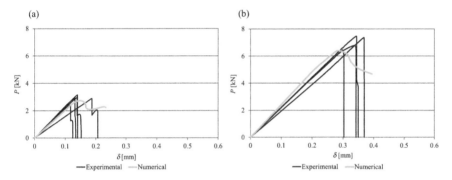

FIG. 11.19 Experimental and numerical P-δ curves for configuration C2 (a) and C3 repair (b) with $\alpha = 45°$ and $c = 5$ mm.

Both figures show the good repeatability of the tests in which concerns the initial stiffness, maximum load (P_{max}) and the failure displacement, which is representative of all tested configurations. The numerical approximations are fairly accurate, but generally show slightly smaller values for the predicted strength than the experiments. Configuration C1 revealed a maximum deviation of 8.1% for the $\alpha = 30°$ repair. For configuration C2, the maximum deviation to the experiments was 20.2% for the repairs with $\alpha = 10°$ and $c = 5$ mm, and for configuration C3 the maximum difference was 15.1%, considering repairs with $\alpha = 10°$ and $c = 5$ mm. As a result, it is assumed that the proposed numerical technique simulates with reasonable accuracy the behaviour of the repairs. Figure 11.20 compares the experimental and numerical P_{max} values for the three tested configurations, including the respective standard deviation of the experiments. For $c = 10$ mm (either configuration C2 or C3), only numerical data is available. For all configurations, an exponential increase of P_{max} with the decrease of α is found. For unreinforced scarf repairs, this behaviour is already documented in the literature (Pinto et al. 2010, Campilho et al. 2011a), and it is related to the exponential increase of the bond length and decrease of normalized σ stresses (Fig. 11.12) with the reduction of α. τ peak stresses also exhibit a reduction at the scarf edges for smaller values of α, which helps to this behaviour (Fig. 11.13). When considering reinforcements, either with configuration C2 or C3, the same tendency is followed, although accompanied by an overall strength improvement due to the crack onset retardation effect of the reinforcements and deviation of the load paths to the outer and inner (when applicable) reinforcement adhesive layers (Fig. 11.14 and Fig. 11.15), despite σ and τ peak stresses at the scarf adhesive layers may increase, depending on the repair configuration (Fig. 11.12 and Fig. 11.13, respectively) (Pinto et al. 2010).

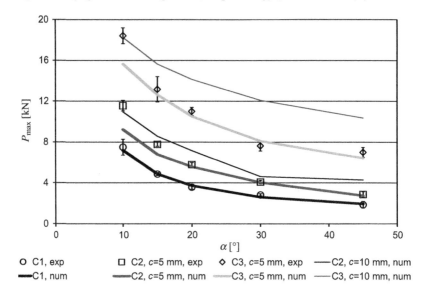

FIG. 11.20 P_{max} as a function of α: comparison between the experimental and numerical results.

From Fig. 11.20 it is observed that for configurations C2 and C3 ($c = 5$ mm) and small values of α the numerical values of P_{max} are slightly out of the experimental range. For higher values of α, the numerical values of P_{max} are within the range of experimental P_{max}. For the $c = 10$ mm C2 and C3 configuration repairs, better results were obtained than for the $c = 5$ mm repairs, because higher values of c give a bigger reinforcement area. Adding to this, τ peak and overall stresses at both outer and inner reinforcement adhesive layers highly diminish (Fig. 11.15), while σ stresses also show small reductions (Fig. 11.14). The highest value of P_{max} (≈ 18 kN) was obtained for the repair with configuration C3, $\alpha = 10°$ and $c = 10$ mm. Figure 11.21 represents the percentile efficiency (η [%]) of the repairs over configuration C1, i.e., without reinforcement, as a function of α, based on the numerical results. Results show that η increases with α up to a repair efficiency above 500% for configuration C3, $\alpha = 45°$ and $c = 10$ mm. The gradual increase of η with α is because of the extremely small bonding areas at the scarf region, which gives the reinforcement an important role in modifying the load path from the scarf to the reinforcement. For small values of α, the maximum values of η are thus much smaller but still to consider.

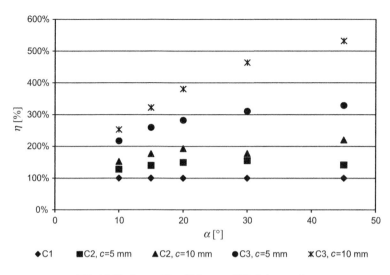

◆C1   ■C2, c=5 mm   ▲C2, c=10 mm   ●C3, c=5 mm   ✕C3, c=10 mm

FIG. 11.21 Percentile efficiency η [%] of the repairs.

11.3.4 Conclusions of the Study

In this case study, the effect of the using external reinforcement (on one or both sides of the structure) for scarf adhesive repairs, including different values of α, was studied experimentally and by CZM modelling. A moderately ductile adhesive was considered (Araldite® 2015). The experimental strength analysis showed that there was a significant increase of the repairs strength using configuration C2 and an even more significant increase for configuration C3. By assessing the influence of α, an exponential strength improvement was found with the decrease of α. For configuration C1, P_{max} improves 270% by reducing α from 45° to 10°. For configuration C2 the strength improves in the order of 302% by reducing α from 45° to 10°. For configuration C3, this improvement is in the order of 163%. By comparing configurations C2 and C3 with C1, a significant improvement was attained. Experimentally, the strength improvement from configuration C1 to C2 ranged between 44% ($\alpha = 30°$) and 62% ($\alpha = 20°$). Between configurations C1 and C3, the improvement was between 145% ($\alpha = 10°$) and 275% ($\alpha = 45°$). A FE analysis was carried out to study the variation of σ and τ stresses at the scarf, outer reinforcement and inner reinforcement adhesive layers, to understand the failure modes and strength tendency. The comparison between the numerical and experimental failure modes revealed an excellent agreement. A CZM-based failure analysis was carried out for strength prediction purposes. It was found that, in the case of configuration C1, the exponential increase of P_{max} with the decrease of α was due to the increase of the bonding area, decrease of normalized σ stresses and τ stress gradients at the scarf edges. For configurations C2 and C3, the same behaviour is due to the crack onset retardation effect of the reinforcements and deviation of the load paths to the outer and inner (when applicable) reinforcement adhesive layers. The CZM-based strength analysis reinforced the exponential increase of P_{max} with the decrease of α. By the comparison of the obtained values with the experiments, the numerical value of P_{max} is always slightly smaller, although the results are comparable. Better results were found in configurations C2 and C3 with $c = 10$ mm rather than with $c = 5$ mm, due to the increase of overlapping area. Adding to this, τ peak and overall stresses at both outer and inner reinforcement adhesive layers highly diminished, while σ stresses also showed small reductions. The highest value of P_{max} (≈ 18 kN) was obtained for configuration C3 with $\alpha = 10°$ and $c = 10$ mm. Numerically, considering both values of c, the value of η surpassed 500% for configuration C3, $\alpha = 45°$ and $c = 10$ mm. In view of the obtained results, it is concluded that the numerical method considered in this work to predict the repairs strength is validated, and these models can be used to predict the repairs strength and for geometric and material repair optimization studies. The obtained results allowed the establishment of design guidelines for repairing, since the obtained results showed clear tendencies of strengthening.

11.4 CONCLUSIONS

The most recent trends in numerical modelling of adhesive joints have been discussed in this chapter. XFEM was excluded from the discussion since another chapter deals specifically with this method. The following conclusions can be drawn. The most relevant strength prediction tools are based on the FE method. In the continuum mechanics approach, the maximum values of stress, strain or strain energy, predicted by the FE analyses, are typically used in the failure criterion where they are compared with the corresponding material allowable values. However, it is known that these maximum predicted values are usually found very near to the singular points of the model (sharp corners or bi-material interfaces). Therefore, their magnitude strongly depends on how well the stress field around the singularity is modelled (i.e., mesh refinement). Additionally, the maximum critical values obtained by the FE analyses are also dependent on the proximity of the critical point from the stress or strain concentrator. In order to overcome this problem, a common approach used by many researchers is to use the same variables (stress, strain or energy) but this time at some arbitrary distance from the point of singularity, where the stress field is clear of any effects from the singular point. The critical distances must be usually calibrated from the FE results and, as a consequence, these critical values obtained can only be used for similar geometric and material configurations. There is usually no physical explanation relating the critical distances with experimental observations. The average plastic density belongs to the same concept, but the use of an value gives a less sensitive result to the mesh size used.

Continuum mechanics assumes that the structure and its material are continuous. Defects or two materials with re-entrant corners obviously violate such an assumption. Cracks are the most common defects in structures, for which the method of fracture mechanics has been developed. Fracture mechanics can be used to predict joint strength or residual strength if there is a crack tip or a known and calibrated singularity. Fracture mechanics is more difficult to apply to strength predictions for joints bonded with ductile adhesives, since G_c is not independent of the joint geometry. This is mainly because the adherends restrict the development of the yield zone in the adhesive bond or they can cause fracture of the adhesive if the material yields.

CZM, which can be categorized under these two lines of analysis, simulate the macroscopic damage along this path by the specification of a traction-separation response between paired nodes on either sides of pre-defined crack paths. The techniques for modelling damage can be divided into either local or continuum approaches. In the local approach, damage is confined to a zero thickness path (2D analyses) or surface (3D analyses). By the continuum approach, damage is modelled over a finite area (2D analyses) or volume (3D analyses). In most of the CZM, the traction–separation relations are such that, with increasing interfacial separation, the traction across the interface reaches a maximum (crack initiation), then it decreases (softening), and finally the crack propagates, permitting a total de-bond. The whole failure response and crack propagation can thus be modelled. A CZM simulates the fracture process, extending the concept of continuum mechanics by including a zone of discontinuity modelled by cohesive zones, thus using both strength and energy parameters to characterize the debonding process. This allows the approach to be of much more general utility than conventional fracture mechanics. The method is also mesh insensitive, provided that enough integration points undergo softening simultaneously. It is similar to criteria where the stress or strain is averaged over a finite area so as to remove the singularity. Studies demonstrated that it is possible to experimentally determine the appropriate cohesive zone parameters of an adhesive bond, and to incorporate them into FE analyses for excellent predictive capabilities. However, CZM present a limitation, as it is necessary to know beforehand the critical zones where damage is prone to occur, and to place the cohesive elements accordingly. Also, for ductile materials, the shape of the traction-separation law must be modified, which may give additional convergence problems.

Damage mechanics has been used to model the progressive damage and failure arbitrarily within a structure. By this technique, the load transfer is locally reduced, globally resulting on a drop of P for a given value of δ applied to the structure. Damage mechanics has been used to numerically model static problems, and constant and variable amplitude fatigue loadings. Compared to fatigue CZM, damage mechanics techniques do not provide a clear distinction between fatigue initiation and propagation phases, although they can give a basis for the predictive analysis. Nonetheless, the evolution of damage prior to macro-crack growth can be simulated. On the other hand, damage modelling with fatigue CZM is restricted to pre-defined crack paths and, in specific applications, damage mechanics may be recommended if the damage is more widespread or the failure path is not known.

From the state-of-the-art description carried through in this chapter, it is clear that numerical methods have been largely exploited for some decades for the strength and fracture prediction of bonded joints by varying approaches, and with formulations tailored for different load scenarios. In the near future, further developments within this scope are expected, and a few research tendencies are proposed that can effectively expand the use of these powerful tools. CZM will definitely be improved, beginning from easier and more efficient calibration tools.

Regarding the analyses capabilities, further research on the residual strength of environmentally degraded joints is required, namely regarding degradation of the adhesive and interfacial regions because of extreme conditions that structures can be subjected to in industrial applications. Rate dependent damage formulations are still very incipient and need exploiting, and fatigue CZM are very recent and need further validations and improvements for a widespread application to different geometries and load conditions.

11.5 REFERENCES

Abaqus® 2013. Documentation. D. Systèmes. Vélizy-Villacoublay.

Adams, R. D. and Peppiatt, N. A. 1974. Stress analysis of adhesive-bonded lap joints. The Journal of Strain Analysis for Engineering Design 9: 185-196.

Adams, R. D. and Harris, J. A. 1987. The influence of local geometry on the strength of adhesive joints. International Journal of Adhesion and Adhesives 7: 69-80.

Adams, R. D. and Mallick, V. 1992. A method for the stress analysis of lap joints. The Journal of Adhesion 38: 199-217.

Adams, R. D., Comyn, J. and Wake, W. C. 1997. Structural Adhesive Joints in Engineering. Chapman & Hall, London.

Ahn, S.-H. and Springer, G. S. 1998. Repair of composite laminates-II: Models. Journal of Composite Materials 32: 1076-1114.

Andersson, T. and Stigh, U. 2004. The stress–elongation relation for an adhesive layer loaded in peel using equilibrium of energetic forces. International Journal of Solids and Structures 41: 413-434.

Assih, J., Li, A. and Delmas, Y. 2001. Strengthened concrete beams by gluing carbon fiber composite sheet: Application of damage theory. Composites in Constructions 127: 623-628.

Baker, A. A., Chester, R. J., Hugo, G. R. and Radtke, T. C. 1999. Scarf repairs to highly strained graphite/epoxy structure. International Journal of Adhesion and Adhesives 19: 161-171.

Banea, M. D., da Silva, L. F. M. and Campilho, R. D. S. G. 2010. Temperature dependence of the fracture toughness of adhesively bonded joints. Journal of Adhesion Science and Technology 24: 2011-2026.

Bikerman, J. J. 1968. The Science of Adhesive Joints. Academic Press, New York.

Bruggi, M., Milani, G. and Taliercio, A. 2014. Simple topology optimization strategy for the FRP reinforcement of masonry walls in two-way bending. Computers & Structures 138: 86-101.

Campilho, R. D. S. G., de Moura, M. F. S. F. and Domingues, J. J. M. S. 2005. Modelling single and double-lap repairs on composite materials. Composites Science and Technology 65: 1948-1958.

Campilho, R. D. S. G., de Moura, M. F. S. F. and Domingues, J. J. M. S. 2007. Stress and failure analyses of scarf repaired CFRP laminates using a cohesive damage model. Journal of Adhesion Science and Technology 21: 855-870.

Campilho, R. D. S. G., de Moura, M. F. S. F. and Domingues, J. J. M. S. 2008. Using a cohesive damage model to predict the tensile behaviour of CFRP single-strap repairs. International Journal of Solids and Structures 45: 1497-1512.

Campilho, R. D. S. G., de Moura, M. F. S. F., Pinto, A. M. G., Morais, J. J. L. and Domingues, J. J. M. S. 2009a. Modelling the tensile fracture behaviour of CFRP scarf repairs. Composites Part B: Engineering 40: 149-157.

Campilho, R. D. S. G., de Moura, M. F. S. F., Ramantani, D. A., Morais, J. J. L. and Domingues, J. J. M. S. 2009b. Buckling behaviour of carbon-epoxy adhesively-bonded scarf repairs. Journal of Adhesion Science and Technology 23: 1493-1513.

Campilho, R. D. S. G., de Moura, M. F. S. F., Ramantani, D. A., Morais, J. J. L. and Domingues, J. J. M. S. 2009c. Tensile behaviour of three-dimensional carbon-epoxy adhesively bonded single- and double-strap repairs. International Journal of Adhesion and Adhesives 29: 678-686.

Campilho, R. D. S. G., de Moura, M. F. S. F., Pinto, A. M. G., Ramantani, D. A., Morais, J. J. L. and Domingues, J. J. M. S. 2011a. Strength prediction of adhesively-bonded scarf repairs in composite structures under bending. Materials Science Forum 636-637: 233-238.

Campilho, R. D. S. G., Banea, M. D., Pinto, A. M. G., da Silva, L. F. M. and de Jesus, A. M. P. 2011b. Strength prediction of single- and double-lap joints by standard and extended finite element modelling. International Journal of Adhesion and Adhesives 31: 363-372.

Campilho, R. D. S. G., Pinto, A. M. G., Banea, M. D., Silva, R. F. and da Silva, L. F. M. 2011c. Strength improvement of adhesively-bonded joints using a reverse-bent geometry. Journal of Adhesion Science and Technology 25: 2351-2368.

Campilho, R. D. S. G., Banea, M. D., Neto, J. A. B. P. and da Silva, L. F. M. 2013. Modelling adhesive joints with cohesive zone models: Effect of the cohesive law shape of the adhesive layer. International Journal of Adhesion and Adhesives 44: 48-56.

Carver, D. R. and Wooley, G. R. 1971. Stress concentration factors for bonded lap joints. Journal of Aircraft 8: 817-820.

Charalambides, M. N., Hardouin, R., Kinloch, A. J. and Matthews, F. L. 1998a. Adhesively-bonded repairs to fibre-composite materials I. Experimental. Composites Part A: Applied Science and Manufacturing 29: 1371-1381.

Charalambides, M. N., Kinloch, A. J. and Matthews, F. L. 1998b. Adhesively-bonded repairs to fibre-composite materials II. Finite element modelling. Composites Part A: Applied Science and Manufacturing 29: 1383-1396.

Chen, Z., Adams, R. D. and da Silva, L. F. M. 2011. Prediction of crack initiation and propagation of adhesive lap joints using an energy failure criterion. Engineering Fracture Mechanics 78: 990-1007.

Choupani, N. 2008. Interfacial mixed-mode fracture characterization of adhesively bonded joints. International Journal of Adhesion and Adhesives 28: 267-282.

Crocombe, A. D. and Adams, R. D. 1981. Influence of the spew fillet and other parameters on the stress distribution in the single lap joint. The Journal of Adhesion 13: 141-155.

Crocombe, A. D. 1989. Global yielding as a failure criterion for bonded joints. International Journal of Adhesion and Adhesives 9: 145-153.

Crocombe, A. D., Hua, Y. X., Loh, W. K., Wahab, M. A. and Ashcroft, I. A. 2006. Predicting the residual strength for environmentally degraded adhesive lap joints. International Journal of Adhesion and Adhesives 26: 325-336.

da Silva, L. F. M., das Neves, P. J. C., Adams, R. D. and Spelt, J. K. 2009a. Analytical models of adhesively bonded joints—Part I: Literature survey. International Journal of Adhesion and Adhesives 29: 319-330.

da Silva, L. F. M., das Neves, P. J. C., Adams, R. D., Wang, A. and Spelt, J. K. 2009b. Analytical models of adhesively bonded joints—Part II: Comparative study. International Journal of Adhesion and Adhesives 29: 331-341.

da Silva, L. F. M. and Campilho, R. D. S. G. 2011. Advances in Numerical Modelling of Adhesive Joints. Springer, Heidelberg.

Dang Hoang, T., Herbelot, C., Imad, A. and Benseddiq, N. 2013. Numerical modelling for prediction of ductile fracture of bolted structure under tension shear loading. Finite Elements in Analysis and Design 67: 56-65.

Das, M., Madenci, E. and Ambur, D. R. 2008. Three-dimensional nonlinear analyses of scarf repair in composite laminates and sandwich panels. Journal of Mechanics of Materials and Structures 3: 1641-1658.

Daudeville, L. and Ladevèze, P. 1993. A damage mechanics tool for laminate delamination. Composite Structures 25: 547-555.

de Moura, M. F. S. F., Gonçalves, J. P. M., Chousal, J. A. G. and Campilho, R. D. S. G. 2008. Cohesive and continuum mixed-mode damage models applied to the simulation of the mechanical behaviour of bonded joints. International Journal of Adhesion and Adhesives 28: 419-426.

Dillard, D. A., Singh, H. K., Pohlit, D. J. and Starbuck, J. M. 2009. Observations of decreased fracture toughness for mixed mode fracture testing of adhesively bonded joints. Journal of Adhesion Science and Technology 23: 1515-1530.

Dugdale, D. S. 1960. Yielding of steel sheets containing slits. Journal of the Mechanics and Physics of Solids 8: 100-104.

Feraren, P. and Jensen, H. M. 2004. Cohesive zone modelling of interface fracture near flaws in adhesive joints. Engineering Fracture Mechanics 71: 2125-2142.

Fernlund, G. and Spelt, J. K. 1991a. Failure load prediction of structural adhesive joints: Part 1: Analytical method. International Journal of Adhesion and Adhesives 11: 213-220.

Fernlund, G. and Spelt, J. K. 1991b. Failure load prediction of structural adhesive joints: Part 2: Experimental study. International Journal of Adhesion and Adhesives 11: 221-227.

Found, M. S. and Friend, M. J. 1995. Evaluation of CFRP panels with scarf repair patches. Composite Structures 32: 115-122.

Fredrickson, B. M., Schoeppner, G. A., Mollenhauer, D. H. and Palazotto, A. N. 2008. Application of three-dimensional spline variational analysis for composite repair. Composite Structures 83: 119-130.

Gleich, D. M., Van Tooren, M. J. L. and de Haan, P. A. J. 2000. Shear and peel stress analysis of an adhesively bonded scarf joint. Journal of Adhesion Science and Technology 14: 879-893.

Gleich, D. M., Van Tooren, M. J. L. and Beukers, A. 2001. Analysis and evaluation of bondline thickness effects on failure load in adhesively bonded structures. Journal of Adhesion Science and Technology 15: 1091-1101.

Goland, M. and Reissner, E. 1944. The stresses in cemented joints. Journal of Applied Mechanics 66: A17-A27.

Greenwood, L. 1969. The strength of a lap joint. pp. 40. In: D. Alner (ed.). Aspects of Adhesion, Volume 5. University of London Press, London.

Groth, H. L. 1988. Stress singularities and fracture at interface corners in bonded joints. International Journal of Adhesion and Adhesives 8: 107-113.

Gunnion, A. J. and Herszberg, I. 2006. Parametric study of scarf joints in composite structures. Composite Structures 75: 364-376.

Hamoush, S., Shivakumar, K., Darwish, F., Sharpe, M. and Swindell, P. 2005. Defective repairs of laminated solid composites. Journal of Composite Materials 39: 2185-2196.

Harman, A. B. and Wang, C. H. 2006. Improved design methods for scarf repairs to highly strained composite aircraft structure. Composite Structures 75: 132-144.

Harris, J. A. and Adams, R. A. 1984. Strength prediction of bonded single lap joints by non-linear finite element methods. International Journal of Adhesion and Adhesives 4: 65-78.

Hart-Smith, L. J. 1973a. Adhesive-bonded single-lap joints. NASA Contract Report, NASA CR-112236.

Hart-Smith, L. J. 1973b. Adhesive-bonded scarf and stepped-lap joints. NASA Langley Report CR 112237, Douglas Aircraft Co.

Hart-Smith, L. J. 1981. Further developments in the design and analysis of adhesive-bonded structural joints. ASTM STP 749: 3-31.

Ikegami, K., Takeshita, T., Matsuo, K. and Sugibayashi, T. 1990. Strength of adhesively bonded scarf joints between glass fibre-reinforced plastics and metals. International Journal of Adhesion and Adhesives 10: 199-206.

Imanaka, M., Hamano, T., Morimoto, A., Ashino, R. and Kimoto, M. 2003. Fatigue damage evaluation of adhesively bonded butt joints with a rubber-modified epoxy adhesive. Journal of Adhesion Science and Technology 17: 981-994.

Kafkalidis, M. S. and Thouless, M. D. 2002. The effects of geometry and material properties on the fracture of single lap-shear joints. International Journal of Solids and Structures 39: 4367-4383.

Karac, A., Blackman, B. R. K., Cooper, V., Kinloch, A. J., Rodriguez Sanchez, S., Teo, W. S. and Ivankovic, A. 2011. Modelling the fracture behaviour of adhesively-bonded joints as a function of test rate. Engineering Fracture Mechanics 78: 973-989.

Kattan, P. I. and Voyiadjis, G. Z. 2005. Damage Mechanics with Finite Elements. Springer, Heidelberg.

Kelly, L. G. 1983. Composite structures repair. Proceedings of the 57th meeting of the AGARD structures and materials panel, Portugal: Vimeiro.

Khoramishad, H., Crocombe, A. D., Katnam, K. B. and Ashcroft, I. A. 2010. Predicting fatigue damage in adhesively bonded joints using a cohesive zone model. International Journal of Fatigue 32: 1146-1158.

Kimiaeifar, A., Toft, H., Lund, E., Thomsen, O. T. and Sørensen, J. D. 2012. Reliability analysis of adhesive bonded scarf joints. Engineering Structures 35: 281-287.

Kimiaeifar, A., Lund, E., Thomsen, O. T. and Sørensen, J. D. 2013. Asymptotic sampling for reliability analysis of adhesive bonded stepped lap composite joints. Engineering Structures 49: 655-663.

Lemaitre, J. and Desmorat, R. 2005. Engineering Damage Mechanics. Springer, Heidelberg.

Liljedahl, C. D. M., Crocombe, A. D., Wahab, M. A. and Ashcroft, I. A. 2006. Damage modelling of adhesively bonded joints. International Journal of Fracture 141: 147-161.

Maiti, S. and Geubelle, P. H. 2005. A cohesive model for fatigue failure of polymers. Engineering Fracture Mechanics 72: 691-708.

Marques, E. A. S. and da Silva, L. F. M. 2008. Joint strength optimization of adhesively bonded patches. The Journal of Adhesion 84: 915-934.

Matthews, T., Ali, M. and Paris, A. J. 2014. Finite element analysis for large displacement J-integral test method for Mode I interlaminar fracture in composite materials. Finite Elements in Analysis and Design 83: 43-48.

Moës, N., Dolbow, J. and Belytschko, T. 1999. A finite element method for crack growth without remeshing. International Journal for Numerical Methods in Engineering 46: 131-150.

Mohammadi, S. 2008. Extended Finite Element Method for Fracture Analysis of Structures. Blackwell Publishing, New Jersey, USA.

Moroni, F. and Pirondi, A. 2011. A procedure for the simulation of fatigue crack growth in adhesively bonded joints based on the cohesive zone model and different mixed-mode propagation criteria. Engineering Fracture Mechanics 78: 1808-1816.

Mortensen, F. and Thomsen, O. T. 1997. Simplified linear and non-linear analysis of stepped and scarfed adhesive-bonded lap-joints between composite laminates. Composite Structures 38: 281-294.

Nazir, S. and Dhanasekar, M. 2014. A non-linear interface element model for thin layer high adhesive mortared masonry. Computers & Structures 144: 23-39.

Objois, A., Assih, J. and Troalen, J. P. 2005. Theoretical method to predict the first microcracks in a scarf joint. The Journal of Adhesion 81: 893-909.

Odi, R. A. and Friend, C. M. 2004. An improved 2D model for bonded composite joints. International Journal of Adhesion and Adhesives 24: 389-405.

Pearson, I. T. and Mottram, J. T. 2012. A finite element modelling methodology for the non-linear stiffness evaluation of adhesively bonded single lap-joints: Part 1. Evaluation of key parameters. Computers & Structures 90–91: 76-88.

Pinto, A. M. G., Campilho, R. D. S. G., de Moura, M. F. S. F. and Mendes, I. R. 2010. Numerical evaluation of three-dimensional scarf repairs in carbon-epoxy structures. International Journal of Adhesion and Adhesives 30: 329-337.

Pocius, A. V. 2002. Adhesion and Adhesives Technology, An Introduction, 2nd Edition. Carl Hanser Gardener Verlag, Munchen.

Raghavan, P. and Ghosh, S. 2005. A continuum damage mechanics model for unidirectional composites undergoing interfacial debonding. Mechanics of Materials 37: 955-979.

Reis, P. N. B., Ferreira, J. A. M. and Antunes, F. 2011. Effect of adherend's rigidity on the shear strength of single lap adhesive joints. International Journal of Adhesion and Adhesives 31: 193-201.

Rice, J. R. 1968. A path independent integral and the approximate analysis of strain concentration by notches and cracks. Journal of Applied Mechanics 35: 379-386.

Roe, K. L. and Siegmund, T. 2003. An irreversible cohesive zone model for interface fatigue crack growth simulation. Engineering Fracture Mechanics 70: 209-232.

Sampaio, E. M., Luiz Bastian, F. and Costa Mattos, H. S. 2004. A simple continuum damage model for adhesively bonded butt joints. Mechanics Research Communications 31: 443-449.

Savage, G. and Oxley, M. 2010. Repair of composite structures on Formula 1 race cars. Engineering Failure Analysis 17: 70-82.

Shenoy, V., Ashcroft, I. A., Critchlow, G. W. and Crocombe, A. D. 2010. Fracture mechanics and damage mechanics based fatigue lifetime prediction of adhesively bonded joints subjected to variable amplitude fatigue. Engineering Fracture Mechanics 77: 1073-1090.

Soutis, C., Duan, D. M. and Goutas, P. 1999. Compressive behaviour of CFRP laminates repaired with adhesively bonded external patches. Composite Structures 45: 289-301.

Teng, J. G., Zhang, S. S., Dai, J. G. and Chen, J. F. 2013. Three-dimensional meso-scale finite element modeling of bonded joints between a near-surface mounted FRP strip and concrete. Computers & Structures 117: 105-117.

Tsai, M. Y. and Morton, J. 1994. An evaluation of analytical and numerical solutions to the single-lap joint. International Journal of Solids and Structures 31: 2537-2563.

Turon, A., Dávila, C. G., Camanho, P. P. and Costa, J. 2007a. An engineering solution for mesh size effects in the simulation of delamination using cohesive zone models. Engineering Fracture Mechanics 74: 1665-1682.

Turon, A., Costa, J., Camanho, P. P. and Dávila, C. G. 2007b. Simulation of delamination in composites under high-cycle fatigue. Composites Part A: Applied Science and Manufacturing 38: 2270-2282.

Tzetzis, D. and Hogg, P. J. 2008. Experimental and finite element analysis on the performance of vacuum-assisted resin infused single scarf repairs. Materials & Design 29: 436-449.

Volkersen, O. 1938. Die nietkraftoerteilung in zubeanspruchten nietverbindungen konstanten loschonquerschnitten. Luftfahrtforschung 15: 41-47.

Voyiadjis, G. Z. and Kattan, P. I. 2005. Damage Mechanics. Marcell Dekker, New York.

Wahab, M. M. A., Ashcroft, I. A., Crocombe, A. D. and Shaw, S. J. 2001. Prediction of fatigue thresholds in adhesively bonded joints using damage mechanics and fracture mechanics. Journal of Adhesion Science and Technology 15: 763-781.

Wang, C. H. and Gunnion, A. J. 2008. On the design methodology of scarf repairs to composite laminates. Composites Science and Technology 68: 35-46.

Whittingham, B., Baker, A. A., Harman, A. and Bitton, D. 2009. Micrographic studies on adhesively bonded scarf repairs to thick composite aircraft structure. Composites Part A: Applied Science and Manufacturing 40: 1419-1432.

Williams, M. L. 1959. The stresses around a fault or crack in dissimilar media. Bulletin of the Seismological Society of America 49: 199-204.

Yang, Q. D., Thouless, M. D. and Ward, S. M. 1999. Numerical simulations of adhesively-bonded beams failing with extensive plastic deformation. Journal of the Mechanics and Physics of Solids 47: 1337-1353.

Yang, Q. D. and Thouless, M. D. 2001. Mixed-mode fracture analyses of plastically-deforming adhesive joints. International Journal of Fracture 110: 175-187.

Zhao, X., Adams, R. D. and da Silva, L. F. M. 2011a. Single lap joints with rounded adherend corners: Stress and strain analysis. Journal of Adhesion Science and Technology 25: 819-836.

Zhao, X., Adams, R. D. and da Silva, L. F. M. 2011b. Single lap joints with rounded adherend corners: Experimental results and strength prediction. Journal of Adhesion Science and Technology 25: 837-856.

Zimmerman, K. and Liu, D. 1995. Geometrical parameters in composite repair. Journal of Composite Materials 29: 1473-1487.

12

Mixed Adhesive Joints for the Aerospace Industry

E.A.S. Marques[1,*], R.D.S.G. Campilho[2],
M. Flaviani[3] and Lucas F.M. da Silva[4]

12.1 INTRODUCTION

In the last few decades, the use of adhesive joints has expanded significantly and currently many technologically advanced industries employ adhesive joints as a high performance joining method. The aerospace industry is among these industries and has had a leading role in the development of construction techniques that exploit the adhesives' ability to join dissimilar materials and to redistribute loads more effectively, which leads to lighter vehicles (da Silva et al. 2011). However, there are several aerospace applications where the use of adhesive joints is very limited, usually due to highly demanding environmental conditions (Pethrick 2015). The joining of ceramic to metals in thermal heat shields is one of those applications. The use of adhesives is a practical solution for this application, but large thermal gradients, highly dissimilar material properties and varied mechanical loads combine to create complex and overwhelming loadings on the joint.

One of the most known and studied practical application of the use of adhesives to handle large temperature gradients is the ceramic to metal bond, used in the now retired US Space Shuttle program. In this application, over 24 000 individual tiles were individually bonded to the surface of the vehicle. The design of such system was a great engineering challenge, as not only should it be reusable and light weight but also extremely reliable. The tiles used initially were made of almost pure silica, combined with a temperature resistant coating (Miller et al. 2013). Due

---

[1] Instituto de Engenharia Mecânica (IDMEC), Rua Dr. Roberto Frias, 4200-465 Porto, Portugal.
[2] Departmento de Engenharia Mecânica, Instituto Superior de Engenharia do Porto, Instituto Politécnico do Porto, Rua Dr. António Bernardino de Almeida, 431, 4200-072 Porto, Portugal.
[3] Università degli studi di Parma –Facoltà di ingegneria, Strada Dell'Università, 12, 43100 Parma, Italy.
[4] Departamento de Engenharia Mecânica, Faculdade de Engenharia, Universidade do Porto, Rua Dr. Roberto Frias, 4200-465 Porto, Portugal.
* Corresponding author: emarques@fe.up.pt

to the ceramic construction and large thickness, this type of tile is necessarily a heavy component and several upgraded and lighter versions were gradually introduced, adding some alumina content and improved coatings. The tiles were attached to the shuttle's aluminium structure using adhesive bonding but required a strain isolation pad bonded between the tile and the underlying structure. This pad limited the vibration damage and accommodated some of the thermal expansion differential between the tiles (Nasa KSC 1989). To improve the bonding surface of the tile, a ceramic slurry was applied to the underside of the tile. Gaps between the tiles are required to allow the relative movements caused by thermal expansion of the metallic and ceramic parts. The adhesive used for bonding the tiles is a room temperature vulcanizing (RTV) silicone, suited for high temperature usage. The adhesive is applied and subjected to vacuum pressure to ensure that there are no voids. RTV silicone screed is also spread on the metallic surface before bonding to fill voids and provide a smooth surface for the bonding process (Nasa KSC 2006). Figure 12.1 depicts the construction of the shuttle tile attachment and the materials used.

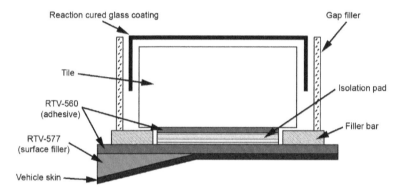

FIG. 12.1 Diagram of shuttle tile attachment.

While the space shuttle is now out of service, the new generation of thermal protection tiles are still being developed. The new generation tiles, known as TUFROC, are manufactured by attaching a ceramic-carbon insulation to a silica-based lower block. The two parts of the tile are joined by combining mechanical methods and a high temperature adhesive (Leiser et al. 2001). It uses an inorganic adhesive, a reaction cured glass applied in a 1.2 mm thick layer. This material uses glass powders mixed with thickeners and is especially resistant to extremely high temperatures and large thermal shocks. It serves not only as an adhesive but also as a non-abrupt transition between thermal gradients. This composite tile can then be attached to a metallic structure using the technique described above for the one-piece tiles (Stewart and Leiser 2008).

Many of the adhesives used for high temperature applications are known to be brittle at low temperatures, while those more suited for low temperatures usually do not provide sufficient strength at high temperatures. To solve this incompatibility, the use of dual adhesive joints was first proposed by Raphael (1966) as a possible solution. This technique is able to reduce the stress concentrations at the ends of the overlap, typical for single-lap joint, and reduce premature joint failures. The concept entails the introduction of a more flexible adhesive at the ends of the overlap, while a stiff adhesive is used in the central section of the joint, less subjected to deformation during loading. Hart-Smith (1973) recognized that the use of mixed adhesive joints could yield improvements in mechanical strength for joints subjected to large temperature gradients. In 2007, da Silva and Adams (2007a) made use of this concept and predicted improvements in the mechanical behaviour of a joint under a large temperature gradient. In their approach, the adhesives to be combined were not only dissimilar in the mechanical properties, but also in their temperature handling capabilities. The stiffer adhesive was also a high temperature adhesive (HTA), responsible for the joint strength when the joint is subjected to heat, while the more flexible adhesive was a low temperature adhesive (LTA), carefully selected to be able to provide strength to the joint under negative temperatures. At high temperatures, the HTA in the middle of the joint retains the strength and transfers the entire load while the LTA is the load bearing component at low-temperatures, making the HTA relatively lightly stressed. At low-temperatures, the load must essentially be supported by the LTA. If its modulus is of the same order as the modulus of the HTA, most of the load will be carried by the LTA. However, if its modulus is much lower than the modulus of the HTA, then there might still be a considerable load in the HTA. Therefore, the geometry and ratios between LTA and HTA must be carefully studied to improve the behaviour over a joint composed only of HTA. Figure 12.2 shows the working principle of such a joint as well as the stress distributions present in the adhesive layer at two extremes of temperature.

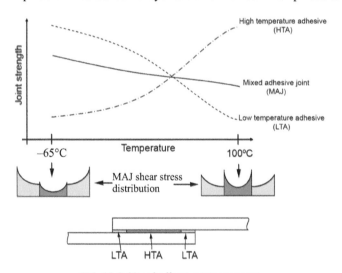

FIG. 12.2 Mixed adhesive joint concept.

It was also demonstrated that this technique is more interesting for practical applications should the adherends be of dissimilar nature. This is mainly due to the volume variation of the adherends under the range of temperatures that the specimen is subjected to. If there is a large disparity in the coefficients of thermal expansion (CTE), using a single HTA will result in early failure due to the inability of the stiff adhesive to accommodate the thermal expansion differential. By introducing a layer of a flexible LTA, the joint can more easily sustain thermal expansion and therefore, should the adhesives be able to withstand it, its temperature range can be expanded. da Silva and Adams (2007b) further presented experimental data that supported these conclusions, proving the concept for a temperature range of –50 to 200°C with titanium and carbon fibre-reinforced plastic (CFRP) adherends.

Marques et al. (2011, 2013) performed a series of experimental studies, bonding ceramic tiles to a metallic adherend using a mixed adhesive joint, combining an RTV silicone with a high temperature epoxy. The joints were tested under shear at room temperature, –65°C and 100°C. With these static tests, mixed adhesive joints were found to have consistent strength at high and low temperatures, while providing a good amount of joint displacement in both cases. Impact tests were also performed and again the mixed adhesive joint was demonstrated as a good alternative to the use of a single adhesive, able to handle large failure loads.

Various geometrical parameters of an adhesive joint can be changed to improve the mechanical behaviour of an adhesive joint, including mixed adhesive joints. A technique commonly employed to reduce stresses is the use of tapered adherends and adhesive layers. This a very common research subject and various authors (Adams et al. 1986, Hildebrand 1994, Rispler et al. 2000, da Silva and Adams 2007c) have employed this technique with good results, always with the aim of lowering the stiffness at the ends of the overlap of a joint and therefore achieve a smoother load transfer.

Another important characteristic that can be controlled in mixed adhesive joints is the ratio between the bonded area of each of the two adhesives on the overlap, a parameter which also causes significant changes in the overall joint strength and behaviour. Srinivas (1975) concluded by the Finite Element Method (FEM) that optimum lengths of stiff and flexible bonds can be chosen to assure the lowest possible stresses in the bondline. Raphael (1966) suggested to select the adhesive ratios so that the stress distributions of both adhesives match. Fitton (2004) improved on this by making a numerical analysis, studying the effect of different adhesive modulus combinations and different geometrical configurations. His studies confirmed that lower shear stresses are obtained when the peak shear stress is equal in both of the adhesives, but he also identified that this concept is not the optimum for achieving strength improvement, instead being more suited for obtaining lower shear stress distributions. Breto et al. (2014) numerically modelled mixed joins with aluminium/composite adherends under shear loads. A simple optimization procedure was performed to determine the optimum discrete grading of properties in a bondline and the ultimate load was improved by around 70%, maximizing the ultimate loading capacity of a single lap joint.

Cohesive zone models (CZM) are a very powerful tool for studying the behaviour of adhesive joints, including mixed adhesive joints. These models are increasingly being used to improve the failure load prediction of FEM models and various authors such as Needleman (1987), Tvergaard and Hutchinson (1992) and Camacho and Ortiz (1996) were among the first to adapt this technique for use in adhesive joints. A CZM is able to represent the fracture process and location, advancing beyond the typical continuum mechanics modelling. It does this by including in the model a series of discontinuities modelled by cohesive elements, which use both strength and energy parameters to simulate the nucleation and advance of a fracture crack (Banea et al. 2014). The relationship between the stresses and displacements is governed by a traction separation law. Figure 12.3 shows the pure-mode (traction or shear) and mixed-mode traction separation laws, where t_n^0 and t_s^0 are the cohesive strengths in tension and shear, respectively, and δ_n^0 and δ_s^0 are the respective strains, and δ_n^f and δ_s^f are the tensile and shear strains at failure, respectively.

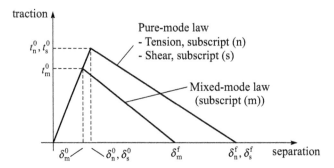

FIG. 12.3 Traction-separation law with linear softening: pure and mixed-mode models.

While the properties for the initial part of the traction-separation law (elastic phase) are relatively simply determined using tensile and shear tests, the second part, relating to the damage of the material, must be determined using fracture mechanics tests (Chaves et al. 2014). The shape of this law can be changed independently of the properties to more adequately fit the mechanical behaviour of the simulated material. The initial elastic portion is always kept linear, but in the literature various shapes for the softening portion of the curve can be found. Needleman (1987) introduced a shape based on more complex functions such as polynomial and exponential laws. Tvergaard and Hutchinson (1992) suggested a trapezoidal model while Liljedahl et al. (2006) proposed the simpler bilinear model. There is however not a consensus regarding the importance of the softening law shape. While some researchers (Tvergaard and Hutchinson 1992, Needleman 1990) have determined that the shape of this portion of the law is not critical for the results accuracy, others have found the opposite effect (Mohammed and Liechti 2000, Chandra et al. 2002). Chandra et al. (2002) published a report and review of the various CZM laws available in the literature. It was concluded that the chosen CZM model for a given application should depend of carefully designed experiments that enable achieving a faithful representation of the failure process. The shape of the CZM law should depend on the inelastic processes occurring in the materials at the micromechanical level. Campilho et al. (2013) addressed the influence of the CZM shape in the strength prediction of composite single-lap joints, considering different adhesives (brittle and ductile) and overlap lengths. Triangular, trapezoidal and linear-exponential shapes were compared. Results showed that the CZM shape is more relevant when considering ductile adhesives, in which case the best results were obtained with the trapezoidal shape law. On the other hand, the results for the brittle adhesive were similar between the three CZM shapes, although the triangular CZM was slightly better.

This work therefore aims to increase the understanding of the mixed adhesive joint capabilities, by mechanically testing metal-ceramic specimens at room temperature and under shear loading and then using this information to allow the construction and validation of FEM models. To more accurately represent the real joint, the models make use of cohesive elements, combining a continuum mechanics approach with a fracture mechanics approach. The cracks can therefore be simulated and matched to the cracks identified in the mechanical testing, thus leading to a validated model that can be used for joint optimization purposes.

12.2 EXPERIMENTAL PROCEDURE

12.2.1 Materials

Success on the use of dual adhesive joints is in large part dependent on the correct selection of the adhesives. The adhesives employed in a dual adhesive joint must be not only compatible but also sufficiently different in properties to complement each other. For this work the adhesives selected were a single part epoxy and a RTV silicone. The selected epoxy is a commercially available stiff and brittle adhesive, suitable for high temperature use produced by Nagase-Chemtex (Osaka, Japan) under the reference XN1244. This adhesive is a one component, high temperature, paste epoxy adhesive, with a glass transition temperature (T_g) of 170°C. Due to its T_g it provides good mechanical properties up to 150°C (Banea et al. 2011a). The cure process is heat activated and requires exposure to a temperature of 140°C during 1h to achieve complete cure. The RTV silicone is also of a commercially available type, produced by ACC Silicones LDT (Bridgewater, UK) under the reference RTV106. This adhesive is very distinct from the XN1244 epoxy by being a very ductile and flexible material, with reduced mechanical strength. The RTV106 silicone also has a much lower T_g (around -130°C), which makes it more insensitive to low temperatures, maintaining a good level of strength while the epoxy becomes extremely brittle.

The curing process of the RTV106 adhesive is very distinct from the curing process of the XN1244 epoxy, by being based on the absorption of humidity from the air (de Buyl 2013). In order to ensure a complete cure, water molecules must diffuse from the surface of the material to the interior. Due to the reduced mobility of water molecules, this cure is a slower process. When thick layers of adhesive are used, curing times as long as 10 days can be required to obtain full cure. The mechanical properties of these two adhesives are presented in Table 12.1 and been fully characterized in the works of Banea et al. where the materials were tested in bulk form (Banea and da Silva 2009, 2010, Banea et al. 2011a), in mode I fracture tests (Banea et al. 2011b, 2011c) and in mode II fracture tests (Banea et al. 2012).

Table 12.1 Mechanical properties of RTV 106 silicone and XN1244 epoxy at room temperature.

Property	RTV106 silicone	XN1244 epoxy
E - Young's modulus (N/mm²)	1.6	5870
G - Shear modulus (N/mm²)	0.86	2150
t_n^0 - Tensile strength (N/mm²)	2.3	68.23
t_s^0 - Shear strength (N/mm²)	1.97	37
G_n^c - Mode I fracture energy (N/mm)	2.73	0.47
G_s^c - Mode II fracture energy (N/mm)	5	2.2

These properties can also be visualized as the traction separation laws, used to run the CZM simulations. Figure 12.4 shows the shape of those laws for both adhesives. Figure 12.4a represents the Mode I (traction) law and Fig. 12.4b represents the Mode II (shear) law.

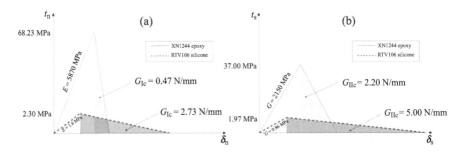

FIG. 12.4 Mode I (a) and mode II (b) triangular traction separation laws for the adhesives.

The metallic adherend of the dual adhesive joint is machined from an aluminium alloy of the 6063 T6 type. The ceramic tile used in this work is made of cordierite, a high temperature resistant ceramic. By being able to withstand large temperature gradients and possessing low thermal conductivity, it is a material suitable for the role of thermal shielding. The characterization of the full cohesive properties of the cordierite material was performed using a reverse method and is described in a following section of this work.

12.2.2 Experimental Specimen Configurations

The specimens studied in this work consists of a ceramic tile bonded to a metallic sheet, as shown in Fig. 12.5. Three specimen configurations were tested in the experimental phase. All three configurations use the same adherends but differ in the adhesive layer used. One configuration used an adhesive layer with RTV silicone adhesive, another used XN1244 epoxy and finally the third combination uses both adhesives simultaneously in a mixed adhesive joint.

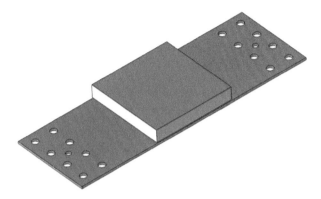

FIG. 12.5 Three-dimensional (3D) drawing of the specimen used.

The ceramic tiles had a dimension $80 \times 80 \times 12.8$ mm^3 and were produced by water-jet cutting of a larger plate. They were bonded to the centre of the aluminium sheet using the before mentioned three configurations of adhesive layers. The area available for bonding is 60×60 mm^2, smaller than the area of the ceramic adherend (80×80 mm^2). A detailed drawing of each configuration and its geometry is shown in Fig. 12.6.

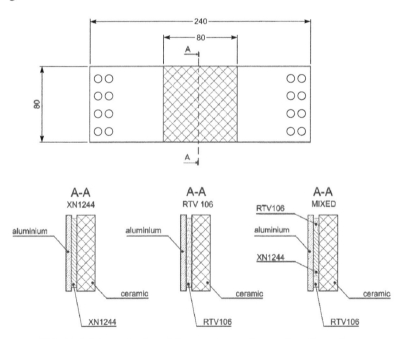

FIG. 12.6 Schematic view of the experimentally tested configurations.

An important parameter to define in a study of mixed adhesive joints is the ratio between the two adhesives. This ratio can be defined using a variety of parameters, but in this work the adhesives' surface area was used to calculate a simple ratio, dividing the area of the silicone layer ($A_{silicone}$) by the area of the epoxy layer (A_{epoxy}). The mixed joint tested here has a ratio of 0.5. Table 12.2 lists the bonded area and ratio (if applicable) of each of the specimens tested.

Table 12.2 Bonded area configurations for experimentally tested specimens.

Bonded area configuration	$A_{silicone}/A_{epoxy}$ ratio
60×60 mm^2 Epoxy	Full Epoxy
40×60 mm^2 Epoxy, 20×60 mm^2 Silicone	0.5 - Mixed joint
60×60 mm^2 Epoxy	Full Silicone

A thick adhesive layer (1 mm) was used to ensure strength of the RTV silicone adhesive. While theoretically a thinner adhesive layer would provide higher joint strength improvements in comparison to a single brittle adhesive (because the load would be more concentrated at the ends of the overlap), in practice the use of the RTV silicone in thin layers results in very weak joint strength due to its extremely low modulus of elasticity. A frame of silicone rubber, 1 mm thick, was cut to constrain the adhesive underneath the ceramic tile and to set the adhesive layer thickness. Two different techniques were used to manufacture the single adhesive joints and the mixed adhesive joints. The latter type, instead of having a full, unobstructed internal square of 60×60 mm^2, has two thin barriers dividing the square into three different bonding areas. This avoids the contact between the two different adhesives. A custom mould (shown in Fig. 12.7) was built to position and restrict the movement of the adherend during the curing process.

FIG. 12.7 Specimens in the mould, ready for bonding.

12.2.3 Testing Procedure

The specimens were tested in a universal testing machine using a custom testing tool, designed to fix the metal adherend while the ceramic tile is pulled away. This introduces a shearing load in the adhesive layer. The testing machine is an INSTRON (Norwood, MA, USA) model 3360 electromechanical testing machine with a load cell of 30 kN. The selected test speed was 1 mm/min and the temperature was 22°C (room temperature). The load-displacement (P-δ) curve was registered during the test allowing a variety of properties regarding the whole specimen to be determined: stiffness by the slope of the curve in the elastic zone, the maximum load, P_{max}, and the maximum displacement, δ_{max}, at failure. Moreover, due to the flexing and rotation of the specimen, an additional displacement measuring equipment was added. Two linear variable differential transformers (LVDT) were placed behind the test machine and supported by a mechanism, which allowed their translation in the vertical and horizontal axis for precise location. Two specific points on the back side of the adherend were selected and kept in contact with the LVDT shafts. As shown in Fig. 12.8, they were located exactly in the medium cross section where the highest displacement of the metallic adherend was expected to occur. The obtained output can be useful to determine the boundary condition of some numerical simulation models.

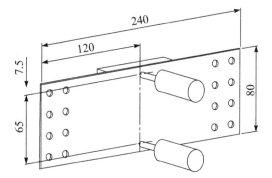

FIG. 12.8 LVDT sensor locations (dimensions in mm).

12.2.4 Testing Tool Compliance Correction

Due to the specimen shape, the testing tool and procedures used for this work are not standardized. During initial testing, the tool and its substantial size and asymmetrical shape was found to introduce additional deformation in the displacement measurements. To quantify the additional deformation a simple test was performed, using a steel dummy specimen with high stiffness. The measured displacement in this calibration test can be then removed from the displacement measured during the experimental test to obtain a P-δ curve without contribution from the displacement derived from the gaps and flexing of the testing tool. The dummy specimen, shown in Fig. 12.9, was constructed with welded steel plates. A simple FEM analysis was carried out to ensure that the stiffness of the specimen was maximum and negligible when compared with the experimental testing results.

FIG. 12.9 Dummy specimen used for compliance measurement.

The dummy specimen was fastened to the testing tool in a manner equivalent of the other experimental specimens. It was then gradually loaded to $P_{max} = 10$ kN and the resultant P-δ curve registered. This curve is therefore a representation of the testing tool displacement for each given load value. This compliance curve was curve-fitted and used to find a displacement per load equation (Fig. 12.10).

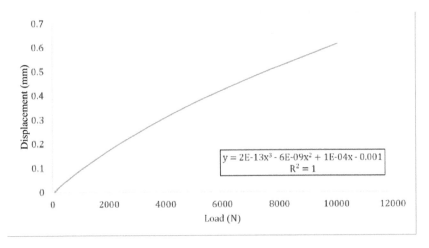

$$y = 2E\text{-}13x^3 - 6E\text{-}09x^2 + 1E\text{-}04x - 0.001$$
$$R^2 = 1$$

FIG. 12.10 Compliance curve of the testing tool and respective compliance equation.

The compliance equation was then used to subtract the excess displacement from the experimental P-δ curve. This process is depicted in Fig. 12.11.

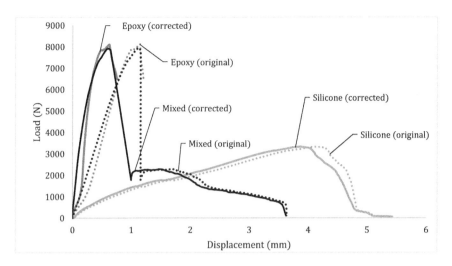

FIG. 12.11 Correction process of experimental P-δ curve using the compliance equation.

12.3 NUMERICAL PROCEDURES

12.3.1 Determination of Ceramic Mechanical Properties

A variety of different failure mechanisms can occur in a bonded joint. In order to predict the failure mode when the failure is located in the ceramic, knowledge of the mechanical properties of the cordierite used is fundamental. Due to the lack of mechanical property values of cordierite in the literature, the mechanical properties of ceramic tile had to be determined. However, direct testing of the mechanical properties of ceramic materials is a complex procedure, which requires careful experimental control and a large number of specimens. Alternatively, and to simplify this process, an inverse method was employed to obtain these properties. As the ceramic tile is loaded in shearing mode, it is most important to study the properties regarding this type of loading. A novel specimen was designed for this purpose. To design this model, a FEM elastic model was first created in order to study the shear loading along a thin central section. Central cuts were designed and optimized to ensure that shear loading was by far the most significant type of loading present. The specimen was then produced by water-jet cutting from a block of cordierite and then tested under compression. The specimen's shape is very similar to a Thick Adherend Shear Testing (TAST) specimen, in which the central portion geometry is designed in a way as to create a pure shear loading when the specimen is compressed between two metal plates. The final specimen, as well as the tool used for its test, can be seen in Fig. 12.12.

After this process, a more complex model, including cohesive elements in the central section (where failure occurs), was built. By means of an inverse method, it was possible to estimate the cohesive properties of the specimen under shear loading conditions, simply by adjusting the properties until there was a good match between the FEM and the experimental P-δ curves as seen in Figure 12.13.

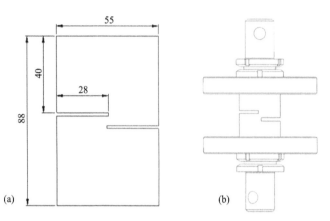

FIG. 12.12 Specimen geometry (dimensions in mm) (a) and testing tool and specimen (b).

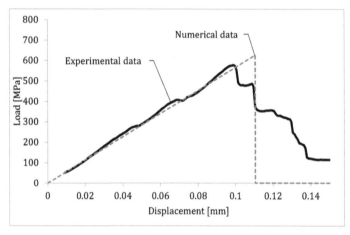

FIG. 12.13 Experimental and numerical P-δ curves for the ceramic specimen.

The properties of the ceramic material obtained via inverse method are given in Table 12.3. These properties were incorporated in cohesive elements as a bi-linear material.

Table 12.3 Mechanical properties of cordierite as determined by experimental/numerical matching.

	Cordierite
E - Young's Modulus (N/mm²)	5200
G - Shear Modulus (N/mm²)	2060
t_n^0 - Tensile Strength (N/mm²)	45
t_s^0 - Shear Strength (N/mm²)	45
G_n^c - Mode I fracture energy (N/mm)	0.5
G_s^c - Mode II fracture energy (N/mm)	0.5

12.3.2 Two-Dimensional Model Construction

12.3.2.1 Shear Loading Specimen Models

FEM models were created for each type of joint configuration: adhesive layer with silicone (silicone joint), adhesive layer with epoxy (epoxy joint) and adhesive layer with dual adhesives (mixed joint). For each of these three configurations, linear elastic and cohesive analysis were separately performed. The initial elastic analysis was performed in order to adjust the boundary conditions. For the cohesive models, thin layers of cohesive elements were placed in the middle of elastic elements and the model was simulated until failure. The boundary conditions for the FEM model were selected by studying each case. The main problem to be solved consisted of reproducing the displacement of the aluminium adherend in the considered cross section. Simple displacement and rotation restrictions, commonly used in two-dimensional (2D) models, were found not to be able to represent the amount of displacement present in the joint during the experimental tests. To accurately represent the displacement of the full joint under load, it was proposed to connect two spring elements with carefully adjusted stiffness in two points of the aluminium adherend. In an Abaqus®/Standard analysis it is possible to define springs that connect points to the ground and exhibit the same linear behaviour independently of field variables. The configuration used for the boundary conditions is shown in Fig. 12.14.

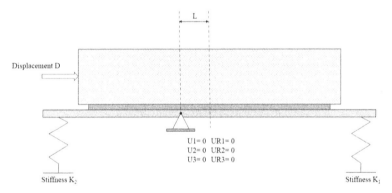

FIG. 12.14 Boundary conditions for the numerical model.

The defined parameters were the position (selection of the points to connect to the ground), the direction (to specify an orientation for the springs using a datum coordinate system) and the spring stiffness. To determine the spring parameters, some additional information about the displacement of the joint during the experimental test was required. As mentioned in the section detailing the experimental procedure, it was decided to instrument the experimental specimens with LDVT sensors (displacement sensors) in order to obtain curves that relate the variation of displacement of the specimen with the applied load. By adjusting the stiffness of the springs, these curves can be matched with the model ensuring that the boundary conditions are adequate. However, the springs by themselves are not able to resist lateral loads; so it was necessary to include another boundary condition to limit the lateral and vertical movement of the specimen. A "pinned" type of boundary condition restricts the x and y axis movement of the specimen while providing a hinge for the springs to actuate on. With these boundary conditions in place and comparing with the LDVT data, the spring stiffness values that were found are presented in Table 12.4.

Table 12.4 Spring parameters and locations for the numerical models.

Parameters	D [mm]	K_1 [N/mm]	K_2 [N/mm]	L [mm]
Silicone	8	80	75	20
Epoxy	2	1	450	7
Mixed	2	1	450	7

After the boundary conditions were satisfactorily determined, the cohesive element models were developed. The main challenge in the construction of these models is the location of the cohesive element layers, which must be located in the areas where failure is expected. Figure 12.15 shows a scheme with the location of each layer for the two main types of specimens.

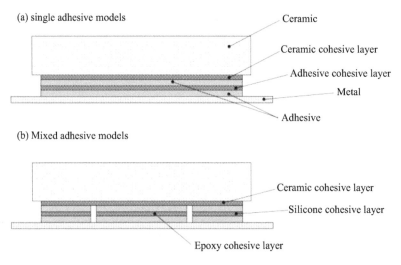

FIG. 12.15 Location of the cohesive layers in the FEM models (some dimensions exaggerated).

The model of the joint containing silicone adhesive has two different layers of cohesive elements (Fig. 12.15a). A layer is introduced in the bottom portion of the ceramic tile and the middle of adhesive layer. The thickness of the cohesive layers is 0.1 mm, much thinner than the actual adhesive layer thickness, which is 1 mm. This low value is necessary to correctly use the cohesive elements as is also the same thickness present in the models and specimens used to determine the ceramic properties. This ensures consistency in the use of cohesive elements. The model of the joint containing epoxy adhesive (also Fig. 12.15a) is similar: a thin layer of cohesive elements was placed in the middle of the adhesive and another layer of cohesive elements was located in the ceramic tile, immediately above the bonding line.

The configuration of the cohesive model for the mixed adhesive joint (Fig. 12.15b) is the most complex as it has four different cohesive element layers. There is a layer for the lower portion of the ceramic, one in the middle of the epoxy adhesive and two additional layers located in the middle of each silicone section. This geometry allows the cracks to propagate in each of the adhesives and also in the ceramic.

12.3.3 Three-Dimensional Model Construction

12.3.3.1 Shear Loading Specimen Models

In addition to the previously presented 2D model, a 3D model was also constructed for the aluminium ceramic joint. The 3D model allows more a more powerful study of the influence of various geometrical parameters on the specimen's behaviour. The modelling of boundary conditions is also simplified, as it is easier to replicate the actual mechanical restrictions found during experimental testing. These characteristics make it a good tool to assess the efficiency of possible enhancements of the specimen's shape. As for the 2D model, cohesive elements are used to replicate the crack propagation in the specimens, providing a simulated P-δ curve and failure load predictions. The cohesive element used for this purpose was of the COH3D8 type, available in the Abaqus® default element library for 3D models. This element allows the use of a triangular traction separation law with linear softening. Due to the extremely computation intensive nature of the 3D cohesive models, some simplification steps were undertaken. Using symmetry considerations, only half of the joint was modelled and the mesh was finely adjusted to reduce the number of elements in non-critical parts of the model. The boundary conditions were similar for all models created and are shown in Fig. 12.16.

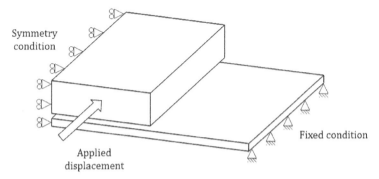

FIG. 12.16 Boundary conditions used in all FEM models.

To perform an initial validation of the FEM models, three different 3D models were created in Abaqus®, representative of each type of joint configuration that was experimentally tested: adhesive layer with silicone (silicone joint), adhesive layer with epoxy (epoxy joint) and adhesive layer with dual adhesives (mixed joint). The models employed cohesive elements, with the aim of achieving a complete modelling of the failure procedure. A cohesive analysis was performed for each of these three configurations. The cohesive models made use of thin layers of cohesive elements, placed in the middle of elastic elements and the behaviour of each model was simulated until failure. The main challenge in the construction of these models is the location of the cohesive element layers, which must be manually placed in the areas where failure is expected. Similarly to the 2D model, the 3D model had two different layers of cohesive elements. A cohesive layer was again introduced in the bottom portion of the ceramic tile and the middle of adhesive layer. The thickness of the cohesive layers was 0.1 mm, much thinner than the actual adhesive layer thickness, which is 1 mm. These dimensions are in accordance with the previously introduced discussion regarding the 2D models. The model geometry and mesh are shown in Fig. 12.17.

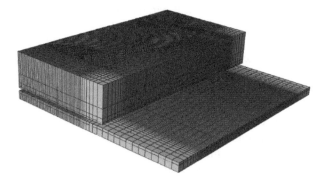

FIG. 12.17 FEM model of mixed adhesive joint.

12.3.3.2 Alternative Specimen Configurations

In an effort to understand the effect of geometrical changes on the studied mixed adhesive joint, several alternative joint configurations were proposed and numerically studied. A description of these configurations are described next.

Square Mixed Joint

To simplify manufacturing, the orientation of the layers of adhesive in the previously described mixed adhesive joints is such that they are optimized to handle loads only in one direction. While this is sufficient to study the mixed joint concept, in real world applications it might be necessary to develop a joint geometry that is able to handle loads in more than one direction. To achieve this purpose, an alternative mixed adhesive joint geometry was modelled using a silicone adhesive layer that encompasses an internal epoxy square. For clarification, a simple comparison between the experimentally tested mixed joint and this square joint is shown in Fig. 12.18.

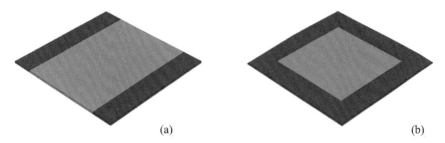

(a)   (b)

FIG. 12.18 Mixed joint experimentally tested (a) and proposed square mixed adhesive joint (b).

In this squared layer, the maximum displacement zones created by any type of side loading on the ceramic tile will always occur in a zone of the adhesive layer that is composed of silicone adhesive. By using the same dimensions of the previously described mixed adhesive joint and adding symmetry, this model has more silicone in the adhesive layer. This translates into a $A_{silicone}/A_{epoxy}$ ratio of 1.25, higher than the 0.5 of the initially proposed mixed joint.

Ramped Joint

A technique that can modify the behaviour of dual adhesive joints consists of the use of a tapered adhesive layer, gradually increasing the amount of adhesive in the overlap ends. A model employing such technique was modelled by FEM and its basic geometry is shown in Fig. 12.19.

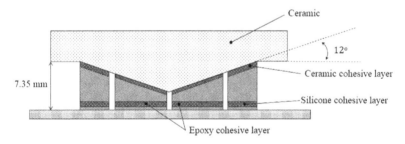

FIG. 12.19 Schematic drawing of the ramp mixed adhesive joint and cohesive layer locations.

This model has an adhesive layer that varies from a 1 mm thickness in the centre of the joint to a 6.35 mm thickness in the joint extremities. This means that the central part of the layer containing epoxy adhesive has the same thickness of the flat joint while the thickness in the silicone section is significantly thicker. The value of the angle was selected to be as big as possible without weakening the ceramic tile. This joint was modelled to have a bonded area ratio equivalent to that of the experimentally tested mixed adhesive joint. The general aspect of the FEM model and its mesh is shown in Fig. 12.20.

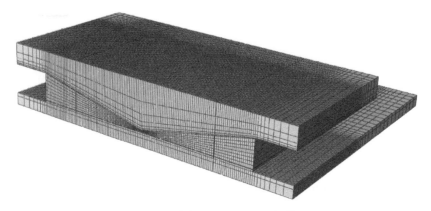

FIG. 12.20 FEM model of the ramp mixed adhesive joint.

Table 12.5 lists the three previously described mixed joint configurations, defining the surface bonded areas and the $A_{silicone}/A_{epoxy}$ ratio for each one.

Table 12.5 Bonded area configurations for square mixed joint.

Configuration reference	Square joint bonded area configuration	$A_{silicone}/A_{epoxy}$ ratio
Initial	40 × 60 mm² Epoxy/20 × 60 mm² Silicone	0.5
Square	40 × 40 mm² Epoxy/2000 mm² Silicone	1.25
Ramp	40 × 60 mm² Epoxy/20 × 60 mm² Silicone	0.5

Ramped Joint (Other Adhesive Ratios)

Besides using the same area ratio (0.5) employed in the previously described analysis, the ramp model was further explored by using five additional configurations to assess the influence of the adhesive distribution on the mechanical behaviour of the specimen. The selected configurations and surface ratios are shown in Table 12.6.

Table 12.6 Bonded area configurations for the ramp mixed joint.

Ramp joint bonded area configurations	$A_{silicone}/A_{epoxy}$ ratio
60 × 60 mm² Epoxy	Full Epoxy
50 × 60 mm² Epoxy/10 × 60 mm² Silicone	0.2
40 × 60 mm² Epoxy/20 × 60 mm² Silicone	0.5
30 × 60 mm² Epoxy/30 × 60 mm² Silicone	1
20 × 60 mm² Epoxy/40 × 60 mm² Silicone	2
60 × 60 mm² Epoxy	Full Silicone

These models range from a specimen containing only epoxy to a specimen containing only silicone and include four other configurations in between.

12.4 RESULTS

12.4.1 Fracture Surfaces

Figure 12.21 shows a typical fracture surface of a specimen containing only silicone adhesive. These specimens exhibited a generally cohesive failure of adhesive layer. A few zones with adhesive failure could be identified but were never substantial in area. The FEM model for this specimen translated the phenomenon correctly. Of the two cohesive layers in the model (ceramic and adhesive) only the layer placed in the adhesive was subjected to failure.

FIG. 12.21 Fracture surface of a specimen containing only silicone.

Figure 12.22 shows the fracture surface of a specimen containing only epoxy adhesive. There is a purely cohesive failure of the ceramic material in the bonded area. This fracture is very near the adhesive layer. Again, the cohesive model used for this purpose was found to also have avoided failure of the adhesive and progressed only in the cohesive layer set in the ceramic material.

FIG. 12.22 Fracture surface of a specimen containing only epoxy.

Figure 12.23 shows the typical fracture surface of a mixed adhesive joint. These specimens simultaneously exhibited cohesive fracture in the ceramic and the silicone layer. The initial crack was found to occur in the ceramic (similarly to the epoxy specimens) and then progressed into a cohesive failure of both silicone portions. This behaviour was also successfully modelled. In the models the crack progresses completely through the cohesive layer installed in the ceramic and then jumps to the cohesive layer installed in the silicone. The cohesive layer for the epoxy is left undamaged.

FIG. 12.23 Fracture surface of a specimen with mixed adhesive layer.

12.4.2 Numerical-Experimental Curve Comparison (Two-Dimensional Models)

The failure load results obtained in experimental testing are shown in Fig. 12.24. To provide some information about the variation of the mechanical properties, experimental results for high temperature and low temperature are also provided. These results were already previously published by the authors.

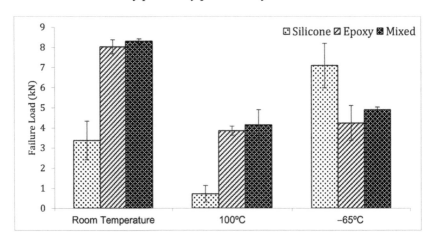

FIG. 12.24 Static testing results.

At room temperature there is a large similarity in results between the mixed adhesive joint and the joint containing only epoxy, while the silicone joint provides smaller failure loads. These results change dramatically with temperature, with the silicone joint being very strong at low temperature and the mixed and epoxy joints being the best at high temperatures.

Results obtained from the experimental work and numerical simulations were compared. Typical experimental curves were considered. The experimental and numerical P-δ curves allowed the study of stiffness, the P_{max} and the displacement of the joints at room temperature. A comparison of these curves can be found in Fig. 12.25.

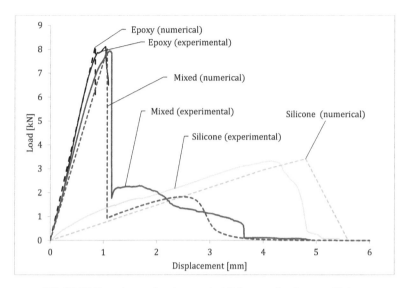

FIG. 12.25 Experimental and numerical P-δ curves for the tested joints.

The experimental curves shown here do not use the calibration factors described previously. This is due to the fact that the calibration process was done in the boundary conditions' adjustment. In both the numerical models and experimental tests the silicone specimens were found to always break cohesively in the silicone layer and the epoxy specimens were found to break in the ceramic layer. The mixed adhesive joint exhibited cohesive fracture in the ceramic near the epoxy layer but, after the test was stopped, the silicone layer was still intact and able to carry a small load, achieving a displacement at failure comparable to the silicone joint and nearly five times larger than that of the epoxy joints. This secondary load increase can be seen in the graph, with a dual behaviour of the mixed adhesive joint, which after reaching P_{max}, it still has a substantial amount of displacement before total failure. Both the FEM model and the experimental tests captured this phenomenon.

A good agreement between experimental and numerical data was generally found. In the initials peaks (after the elastic phase) a small error can be found, as the simulation curves cannot match the slight reduction in stiffness near P_{max}. This is due to the triangular CZM law adopted in Abaqus®, which does not allow the displacement to increase with constant load. A trapezoidal CZM would be able to more precisely simulate this part of the graph, albeit with increased complexity. The other main divergence between the CZM and the experimental data can be found in the curves relating to the mixed adhesive joint, immediately after the first peak. The first peak corresponds to the failure of the ceramic tile, while the second peak is due to the load being transferred to the still intact silicone layers. This second peak is very hard to model, because in the experimental specimens there is significant drag caused by the interference of the broken (and now sliding) materials against the adhesive connections still remaining in place. The modelled curve can therefore roughly match the loads and displacements but will not able to accurately simulate the unstable nature of the load progression.

12.4.3 Numerical-Experimental Curve Comparison (Three-Dimensional Models)

As performed for the 2D models, the 3D models were initially used to validate the properties, boundary conditions and simulation techniques employed in this work, being compared against experimental data. The failure propagation on a specimen is an aspect that can be studied using these models. Figure 12.26 shows an example of such process, depicting the three steps in propagation of the crack in a mixed adhesive joint.

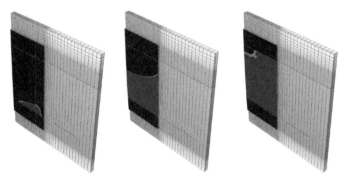

FIG. 12.26 Crack progression in a mixed adhesive joint (cracked areas shown in red).

The crack initiates in the central part of the mixed joint (containing epoxy), in the interface of the ceramic material. It is only after this central section suffers damage that the outer regions, containing silicone, begin to fail. As done for the 2D models, the P-δ curves can be used to compare the experimental and numerical results. Figure 12.27 shows the P-δ curves, comparing the results obtained with the 3D models with the calibrated experimental curves.

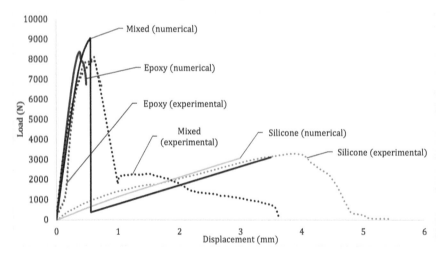

FIG. 12.27 Experimental and numerical P-δ curves of the specimens used for the FEM model validation procedure.

The numerical results obtained with the 3D CZM can be described as having a good agreement with the experimental data, especially in which regards to P_{max}. There is also a good agreement with the experimental specimen stiffness, demonstrating the importance of the previously described calibration procedure. There was however difficulty in accurately modelling the behaviour of the ductile silicone adhesive. In the silicone only model and the mixed adhesive model it became impossible to model the last portion of the joints' mechanical behaviour. The large displacements and relatively small loads involved are not especially suited to the triangular cohesive elements used, again suggesting that the use of a trapezoidal or exponential cohesive element law could probably yield improvements. As described for the 2D models, the failure mode of the mixed adhesive is relatively complex, which leads to introduction of friction after the initial failure of the epoxy layer and creates further problems in the correct modelling of this phase of the joints behaviour. Figure 12.28 shows a comparison between three different FEM configurations. Numerical P-δ curves are shown for the original mixed joint, the squared joint and a simple ramped joint.

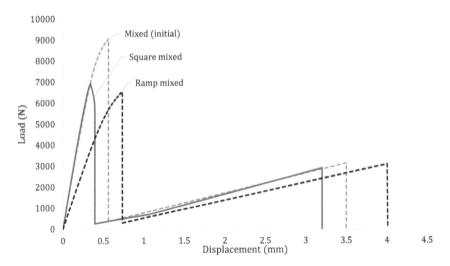

FIG. 12.28 Numerical P-δ curves for the three main mixed joint configurations tested.

This data demonstrates that the initial simpler mixed configuration is clearly stronger than the other two proposed configurations. The symmetrical squared joint exhibits a large reduction in joint strength, mainly due to the reduced amount of epoxy present in the central section of the overlap, now partially replaced by a border of silicone. However, the overall stiffness of this type of joint is equivalent to that of the initial mixed joint. While the ramped joint exhibits substantially lower failure load when compared to the standard mixed joint, the failure occurs at a larger displacement. In this case, the difference is the ramp geometry and the large amount of flexibility that this configuration introduces on the joint. While this is not directly translated into an increase of peak load, the decreased stiffness is beneficial to a joint that will be subjected to large thermal gradients. All the joints exhibit a two phase failure in which, after an initial peak load and failure in the epoxy section, the silicone is still able to sustain some load. Figure 12.29 shows the numerical P-δ curves of the ramp models simulated with different adhesive ratios. To allow for comparison, a numerical P-δ curve of the initial (non-ramped) mixed joint is also presented.

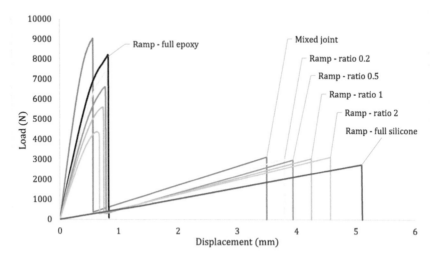

FIG. 12.29 Numerical P-δ curves for the ramped mixed adhesive joints with varying adhesive ratios.

Figure 12.30 shows the same data, but studies the influence of the epoxy amount on the initial specimen stiffness (the slope of the initial part of the P-δ curve) and the failure load.

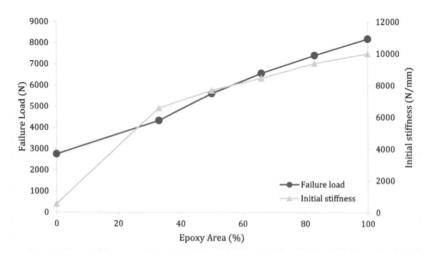

FIG. 12.30 Numerical P-δ curves for the ramped mixed adhesive joints with varying adhesive ratios.

It can be seen that there is an almost linear progression of the failure load with an increase in the epoxy area. The specimen stiffness, however, does not follow this tendency, with a large increase in the stiffness as the epoxy content increases. This data shows that, by using the proposed ramp configuration, there is no ratio of adhesive that is able to achieve a strength as high as the flat standard mixed adhesive joint. There are, however, significant changes in the joint initial stiffness. By gradually increasing the amount of silicone in the adhesive layer, the stiffness of the joint is reduced. Counter intuitively, this reduction of stiffness is also accompanied by a reduction in the displacement at which P_{max} occurs. This is caused by the progressively smaller area of epoxy that sustains this initial load peak. Additionally, as the epoxy sections reduce in size the more they are restricted to the thinner central section of the adhesive layer, which also contributes to this early peak load.

12.5 CONCLUSIONS

In this work, numerical simulations were developed to model the mechanical behaviour of ceramic-metal joints. Joints with a high strength/high temperature epoxy (XN1244) and a RTV silicone (RTV106) were manufactured and tested in a specially designed apparatus in order to study their behaviour and provide data to validate the numerical models. In an attempt to explore the synergetic advantages of a combination between these two materials, mixed adhesive joints were produced.

In the experimental procedure presented in this work, RTV106 silicone joints were found to exhibit low strength when compared with the other configurations studied. In fact, P_{max} for the silicone was 3500 N, while the other joints reached around 8000 N. On the other hand, the maximum displacement measured for the silicone joints was about 5 times larger than the displacement of the epoxy joint. In the mixed joint, P_{max} was identical to the epoxy adhesive joints, but with a slightly higher deformation. All joints failed cohesively during these tests. The joints using only RTV106 silicone had cohesive failure in the adhesive, while the joints that used XN1244 had cohesive failure in the ceramic adherend. In the mixed joint, even after the ceramic near the epoxy breaks, some amount of load is still absorbed, as the silicone layer and the ceramic next to it is still reasonably intact. The brittle nature of the XN1244 epoxy joints was partially corrected with the addition of RTV106 to the joints, resulting in a stronger and safer joint.

For the numerical simulations, the first step consisted in the determination of the mechanical properties of the ceramic and adhesives for implementation in the cohesive models. The ceramic parameters were successfully determined by means of an inverse method. A novel specimen was designed to assess the cohesive properties of the ceramic and the mechanical behaviour of this specimen was compared with a cohesive model. The numerical models presented in this model are divided into 2D and 3D models. The 2D models were studied for their simplicity and low computation costs while the 3D models were explored due to their ability to more exactly model the actual specimens. In the 2D models, the boundary conditions have to be carefully selected to avoid erroneous results. In these models, the boundary conditions were mainly represented by spring elements whose stiffness was determined by comparing the movement of the actual joint during the experimental work with the movement of the simulated joints. With the properties and the boundary conditions satisfactorily defined, a cohesive 2D model was implemented for each type of joint used. The P-δ curves obtained with these FEM models were found to be in good agreement with the experimental results, thus validating the simulation procedures. Due to the limitations on geometrical changes posed by the use of 2D models, 3D models were then produced. As performed for the 2D models, the experimental data was calibrated and used to validate 3D FEM models of joints containing different type of layers (mixed and single adhesive). After the successful validation of these models, different geometrical configurations were proposed and their merits studied. The first comparison was between the experimentally tested mixed adhesive joint configuration and two alternative configurations: a symmetrical square configuration, with silicone fully surrounding the epoxy central section, and a ramped configuration, with a tapered ceramic adherend and a tapered adhesive layer. The symmetrical and tapered joints were found to have lower failure loads than the standard mixed joint but the tapered joint exhibited significantly lower stiffness, which is a beneficial characteristic for operation under large thermal gradients. The symmetrical joint presented the same stiffness and lower strength when compared with the standard mixed joint. This was expected and represents the trade-off that must be made to ensure that this type of joint can provide improvements in more than one direction. Lastly, the study of different adhesive ratios on the ramped joint led to the conclusion that an increase in silicone adhesive content translates into a consistent reduction in peak loads and joint stiffness.

12.6 ACKNOWLEDGMENTS

FCT (Fundação da Ciência e Tecnologia) is thanked for financing the project (PTDC/ EME-PME/67022/2006–P0716) and also for the PhD grant with the reference SFRH/BD/71794/2010. Nagase-Chemtex is also gratefully acknowledged for free supply of adhesive.

12.7 REFERENCES

Adams, R. D., Atkins, R. W., Harris, J. A. and Kinloch, A. J. 1986. Stress-analysis and failure properties of carbon-fiber-reinforced-plastic steel double-lap joints. The Journal of Adhesion 20: 29-53.
Banea, M. D. and da Silva, L. F. M. 2009. Mechanical characterization of flexible adhesives. The Journal of Adhesion 85: 261-285.
Banea, M. D. and da Silva, L. F. M. 2010. Static and fatigue behaviour of room temperature vulcanizing silicone adhesives for high temperature aerospace applications. Materialwissenschaft und Werkstofftechnik 41: 325-335.
Banea, M. D., de Sousa, F. S. M., da Silva, L. F. M., Campilho, R. D. S. G. and de Bastos, A. M. P. 2011a. Effects of temperature and loading rate on the mechanical properties of a high temperature epoxy adhesive. Journal of Adhesion Science and Technology 25: 2461-2574.

Banea, M. D., da Silva, L. F. M. and Campilho, R. D. S. G. 2011b. Mode I fracture toughness of adhesively bonded joints as a function of temperature: Experimental and numerical study. International Journal of Adhesion and Adhesives 31: 273-279.

Banea, M. D., da Silva, L. F. M. and Campilho, R. D. S. G. 2011c. Temperature dependence of the fracture toughness of adhesively bonded joints. Journal of Adhesion Science and Technology 24: 2011-2026.

Banea, M. D., da Silva, L. F. M. and Campilho, R. D. S. G. 2012. Mode II fracture toughness of adhesively bonded joints as a function of temperature: Experimental and numerical study. The Journal of Adhesion 88: 534-551.

Banea, M. D., da Silva, L. F. M. and Campilho, R. D. S. G. 2014. Effect of temperature on the shear strength of aluminium single lap bonded joints for high temperature applications. Journal of Adhesion Science and Technology 28: 1367-1381.

Breto, R., Chiminelli, A., Duvivier, E., Lizaranzu, M. and Jiménez, M. A. 2014. Functionally graded bond-lines for metal/composite joints. European Conference on Composite Materials, Seville, Spain.

Camacho, G. T. and Ortiz, M. 1996. Computational modelling of impact damage in brittle materials. International Journal of Solids and Structures 33: 2899-2938.

Campilho, R. D. S. G., Banea, M. D., Neto, J. A. B. P. and da Silva, L. F. M. 2013. Modelling adhesive joints with cohesive zone models: Effect of the cohesive law shape of the adhesive layer. International Journal of Adhesion and Adhesives 44: 48-56.

Chandra, N., Li, H., Shet, C. and Ghonem, H. 2002. Some issues in the application of cohesive zone models for metal–ceramic interfaces. International Journal of Solids and Structures 39: 2827-2855.

Chaves, F. J. P., da Silva L. F. M., de Moura, M. F. S. F., Dillard, D. A. and Esteves, V. H. C. 2014. Fracture mechanics tests in adhesively bonded joints: A literature review. The Journal of Adhesion 90: 955-992.

da Silva, L. F. M. and Adams, R. D. 2007a. Joint strength predictions for adhesive joints to be used over a wide temperature range. International Journal of Adhesion and Adhesives 27: 362-379.

da Silva, L. F. M. and Adams, R. D. 2007b. Adhesive joints at high and low temperatures using similar and dissimilar adherends and dual adhesives. International Journal of Adhesion and Adhesives 27: 216-226.

da Silva, L. F. M. and Adams, R. D. 2007c. Techniques to reduce peel stresses in adhesive joints with composites. International Journal of Adhesion and Adhesives 27: 227-235.

da Silva, L. F. M., Öchsner, A. and Adams, R. D. 2011. Handbook of Adhesion Technology. Springer-Verlag, Berlin Heidelberg.

de Buyl, F. 2013. A generalized cure model for one-part room temperature vulcanizing sealants and adhesives. Journal of Adhesion Science and Technology 27: 551-565.

Fitton, M. D. 2004. Multi-modulus adhesive bonding of advanced composite materials. Ph.D. Thesis, Oxford Brookes University, UK.

Hart-Smith, L. J. 1973. Adhesive-bonded double lap joints. Nasa Report CR-112235.

Hildebrand, M. 1994. Non-linear analysis and optimization of adhesively bonded single lap joints between fibre-reinforced plastics and metals. International Journal of Adhesion and Adhesives 14: 261-267.

Leiser, D., Hsu, M. and Chen, T. 2001. Washington DC, Refractory Oxidative-Resistant ceramic carbon Insulation, US Patent No. 6,225,248, filed 1 May, 2001.

Liljedahl, C., Crocombe, A., Wahab, M. and Ashcroft, I. 2006. Damage modelling of adhesively bonded joints. International Journal of Fracture 141: 147-161.

Marques, E. A. S., Magalhães, D. N. M. and da Silva, L. F. M. 2011. Experimental study of silicone-epoxy dual adhesive joints for high temperature aerospace applications. Materialwissenschaft und Werkstofftechnik 42: 471-477.

Marques, E. A. S., da Silva, L. F. M. and Sato, C. 2013. Testing of dual adhesive ceramic-metal joints for aerospace applications. pp. 170-190. *In*: S. Kumar and K. L. Mittal (eds.). Advances in Modelling and Design of Adhesively Bonded Systems. Scrivener Publishing, Beverley, MA.

Miller, J. E., Bohl, W. E., Christiansen, E. L. and Davis, B. A. 2013. Ballistic performance of porous-ceramic, thermal protection systems. Procedia Engineering 58: 584-593.

Mohammed, I. and Liechti, K. M. 2000. Cohesive zone modeling of crack nucleation at bimaterial corners. Journal of the Mechanics and Physics of Solids 48: 735-764.

Needleman, A. 1987. A continuum model for void nucleation by inclusion debonding. Journal of Applied Mechanics 54: 525-531.

Needleman, A. 1990. An analysis of tensile decohesion along an interface. Journal of the Mechanics and Physics of Solids 38: 289-324.

Orbiter Thermal Protection System NASA KSC Release No. 11-89. 1989. Kennedy Space Center, FL.

Orbiter Thermal Protection System NASA FS-2004-09-014-KSC. 2006. Kennedy Space Center, FL.

Pethrick, R. A. 2015. Design and ageing of adhesives for structural adhesive bonding – A review. Proceedings of the Institution of Mechanical Engineers, Part L: Journal of Materials: Design and Applications 229: 349-379.

Raphael, C. 1966. Variable adhesive bonded joints. Applied Polymer Symposia 3: 99-108.

Rispler, A. R., Tong, L., Steven, G. P. and Wisnom, M. R. 2000. Shape optimisation of adhesive fillets. International Journal of Adhesion and Adhesives 20: 221-231.

Srinivas, S. 1975. Analysis of bonded joints, NASA TN D-7855.

Stewart, D. and Leiser, D. 2008. Washington DC, Toughened Uni-piece fibrous reinforced oxidization-resistant composite US patent No. 7381459, filed 3 June, 2008.

Tvergaard, V. and Hutchinson, J. W. 1992. The relation between crack growth resistance and fracture process parameters in elastic–plastic solids. Journal of the Mechanics and Physics of Solids 40: 1377-1397.

Index

Printed and bound by CPI Group (UK) Ltd, Croydon, CR0 4YY

01/11/2024

01782622-0011